W9-DIU-143

SOLID-STATE ELECTRONICS CONCEPTS

SOLID-STATE ELECTRONICS CONCEPTS

JOHN I. MATTHEWS

Department of Industry and Technology
Chico State College, California

McGRAW-HILL BOOK COMPANY

New York □ San Francisco □ St. Louis □ Düsseldorf
Johannesburg □ Kuala Lumpur □ London □ Mexico
Montreal □ New Delhi □ Panama □ Rio de Janeiro
Singapore □ Sydney □ Toronto

To my ever-patient
wife and family

This book was set in Times Roman by Applied Typographic Systems and printed on permanent paper and bound by Von Hoffmann Press, Inc. The designer was Janet Bollow; the drawings were done by Ayxa Art. The editors were Alan W. Lowe and Eva Marie Strock. Charles A. Goehring supervised production.

SOLID-STATE
ELECTRONICS
CONCEPTS

Printed in the United States of America.

Library of Congress catalog card number: 70-167496

1234567890 VHVH 798765432

07-040960-9

CONTENTS

PREFACE

Our technological knowledge is increasing at such an enormous rate that few people are able to keep up to date in any field. This is especially true in electronics. However, with a thorough grasp of the basic concepts of electronics, your chances of remaining current are considerably better. Once the fundamental principles are clearly understood, they may be extended to any application. *Solid-state Electronics Concepts* and its accompanying lab manual present contemporary solid-state circuit design in terms of the common conceptual basis of all solid-state devices. Although there is currently emphasis on the microsized integrated circuit, it is based on the same design and circuit concepts presented here. Thus, even though the state of the art is advancing rapidly, this conceptual approach should continue to apply. In some instances it is necessary to clear up common misconceptions that tend to be misleading in the study of semiconductor theory. This systematic approach is the result of many years of teaching electronics at all levels: high school, adult education, community college, state college, and university.

This book was planned for a one-semester course on semiconductors and basic applications. Part One is intended primarily as a review of an introductory course in dc and ac theory and instrumentation. However, it covers all the basic concepts needed to understand semiconductor systems. The treatment uses only basic algebra and a few concepts of trigonometry. Part Two develops the semiconductor devices systematically in terms of their underlying concepts and typical circuit-design considerations. Each experimental circuit is designed to be functional and is planned for easily available components and instruments.

Grateful acknowledgment must go to those colleagues who have helped in their review of the materials in this effort. However, special thanks are due to Robert S. Elmore of Santa Rosa, who ran a two-year pilot study which established that high school second-year electronics students could learn solid-state electronics using this instructional material. Additional reviews which were performed indicated that this presentation is ideal for technician programs at the two-year college level.

JOHN I. MATTHEWS, Ph.D.

SOLID-STATE ELECTRONICS CONCEPTS

PART ONE

CHAPTER 1

OBJECTIVES

1. To establish a point of reference for the study of solid-state electronics
2. To describe the place of solid-state electronics in history
3. To discuss the impact of electronics on our social and industrial lives
4. To relate electronics to its parents, physics and chemistry
5. To examine the advantages of a conceptual approach to the study of electronics

MAN AND ELECTRONICS

Electronics is an exciting field of study that can open up a whole new world, a world in which the energy and materials that surround us seem almost alive. Our knowledge in this momentous field is expanding so rapidly and in so many directions that it constantly presents new challenges to the imagination. Nevertheless, even serious students of electronics sometimes only scratch the surface. They seem almost afraid that electronics is too difficult a subject to be fun.

Electronics is in fact based on entirely logical principles and concepts. It need not be masked by heavy mathematics and explained in complex and highly scientific terms. A mathematical and scientific background is helpful, to be sure, but all that is really required is the ability to grasp the basic principles and follow them through to their logical conclusions.

On this basis, we shall approach electronics from the standpoint of its fundamental concepts—the basic ideas or principles that underlie all of solid-state electronics, the world of diodes, transistors, and integrated circuits.

ELECTRONICS AND THE SOCIAL ENVIRONMENT

Throughout history man has proceeded from one revolution to the next. The Stone Age was followed by the Iron Age, the Bronze Age, and so on through time to the Age of Miracles. The Industrial Revolution, when man really began to rely on machines for his survival, is still not complete. Each new development has opened up possibilities for still further advances. For example, the advent of artificial light literally loosed man from his environment. For the first time he was freed from a schedule based on the daylight hours. With the evolution of the electric light bulb into the vacuum tube, he was able to communicate well beyond his physical reach. Television brought the entertainment stage into the living room and provided immediate access to world news. It also provided a constant babysitter—in time with the brilliance of living color.

The impact of electronics on industry was in a sense a revolution of its own. By the late 1940s the sheer volume of electronic equipment being manufactured each year was enough to boost the economy by millions of dollars. The first of the giant electronic computers had become a reality at the University of Pennsylvania. It took up a vast amount of space although it had only a few words of memory.

In the next few years the transistor became important in industry. The military saw in its performance and small size many advantages that threatened the vacuum tube with near extinction. The race for space was creating a demand for tremendous amounts of electronic equipment for such varied applications as remote control, data transmission, radar, sonar, computer control, and numerical control. By the time Sputnik was launched by the Soviet space program, the United States was well into the era of electronic control and miniaturization.

In the 1960s companies such as Fairchild Semiconductor began producing the tiny integrated circuits to handle work previously done by huge vacuum-tube systems. Equipment manufacturers were no longer thinking in terms of radios and television sets. Instead they were concerned with enormous jobs of data handling, dealing with millions of bits of information per second. The space industry alone was consuming billions of dollars worth of electronic devices and systems each year.

Since then the vacuum tube has been replaced by a host of discrete solid-state devices whose names alone are almost endless—the junction diode, the Zener diode, the tunnel diode, the Gunn diode, the impatt diode, the light-emitting diode, the photo-diode, the transistor, the unijunction transistor, the field-effect transistor (FET), the metal oxide field-effect transistor (MOS-FET), the photo-transistor, the photo-field-effect transistor (photo-FET), the silicon-controlled rectifier (SCR), the silicon-controlled switch (SCS), to mention only a few. With the development of new manufacturing techniques using diffusion and ion implantation it is now possible to produce by large-scale integration circuits containing thousands of transistors, diodes, and resistors. These tiny integrated circuits are so complex that only a computer can keep track of where the interconnecting conductors must go. In fact, in many cases a computerized numerical-control graphical plotter is needed just to lay out the metalizing patterns for their manufacture.

Industry as we now know it could not even handle the data, much less the work, without electronic control of the whole process.

However, the impact of electronics is by no means limited to industry. The fact that mountains of data can be stored and processed electronically has paved the way for developments in other areas that would have been impossible with hand computation. A multitude of electrical and electronic labor-saving devices have released man from countless hours of drudgery. At the same time they have made him increasingly dependent on machinery in every aspect of his life. Whole cities can be immobilized by an electric power failure. For such basic necessities as food man relies on electric refrigeration, electronically controlled food packaging, and even electric can openers. Almost an entire generation now reaching middle age has never known the rigors of physical labor. Children overexposed to the advantages of television may well spend less time in any kind of physical activity—a serious threat to their health. Some, in fact, would have no idea how to occupy their time and thoughts without television, a phonograph, or a transistor radio.

It has become painfully apparent that man's social evolution has not kept pace with his rapidly expanding technology. The results have been that in many respects he is corrupting his own surroundings. Nevertheless, if man is ever to straighten out the mess he has generated, it will be through the very instruments he has created. The problems of society have reached a point where they can no longer be resolved by traditional means. It is only through electronics that men now communicate with each other from opposite sides of the world, and even the planes that bring them face to face rely on the aid of electronic systems. The electronic computer provides a means of spreading and applying increased knowledge in all fields. For example, with the vast amounts of data that can be stored in a computer memory bank, information in all fields of medicine can be pooled so that every doctor—and hence every patient—has access to the latest advances in any area of specialization.

THE CONCEPTUAL APPROACH TO ELECTRONICS

Man is at this point dependent on electronics for his very survival. There is probably no area of present-day living that is not affected, either directly or indirectly, by some phase of this rapidly growing field. Electronics now involves so many areas that no one person could possibly keep abreast of all the new information and its applications. However, he can acquire the fundamental principles in a way that enables him to relate new information in any area to what he already knows. Then, with a firm foundation and a grasp of electrical and solid-state concepts, he is equipped to proceed in any direction his interest carries him.

A *concept* is the basic idea or principle from which other relationships may be developed. It is the least common denominator, without which problems cannot be solved. It makes up the framework into which other ideas must be fitted in order to develop a structural whole. It is the set of rules that relates one fact or application to another.

In electronics an understanding of basic concepts means the ability to isolate into a sequence those ideas or basic bits of knowledge needed to understand a complete system. For example, consider the systems diagram for a common radio shown in Fig. 1-1. A *system* may be a complete project, a complete structure, a complete thought—in this case it is a radio. Note that it is made up of several subsystems, each of which performs a specific function in terms of overall operation. These subsystems are, in turn, made up of various *components*. In some cases different subsystems contain the same types of components. The IF (intermediate-frequency) amplifier and the AF (audio-frequency) amplifier both contain half-Watt resistors. However, in order to understand how each of these subsystems works you must understand what a half-Watt resistor does. In other words, you must first learn something about the basic concepts of electricity. Otherwise, even though you may eventually figure out how the system works, you may never understand *why* it works, which is after all the most interesting thing about it.

In any system there is generally a choice among several subsystems to perform a certain function in the overall operation. Some-

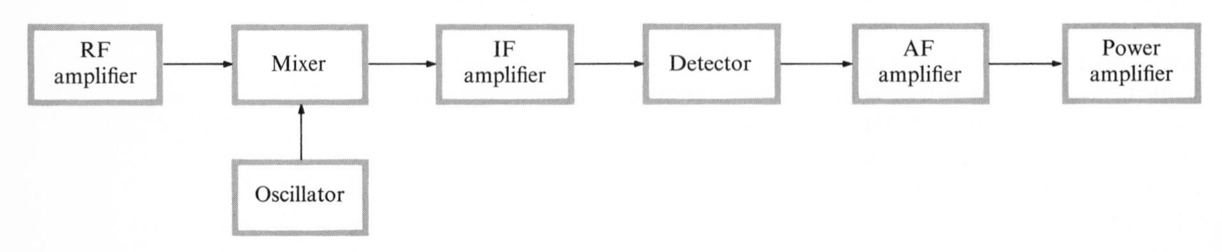

Fig. 1-1 The system diagram of a superheterodyne radio showing the various subsystems involved in its operation.

times a ready-made unit is available. A number of fairly complex electronic circuits are manufactured in integrated-circuit form. In many cases such ready-made subsystems must be modified for a particular application. Frequently the entire circuitry must be designed and developed from scratch. This, of course, requires a thorough grasp of the basic concepts that underlie its operation. This must be the starting point. However, once this foundation has been acquired, the principles may be extended to any application. For example, the transformers in the IF amplifier and a transformer-coupled power amplifier are not the same, but both are based on the concept of inductance. With a clear understanding of inductance and the basis of transformer action, the design of an IF amplifier may be generalized to any transformer-coupled amplifier. The same rules that apply to voltage in a radio also apply to voltage in a space probe. The multivibrator in a television set operates like the multivibrator in a computer. In other words, it is necessary to start at the beginning only once.

For these relationships to be apparent, however, we must approach the underlying concepts in their most fundamental form. For example, the loose definition of a voltage as a "force" is close enough for some applications, but for a real understanding of exactly how voltage relates to current and resistance we must start with a much clearer concept of voltage. In order to understand current flow in a transistor we must start with some basic principles of chemistry. Electronics draws very strongly, in fact, on concepts of both physics and chemistry.

ELECTRONICS AND THE BASIC SCIENCES

It has been said that physics is the father of electronics. If this is true, then certainly chemistry was its mother. A few illustrations should demonstrate this reasoning. *Physics* is the study of energy in its many forms—motion, light, sound, electricity, and heat, to name a few. Of course, it is a fine line that separates physics from chemistry. *Chemistry* is the study of matter and its interactions. To properly understand the concepts of chemistry—the atom, the molecule, the electrons and protons in motion, ionization, and so on—we must consider the various amounts and kinds of energy that are involved. *Solid-state electronics*, then, falls in the realm of chemical-physics, or perhaps physical chemistry.

From either viewpoint, electronics draws very strongly on the concepts of both physics and chemistry. This does not mean that rigorous courses in physics and chemistry are necessary in order to appreciate or to understand electronics. Although they would be of great value, those concepts of physics and chemistry that are involved in the understanding of electronics can be acquired as needed. In fact, the need and application may actually make the physics and chemistry easier to learn.

By the same token, there is no way to escape some amount of mathematics in the study of electronics. However, it is the applications that are of importance in solid-state electronics. Thus, an open mind and application of the simple rules of arithmetic sprinkled lightly with some algebra and a bit of trigonometry make electronics a lot more understandable than most people believe. The necessary refinements are usually picked up along with the material being studied at the time.

One final point is that much of the effort in learning a subject such as electronics is in learning the language. The terms, at least many of them, are new or have more specialized meanings than they do in general usage. However, as with any other language, the language of electronics is best learned through use.

SUMMARY OF CONCEPTS

1. Electronics and its applications are the outgrowth of basic concepts from mathematics, physics, and chemistry followed through to logical conclusions.

2. Solid-state electronics is in a state of rapid expansion that will make the vacuum tube a museum piece in the near future.

3. Solid-state electronics has revolutionized and miniaturized our systems of communication and controls for transportation, data handling, and most other areas of environment.

4. Modern social and industrial life is so complex that only electronic data handling systems can function fast enough to stay ahead and find problem solutions.

5. Without the concepts of mathematics to relate the concepts of atomic structure and matter from chemistry to the concepts of energy and mechanics from physics, solid-state electronics might not have been developed.

6. The conceptual approach to learning allows one to master the basic principles in a field then generalize these concepts into well-structured and related applications without starting at the beginning each time.

7. Facts and applications are valuable, but unless they are put into a related structure they may find little use.

8. A completed structure, or system, is generally made up of subsystems that are in turn made from other subsystems or components. The system is a logical outgrowth or application of several individual components and subsystems.

GLOSSARY

Chemistry The study of matter, its composition, and its reactions.

Component An individual part that combines with other parts to make up a subsystem or system.

Concept A basic principle, thought, or idea from which more complex ideas can be generalized or developed.

Integrated circuit A tiny electronic system containing many transistors, diodes, and resistors.

Physics The study of energy in its many forms.

Principle A basic truth or law: a concept.

Solid-state electronics That branch of electronics which concerns active devices made from solid crystalline materials, such as silicon, rather than vacuum tubes.

Subsystem A functional building block of a complete operational system.

System A group of units or subsystems so combined as to form a whole organized body.

Technology The applied form of the basic sciences involving various functions of engineering and the industrial arts.

CHAPTER 2

OBJECTIVES

1. To examine the common electrical units in terms of basic principles and concepts
2. To review dc theory in a conceptual frame of reference
3. To examine each principle in terms of its actual meaning
4. To explore the relationships of the various electrical "laws"
5. To relate complex dc circuits to their simple equivalent circuits

CONCEPTS IN DIRECT-CURRENT THEORY

Most students approaching the study of solid-state electronics have some familiarity with direct current and Ohm's law. Generally, however, such knowledge is on the functional "how" level, rather than the more conceptual "why" level. This failure to understand the actual meaning of the fundamental units used in electronics may not become apparent until the serious student proceeds to the study of solid-state theory.

In order to clarify any fuzzy notions, this chapter provides a review of the basic concepts of electricity. It is not intended as a full introduction to electricity; rather, it is designed to be used for reference as needed.

THE BASIC UNITS OF FORCE, WORK, AND POWER

Man has always been concerned with his surroundings—chiefly, of course, for reasons of survival. This concern led to the development of the wheel, an outgrowth of the more primitive lever, which is a means of multiplying force. After many thousands of years man learned that all the motion around him could be described in terms of forces. Nevertheless, countless electronics students still do not have a functional understanding of how force relates to electricity.

FORCE: THE NEWTON

The term that is perhaps most overworked and least understood is *electromotive force* (emf), which is supposed to mean *voltage*. This usage often leads to the assumption that voltage must be force. To be sure, voltage involves force, but it is not in itself force.

A little attention to the laws of motion readily shows that force is also basic to other units. The system of units we shall use is the standard *meter-kilogram-second* (*mks*) *system*. These units are set up so that all numbers may be expressed as powers of 10.

Sir Isaac Newton, after whom the unit of force is named, explained that to change the *momentum* of a body in motion requires some force. Momentum, of course, is that property of a body, or mass, moving at some velocity, for example, a train bearing down the tracks toward us. A hummingbird might be bearing down on us at the same velocity, but with considerably less momentum because of its lesser mass. The results of the impact in the two cases would be enormously different.

Since a change in momentum is generally due to a change in velocity over some period of time, force is sometimes defined in terms of acceleration rather than momentum. Acceleration, you will recall, is a change in velocity per unit of time. Thus force is expressed mathematically as

$$\begin{aligned}\text{Force} &= \text{mass} \times \text{acceleration}\\ &= \text{kilogram} \times \frac{\text{meter/second}}{\text{second}}\\ &= \frac{\text{kilogram meter}}{\text{second}^2} = \text{Newton (N)} \qquad (2\text{-}1)\end{aligned}$$

Thus one Newton is the amount of force needed to move one kilogram of mass a distance of one meter per second per second.

WORK: THE NEWTON METER

Our present concern is the electron and its motion. As tiny as it is, the electron has mass and is capable of changing velocity or being accelerated when it absorbs energy. The absorption of energy involves a form of *work*.

Work has sometimes been defined as some action that makes you tired. More accurately, work is performed only when some object having *mass is forced through some distance*. The unit of work is the Newton meter (Nm), defined in mks units as

$$\begin{aligned}\text{Work} &= \text{force} \times \text{distance}\\ &= \frac{\text{kilogram meter}}{\text{second}^2} \times \text{meter}\\ &= \frac{\text{kilogram meter}^2}{\text{second}^2} = \text{Newton meter} \qquad (2\text{-}2)\end{aligned}$$

ENERGY: THE JOULE

When work is discussed in connection with electricity, the unit commonly used is the *Joule* (J) rather than the Newton meter. The terms are synonymous; they describe exactly the same quantity. However, it is more convenient to think of potential energy in terms of the Joule.

Potential energy is energy that is "stored" but available to do work; or, stated another way, it is work available. An example is the energy available in a hammer as it is poised over a carpenter's head. The hammer has potential energy equivalent to the force that is required to move it from this position to the object to be struck. When the carpenter begins his swing downward, this stored or potential energy becomes kinetic energy. *Kinetic energy* is the *energy due to motion*.

In both these cases the unit of energy is the Newton meter. Suppose the potential energy is the energy stored in a battery used to power some circuit—that is, to give the electrons the kinetic energy to move through the circuit. The unit in this case could be the Newton meter, but it is usually the Joule.

POWER: JOULES PER SECOND

Many discussions of electricity leave the false impression that voltage is a force which is work or perhaps some form of power. It is apparent from the foregoing discussion that this is not so. When work is done in moving an electron over some distance through a circuit, the motion obviously involves force and takes time, whether the time is a millionth of a second or an hour. In other words, it is only when we talk about the number of Joules per second—that is, the *work done per unit of time*, that we are talking about power:

$$\text{Power} = \frac{\text{force} \times \text{distance}}{\text{time}}$$

$$= \frac{\text{Joules}}{\text{second}} \qquad (2\text{-}3)$$

The electrical equivalent of one Joule per second is the *Watt* (W).

ELECTRICAL UNITS

Once the basic concepts of force, work, energy, and power have been grasped, it is easy to relate the mechanical world to the electrical world. Basic to the study of electronics is an understanding of the electron in motion. The electron has an apparent mass of about 9×10^{-30} kg. This is about like describing the weight of a fly's wing in terms of the mass of the moon. If we simply move the decimal point 30 places to the left, we see that it is certainly small. The point, however, is that it does have mass. It has other interesting properties as well.

ELECTRIC CHARGE: THE COULOMB

One important property of the electron is that it has an electrically *negative charge*. Each electron has one negative charge. It is difficult to measure even the effect of a single charge, let alone the charge itself. Long before the actual charge on the electron was measured accurately, a functional basic unit for charge was established, the *Coulomb* (C). Charles A. Coulomb (1736–1806), a French physicist, showed that like charges repel each other and unlike charges attract each other. The Coulomb, the unit of electric charge named in his honor, represents the *collective charge found on* 6.25×10^{18} *electrons.*

The number of electrons in one Coulomb of charge is important, because this number is the basis for the unit of electric current.

CURRENT: THE AMPERE

The basic unit of electric current is the *Ampere* (A), named in honor of André M. Ampère (1775–1836). This unit is closely related to the Coulomb; the Ampere is defined as *one Coulomb of charge moving past a point in one second.* That is

$$\text{Current} = \frac{\text{no. of charges}}{\text{second}} = \frac{6.25 \times 10^{18} \text{ electrons}}{\text{second}}$$

$$= \frac{\text{Coulombs}}{\text{second}} = \text{Ampere} \qquad (2\text{-}4)$$

Current may be either *electron flow* or *ion* (charged-atom) *movement*. In the case of copper wire it is electron flow. It is therefore a *motion of negative charges*. From Coulomb's discovery, this means that electrons must move in the direction of the opposite charge—in other words, toward a positive charge. *Current*, then, is the *rate of flow of electrons moving toward a positive charge*.

POTENTIAL DIFFERENCE: THE VOLT

In order to move electrons around a circuit, work must be done on these charges. Stated another way, a *potential-energy difference*, or voltage, between two points is needed to move a group of charges from one point to another. The *Volt* (V), named for Alessandro Volta (1745–1827), is the unit of *potential difference required by one Coulomb of charge as it uses one Joule of energy in moving from one point to another*. That is,

$$\text{Potential difference} = \frac{\text{potential energy}}{\text{no. of charges}}$$

$$= \frac{\text{Joule}}{\text{Coulomb}} = \text{Volt} \qquad (2\text{-}5)$$

A simple comparison of units at this point should indicate the relationship between voltage and force. Force is expressed in Newtons, and the voltage is expressed in Volts. Note that

$$\text{Force} = \frac{\text{kilogram meter}}{\text{second}^2} = \text{Newton} \qquad (2\text{-}6)$$

$$\text{Voltage} = \frac{\text{work}}{\text{Coulomb}} = \frac{\text{Joule}}{\text{Coulomb}} = \frac{\text{Newton meter}}{\text{Coulomb}}$$

$$= \frac{\text{kilogram meter}^2/\text{second}^2}{\text{Coulomb}} = \text{Volt} \qquad (2\text{-}7)$$

ELECTRIC POWER: THE WATT

Any time that electrons move through a conductor, energy is required and used. Actually, the energy *changes from one form to another*, usually to heat or light. The rate at which the energy is used is called power, as discussed above. In other words, *the change in energy per unit of time* represents the power consumed in moving the charges. As a result of this change in energy across a circuit as power is consumed, the potential difference is often referred to as *voltage drop*.

Technically speaking, there is a voltage that can be measured across any circuit which consumes power. This means that power, or Joules per second, must be the product of potential difference and current. That is, since voltage is Joules per Coulomb and current is Coulombs per second,

$$\text{Power} = \frac{\text{Joule}}{\text{Coulomb}} \times \frac{\text{Coulomb}}{\text{second}} = \frac{\text{Joule}}{\text{second}}$$
$$= \text{Volt} \times \text{Ampere} = \text{Watt} \qquad (2\text{-}8)$$

From this relationship it can be seen that the amount of power consumed by any conductor will increase if either the voltage or the current or both increase.

RESISTANCE: THE OHM

Perhaps the most important contribution to the field of electricity was the relationship introduced by George S. Ohm (1787–1854). He showed that for a given potential difference, or voltage, the amount of current that can flow is *limited* by the opposition, or *resistance*, of the circuit. The unit of resistance in use still bears his name, the Ohm (Ω).

The Ohm is defined as *the amount of resistance that will limit the current to one Ampere when a potential difference of one Volt is applied*. That is,

$$\text{Voltage} = \text{current} \times \text{resistance}$$

Therefore

$$\text{Resistance} = \frac{\text{voltage}}{\text{current}}$$
$$= \frac{\text{Volt}}{\text{Ampere}} = \text{Ohm} \qquad (2\text{-}9)$$

CONDUCTANCE: THE SIEMENS (MHO)

Although Ohm was correct in his assumption that resistance limits current in a circuit, there is another view. If a conductor could be observed from its atomic level, it might be said that the amount of current that drifts through a circuit is due to its *conductance*, rather than its resistance. This is merely another way of looking at the same thing and does not affect the validity of Ohm's law.

Perhaps the conductance idea is more useful than the resistance concept, since current will be easily conducted through a good conductor but not through a highly resistive material such as glass or rubber, which has very little conductance. We shall refer to both resistance and conductance, since each term has its own use. The unit of conductance is the *Siemens* (S), formerly called the *mho*, since it is simply the inverse, or *reciprocal*, of the resistance. Thus

$$\text{Conductance} = \frac{1}{\text{resistance}} = \frac{1}{\text{Ohm}} = \text{Siemens} \qquad (2\text{-}10)$$

Current may also be described in terms of conductance, then, as

$$\text{Current} = \text{voltage} \times \text{conductance} = \text{Volt} \times \text{Siemens} = \text{Ampere} \qquad (2\text{-}11)$$

That is, *one volt of potential difference across one Siemens of conductance will cause one Ampere of current to pass through the conductance.*

This alternative concept is quite useful in working with parallel circuits, since we must work with currents that are dividing.

CAPACITANCE: THE FARAD

An interesting phenomenon concerning the control of electric charges is that of *capacitance*. When two conductor surfaces are placed very near to each other but are separated by some insulating material, an electric charge may be stored on the surfaces. The amount of charge that may be stored depends on the capacitance of the system and the potential difference, or voltage, across it.

Two pieces of aluminum foil separated by a thin sheet of plastic may be used to construct a crude capacitor. If a second sheet of plastic is used, the foil-plastic package may be rolled tightly into a small convenient size. The capacitance of the capacitor so constructed depends on the surface area between the two pieces of foil and the insulating properties of the *dielectric*, the plastic separating the two plates.

The unit of capacitance is the *Farad* (F), the capacitance needed to store *one Coulomb of charge with a potential difference of one Volt*. Since the amount of charge stored in the capacitor depends on the capacitance and the potential difference,

$$\text{Charge} = \text{capacitance} \times \text{voltage}$$

the relationship becomes

$$\text{Capacitance} = \frac{\text{charge}}{\text{voltage}} = \frac{\text{Coulomb}}{\text{Volt}} = \text{Farad} \qquad (2\text{-}12)$$

In practice the Farad is much too large for convenient use. Hence capacitance is usually expressed in *microfarads* (μF), which means $1/1{,}000{,}000 = 1 \times 10^{-6}$ F; or nanofarads (nF), $1/1{,}000{,}000{,}000 = 1 \times 10^{-9}$ F, and even picofarads (pF), $1/1{,}000{,}000{,}000{,}000 = 1 \times 10^{-12}$ F. Most capacitors are in the microfarad range. Many have mfd stamped on the case; this means microfarad, not millifarad.

The unit of capacitance was named in honor of Michael Faraday (1791–1867) for his great contributions to the study of electrical phenomena.

INDUCTANCE: THE HENRY

The final basic electrical unit is the *Henry* (H), named after the American physicist Joseph Henry (1797–1878). The Henry is the unit of *inductance*, a property of every coil of wire. Inductance is that property of a coil of wire which allows it to develop an *induced voltage* when a current increases or decreases through it. Thus it is the *change* in the current that is of interest.

For a voltage to be induced in a coil two factors are required: a magnetic field and a changing current. If the coil moves through the magnetic field, a current is generated in the wire, and so a voltage is induced between the ends of the wire. If a current changes through the wire, a changing magnetic field is generated, and of course the voltage changes too. In other words,

$$\text{Induced voltage} = \frac{\text{inductance} \times \text{current change}}{\text{second}}$$

or

$$\text{Inductance} = \frac{\text{induced voltage}}{\text{current change/second}} = \text{Henry} \qquad (2\text{-}13)$$

That is, *one Henry is the amount of inductance present in a coil when a change in current of one Ampere per second induces a voltage or potential difference of one Volt across the conductor or coil.*

OHM'S LAW

The basic laws governing the motion of electrons through a conductor were formulated over 100 years ago. Probably the most significant one for the general study of electronics is *Ohm's law* concerning the relationship of voltage, current, and resistance. In its most basic form Ohm's law states that the voltage measured across any conductor having resistance is proportional to the current passing through the conductor (resistor) and to the value of the resistance. That is, *voltage equals current times resistance*,

$$V = IR \qquad (2\text{-}14)$$

There are several forms of Ohm's law. However, once the basic form of Eq. (2-14) has been learned, the others may be quickly worked out or recalled. If they are merely memorized, they are usually reproduced in some faulty form. Dividing Eq. (2-14) through by R yields the second form,

$$I = \frac{V}{R} \qquad (2\text{-}15)$$

and dividing Eq. (2-14) through by I yields a third form,

$$R = \frac{V}{I} \qquad (2\text{-}16)$$

Note that V, I, and R are mathematical symbols that stand for voltage, current, and resistance, respectively. Do not confuse them with the *units* in which their values are expressed—V (Volts), A (Amperes), and Ω (Ohms).

Recall that conductance is the reciprocal of resistance. Thus conductance, which is denoted by G, may be expressed as

$$G = \frac{1}{R} \qquad (2\text{-}17)$$

Then, if we substitute G for R in Eq. (2-15), we may also express current as

$$I = V\frac{1}{R} = VG \qquad (2\text{-}18)$$

SERIES CIRCUITS

Let us now consider how the fundamental concepts and units of electricity apply to the *series circuit*. The series circuit is just what its name implies. A complete circuit is made up of a series of resistances connected end to end, as shown in Fig. 2-1. Note that with this arrangement the same current must pass through each resistance in the circuit. If 1 A of current passes through resistance R_1, then 1 A of current must also pass through R_2 and R_3.

The source of the electron current is the battery, which supplies electrons to the circuit from its negative terminal. As the current drifts through the resistances, this polarity direction is maintained. That is, the voltage measured across each resistance in the series will be negative to positive, with *the more positive end of the resistance closest to the positive terminal of the battery*. In other words, the positive end of the resistance attracts the electrons from the negative end.

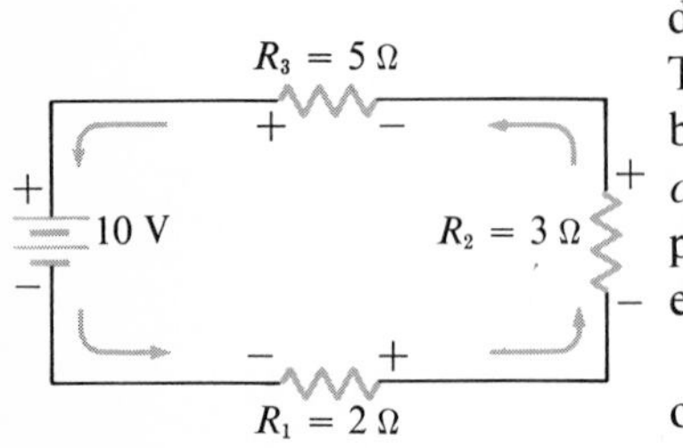

FIG. 2-1 A simple series circuit.

From Ohm's law, the source voltage must be equal to the product of the current and the total resistance of the circuit ($V = IR$). If the supply voltage is 10 V and the current is 1A, the total resistance must be 10 Ω. Simple addition shows that the sum of the three resistances, 2 Ω, 3 Ω, and 5 Ω, is 10 Ω. In other words, the total resistance R_T of a series circuit is simply the sum of the values of the individual resistances in the circuit,

$$R_T = R_1 + R_2 + R_3 + \cdots \tag{2-19}$$

KIRCHHOFF'S VOLTAGE LAW

One of the outgrowths of Eq. (2-19) and Ohm's law is *Kirchhoff's voltage law*. Note that the same current passes through each resistance in Fig. 2-1. Hence we may modify Eq. (2-19) slightly by multiplying both sides by the current I,

$$IR_T = IR_1 + IR_2 + IR_3.$$

Then, since $V = IR$, the *supply voltage* V_S is

$$V_S = V_1 + V_2 + V_3 \tag{2-20}$$

In other words, for series circuits *the sum of the voltages around the circuit external to the supply voltage is equal in magnitude to the supply voltage.*

Since the circuit of Fig. 2-1 has 1 A of current flow, we may use these values to verify Kirchhoff's voltage law by Ohm's law:

$$\begin{aligned} V_S &= V_1 + V_2 + V_3 \\ &= IR_1 + IR_2 + IR_3 \\ &= 1\text{ A} \times 2\ \Omega + 1\text{ A} \times 3\ \Omega + 1\text{ A} \times 5\ \Omega \\ &= 2\text{ V} + 3\text{ V} + 5\text{ V} = 10\text{ V} \end{aligned} \tag{2-21}$$

Note in Fig. 2-1 that the polarity of the supply voltage is opposite in polarity to the individual voltages. The sum of voltages around the entire circuit must therefore equal *zero*. Otherwise the circuit could not be in balance, and there would be voltage left over somewhere.

Ohm's law and Kirchhoff's voltage law can be verified in a rather simple experiment. Select three resistors with the values indicated in Fig. 2-2. Measure their actual values with an ohmmeter and record these measurements. The amount of current in the circuit will depend on the actual values of the resistors, not on their color-code values.

Connect the circuit shown in Fig. 2-2*a*. Be careful to place the milliammeter in the circuit with the polarity indicated; otherwise the meter will attempt to move in the wrong direction and may be damaged. Measure the total resistance of the circuit, including the resistance of the milliammeter itself, and record this measurement. Now add up the individual resistances you recorded before. How does this sum compare with your measured total resistance? What must the resistance of the meter be?

Now connect the circuit of Fig. 2-2*b*, with the proper polarity of the battery carefully noted. Measure and record the amount of current used in the circuit. Measure the voltage of the power supply accurately and record this value. Next measure and record the voltage across each resistance, watching polarity as you move the voltmeter from one resistor to the next. Be careful that the voltages you record are put down for the proper resistances. With the measured values of supply voltage and total current flow, compute the total resistance of the circuit from Ohm's law. Compare this computed value with the values you have already measured or computed. How do you account for any differences?

Using the individual resistance values just measured and the measured current value, compute the voltage across each resistor. How do your computed values compare with your measured values? Add the three measured voltages and compare this sum to the measured value of the supply voltage. How do these two values compare?

Do you agree with Ohm's law? Do you agree that Kirchhoff's voltage law holds true?

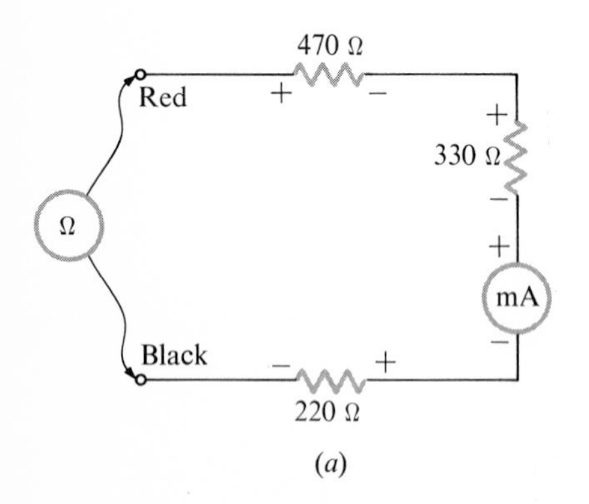

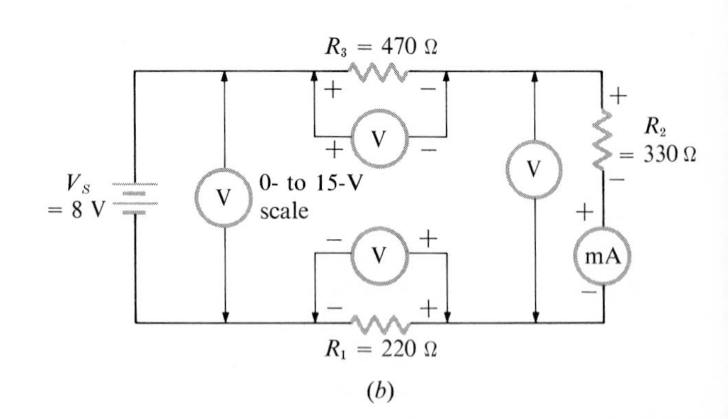

FIG. 2-2 (*a*) Experiments to verify Ohm's law and (*b*) Kirchhoff's voltage law.

WATT'S POWER LAW

As emphasized previously, work is done in moving current through a circuit, and the rate at which this work is done is defined as power. In discussing electric circuits the relationships of power to voltage and current is sometimes referred to as *Watt's law*. Like Ohm's law, Watt's law has several forms. The basic form states that *power equals voltage times current*,

$$P = VI \tag{2-22}$$

However, since $V = IR$, we may also write this as

$$P = (IR)I = I^2R \tag{2-23}$$

Since $I = V/R$, substitution in Eq. (2-22) provides another relationship,

$$P = V\frac{V}{R} = \frac{V^2}{R} \tag{2-24}$$

The implications of the power law are many, but perhaps the most familiar is based on Eq. (2-23). Note that the power consumed in a resistance increases as the *square* of the current used. Since power is usually dissipated in the form of heat, resistors may become quite hot when much current passes through them. For this reason power companies generally transmit their power at a high voltage and low current instead of low voltage and high current. This also allows them to save a great deal of money by using smaller conductors.

At times we must find out the power when the current is unknown. If it is possible to measure the resistance and the voltage applied to the circuit, the power may be computed from Eq. (2-24).

To verify Watt's power law use the three forms to compute the power consumed by each of the resistors in Fig. 2-2. Remember that voltage, current, and power must be computed in basic units. Thus, when current is expressed in milliamperes (mA), it must be converted back into Amperes. The prefix *milli-*, which means 1/1,000, or 1×10^{-3}, should not be confused with *micro-*, which means 1/1,000,000 or 1×10^{-6}. By the same token, resistance is often given in *kilohms* (kΩ), which means 1,000, or $1 \times 10^3\ \Omega$, and sometimes *megohms* (MΩ), or 1,000,000, or $1 \times 10^6\ \Omega$. In such cases it must be converted back into Ohms. For example, if 5.0 mA of current passes through 5 kΩ of resistance, the power consumed is, from Eq. (2-23),

$$\begin{aligned} P &= I^2R \\ &= (5.0\ \text{mA})^2 \times 5\ \text{k}\Omega \\ &= (0.005 \times 0.005) \times 5{,}000 = 0.125\ \text{W} \end{aligned}$$

VOLTAGE-DIVIDER ACTION

When resistances are used in series, the voltage of the supply is divided up around the circuit in proportion to each of the individual resistances. This follows from Kirchhoff's voltage law. Because the current in a series circuit is constant—that is, it is the same through each resistance—with proper choice of resistance values a large voltage can be made to provide a low voltage by means of the *divider action* of a series circuit.

For example, if a radio requires 9 V and only a 12-V supply is available, a voltage divider can be used to provide the proper 9 V, as shown in Fig. 2-3. Note that the radio is in effect a resistance that allows 20 mA of current to pass when the voltage is 9 V. This means that its resistance is 9 V/0.02 A = 450 Ω. From Ohm's law, the value of the series resistance may be computed as

$$R_{\text{series}} = \frac{3\ \text{V}}{0.02\ \text{A}} = 150\ \Omega$$

Thus two resistors in series, 150 and 450 Ω, act to divide 12 V into 3 and 9 V, respectively.

OTHER CURRENT LIMITERS

If Kirchhoff's voltage law and Ohm's law are to hold, then any circuit component that limits current has resistance and must be taken into account. Examples of such components are coils, leaky capacitors, diodes, meters, and even other batteries. In approaching a complex

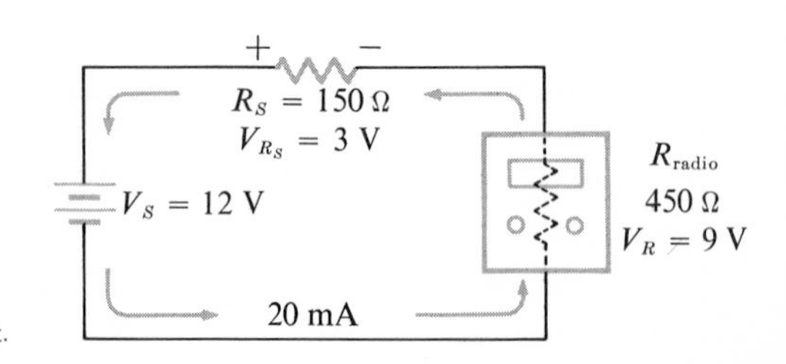

FIG. 2-3 The voltage-divider action of the series circuit.

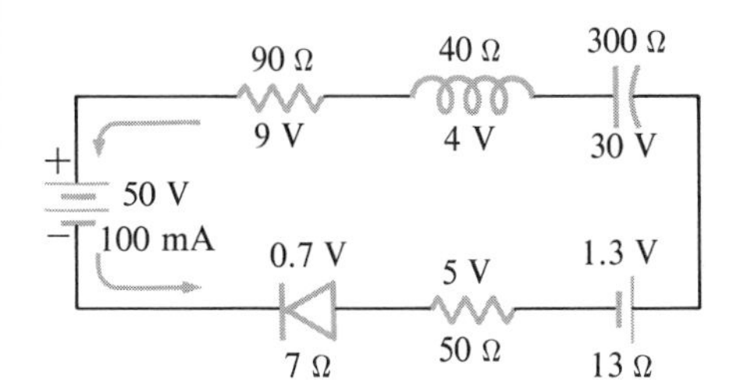

FIG. 2-4 A multiple-component series circuit containing resistances, a coil, a leaky capacitor, a diode, and a second voltage source.

circuit such as that in Fig. 2-4, let us first consider the voltage division. Since the −1.3 V of the cell is opposite in polarity from the main supply voltage, it subtracts from this voltage, leaving 48.7 V to be divided among the remaining circuit components. The silicon diode, when it is conducting properly, has about 0.6 to 0.7 V across it in most circuits. The capacitor in this case is very leaky, since it passes 100 mA of current; a good capacitor should pass almost none.

The resistance of each component is calculated from Ohm's law just as before:

$$R_C = 30\ \text{V}/0.1\ \text{A} = 300\ \Omega$$

$$R_D = 0.7\ \text{V}/0.1\ \text{A} = 7\ \Omega$$

$$R_L = 4\ \text{V}/0.1\ \text{A} = 40\ \Omega$$

The battery connected in reverse has a resistance in this direction of

$$1.3\ \text{V}/0.1\ \text{A} = 13\ \Omega$$

The sum of the resistances in the circuit is

$$90 + 40 + 300 + 13 + 50 + 7 = 500$$

Thus with a current of 0.1 A the supply voltage is $0.1\ \text{A} \times 500\ \Omega = 50\ \text{V}$. When both voltage sources are included in the circuit the sum of the voltages is zero,

$$+50 - 9 - 4 - 30 - 1.3 - 5 - 0.7 = 0\ \text{V}$$

which agrees with Kirchhoff's voltage law as well as Ohm's law.

PARALLEL CIRCUITS

When the current in a circuit has more than one possible path, the circuit is called a *parallel circuit*, since the paths are in effect parallel.

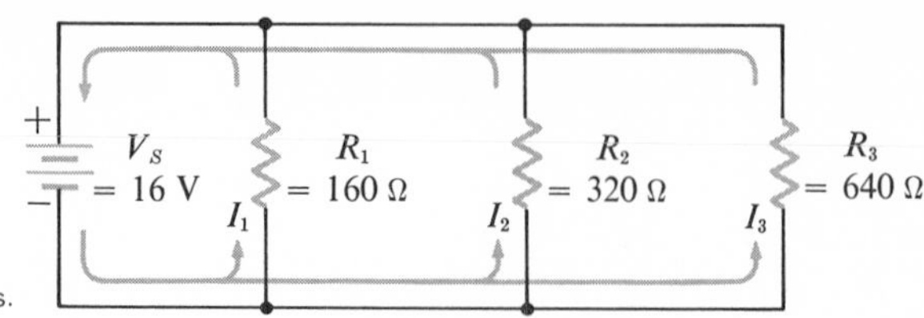

FIG. 2-5 A parallel circuit showing current paths.

As shown in Fig. 2-5, the supply voltage is the same for all the resistances in the circuit. However, the current divides, with some passing through each branch. For this reason the parallel circuit is sometimes called a *current divider.*

KIRCHHOFF'S CURRENT LAW

Each resistance in Fig. 2-5 has 16 V applied across it. From Ohm's law, the total current I_T through R_1, R_2, and R_3 is given by $I_T = V_S/R_T$. Then, since

$$I_1 = \frac{V_S}{R_1} = \frac{16\ \text{V}}{160\ \Omega} = 0.1\ \text{A}$$

$$I_2 = \frac{V_S}{R_2} = \frac{16\ \text{V}}{320\ \Omega} = 0.05\ \text{A}$$

$$I_3 = \frac{V_S}{R_3} = \frac{16\ \text{V}}{640\ \Omega} = 0.025\ \text{A}$$

we have

$$\begin{aligned} I_T &= I_1 + I_2 + I_3 \\ &= 0.1 + 0.05 + 0.025 = 0.175\ \text{A} \end{aligned}$$

When the total current and the supply voltage are known, the total resistance may be computed as

$$\begin{aligned} R_T &= \frac{V_S}{I_T} \\ &= \frac{16\ \text{V}}{0.175\ \text{A}} = 91.4\ \Omega \end{aligned}$$

This is a reasonable value for the total parallel circuit, since the smallest resistance in the circuit is 160 Ω. In addition to this resistor, there are two other paths through which current may pass. Thus the total resistance of a parallel circuit is always less than the smallest resistance branch of the circuit.

Note in Fig. 2-5 that the 0.175 A of current leaves the power supply, divides to pass through each of the branches, and returns to the power supply. According to *Kirchhoff's current law, the sum of the currents leaving a junction must equal the current entering that junction*,

$$I_T = I_1 + I_2 + I_3 \qquad (2\text{-}25)$$

Since the voltage is constant we have

$$I_T R_T = I_1 R_1 + I_2 R_2 + I_3 R_3 \qquad (2\text{-}26)$$

Thus

$$\frac{V_S}{R_T} = \frac{V_S}{R_1} + \frac{V_S}{R_2} + \frac{V_S}{R_3} \qquad (2\text{-}27)$$

and

$$\frac{1}{R_T} = \frac{1}{R_1} + \frac{1}{R_2} + \frac{1}{R_3} \qquad (2\text{-}28)$$

Equation (2-28) is quite straightforward, but it is one with which many students make mistakes. This is primarily because they do not complete the problem. A simple change in form helps somewhat and leads to

$$R_T = \frac{1}{1/R_1 + 1/R_2 + 1/R_3} \qquad (2\text{-}29)$$

For simplicity's sake, since $1/R$ is the same as conductance G, we can substitute in Eq. (2-29) to obtain

$$R_T = \frac{1}{G_1 + G_2 + G_3} \qquad (2\text{-}30)$$

An example will illustrate the usefulness of Eq. (2-30). Note that all fractions are handled in decimal form. First convert every resistance value in Fig. 2-6 to its conductance by dividing into 1. If you are using a slide rule, set the hairline or cursor over the number on the C scale and read the reciprocal value on the C-I or C-Inverted scale. The first resistance in Fig. 2-6, $R_1 = 4.7\ \text{k}\Omega$, has a conductance of $1/4{,}700 = 0.000213$ S (recall that conductance is measured in Siemens). If scientific notation is used, then $1/4.7 \times 10^3 = 0.123 \times 10^{-3}$; this holds all the zeros as a power of 10 and is far less likely to lead to an error. The second resistance, 10 kΩ, has a conductance of $1/10{,}000 = 0.1 \times 10^{-3}$ S. The third resistance has a conductance of $1/7{,}500 = 0.133 \times 10^{-3}$ S.

It is a simple matter to add the three conductances and to divide their sum into 1. However, in order to add with scientific notation the powers of 10 must be the same. Therefore to add 0.213×10^{-3} and 1.00×10^{-4} we must first convert 1.00×10^{-4} to 0.10×10^{-3}. We then have

$$G_T = G_1 + G_2 + G_3$$
$$= 0.213 \times 10^{-3} + 0.100 \times 10^{-3} + 0.133 \times 10^{-3} = 0.446 \times 10^{-3}\ \text{S}$$

or

$$R_T = \frac{1}{G_T}$$
$$= \frac{1}{0.446 \times 10^{-3}} = 2.24\ \text{k}\Omega$$

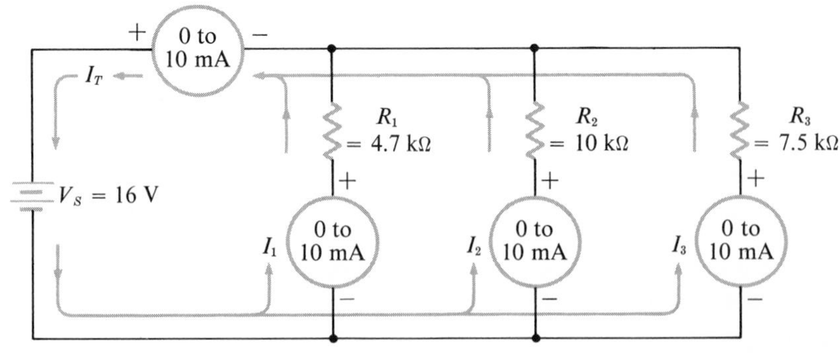

FIG. 2-6 An experimental parallel circuit.

The total parallel sum of 2.24 kΩ may be verified by determining the total current and dividing that value into the constant voltage of 16 V. Thus

$$I_T = I_1 + I_2 + I_3$$
$$= 3.4 + 1.6 + 2.13 = 7.13 \text{ mA}$$

and

$$R_T = \frac{V_S}{I_T}$$
$$= \frac{16 \text{ V}}{7.13 \text{ mA}} = 2.24 \text{ k}\Omega$$

In solving parallel-circuit problems use whichever of these methods is most convenient. When no voltage is given, one may be assumed in order to solve the problem by the current method without introducing error. For example, if the circuit in Fig. 2-6 had not had an indicated supply voltage of 16 V, 10 V could have been assumed. The currents could then have been computed through each resistance, and their sum would have given the total current for the assumed 10-V supply. Then the total resistance could have been determined from Ohm's law.

It will be helpful in visualizing this example to construct the circuit of Fig. 2-6 and check it out for total resistance and the branch and total current values. After the circuit is constructed, but before the power supply is attached, measure the total resistance with an ohmmeter. Remove the ohmmeter from the circuit and record the total resistance.

Next attach the power supply and measure the total current and the individual branch currents. Compare the values obtained experimentally with those computed mathematically. Be certain to take into consideration the *actual* values of the resistances used, and not the color-coded values.

SERIES-PARALLEL CIRCUITS

Most circuits used in electronics are not simple series or parallel circuits. Instead they are combinations of the two. However, the concepts and techniques already explained for handling them apply readily to the more complex series-parallel combination circuits. This is because most series-parallel circuits can be redrawn so that all parallel resistances are located and can be solved for their equivalent

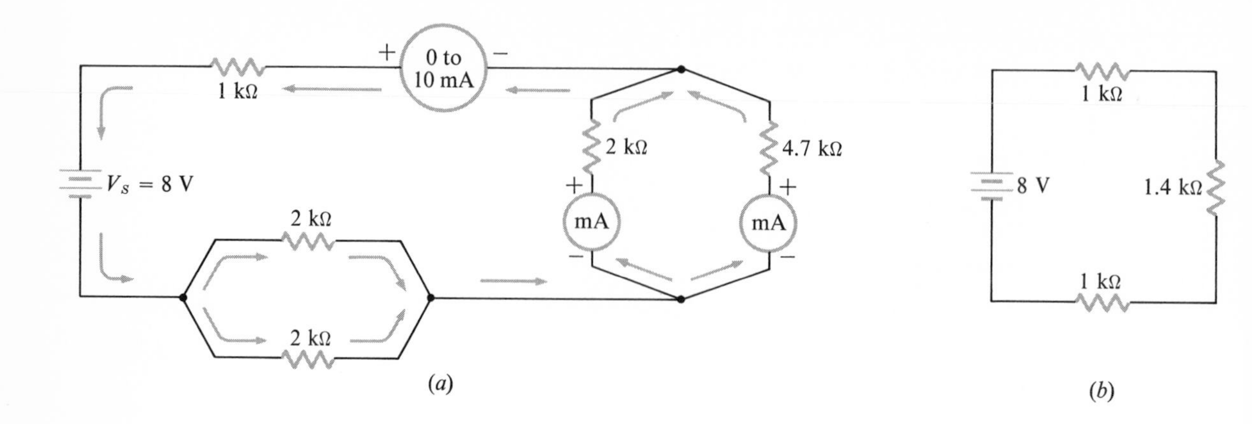

Fig. 2-7 (*a*) The series-parallel circuit and (*b*) its series equivalent.

values; then the circuit may be treated like an ordinary *series circuit.* Figure 2-7 shows a simple example of this basic principle.

Note in Fig. 2-7*a* that there are two sets of parallel resistances. The first set contains two 2-kΩ resistances and the second set contains one 2-kΩ and one 4.7-kΩ resistance. When these parallel combinations are reduced to equivalent resistances, the circuit will appear as in Fig. 2-7*b*.

When two resistances in parallel are of equal value, the current divides equally, with half going through each resistance. The total resistance of that parallel combination is then equal to one-half the value of one resistance. When the resistances are not equal, the total must be computed from Ohm's law or Kirchhoff's law.

When only two resistances are in parallel, Eq. (2-29) reduces to a simple but very useful form,

$$R_T = \frac{1}{1/R_1 + 1/R_2} = \frac{1}{R_2/R_1R_2 + R_1/R_1R_2}$$

$$= \frac{1}{(R_1 + R_2)/R_1R_2} = \frac{R_1R_2}{R_1 + R_2} \tag{2-31}$$

Therefore

$$R_T = \frac{2\ \text{k}\Omega \times 4.7\ \text{k}\Omega}{2\ \text{k}\Omega + 4.7\ \text{k}\Omega} = \frac{9.4 \times 10^{-6}}{6.7 \times 10^{-3}} = 1.4 \times 10^3 = 1.4\ \text{k}\Omega$$

THE THÉVENIN EQUIVALENT CIRCUIT

Frequently series-parallel circuits are quite complex, and the normal methods used for their solution become long and cumbersome. It is possible to reduce the whole circuit quickly to an equivalent circuit by making use of Thévenin's theorem, which states that a circuit should be considered as a black box. Inside the black box there *appears* to be a source voltage and a resistance in series with it. What the actual network of resistance looks like or how big the actual voltage source is makes no difference in most cases.

If, for example, a complex circuit is supplying power to some load such as a speaker or a motor, the only points in terms of power transfer to the load are:

1. What is the output voltage with the load disconnected?
2. What is the internal resistance of the black box?
3. What is the output voltage with the load connected?
4. How much current passes through the load resistance?

The equivalent-circuit development of Fig. 2-8 illustrates the main aspects of Thévenin's theorem. The current from the power supply must divide, with part of it passing through the branch containing the load resistance, and the remainder passing through the other branch. If the circuit is redrawn as in Fig. 2-8*b*, the equivalent circuit begins to develop, since the load resistance and the voltage supply are outside the box. If the load resistance is removed, the resistance as viewed back into the load terminals is known as the *internal resistance* or the *Thévenin resistance*.

Measurement of the Thévenin resistance The Thévenin resistance inside the black-box circuit of Fig. 2-8 may be computed or measured easily. If the parts and power supply are available, connect the circuit

FIG. 2-8 (*a*) Development of a Thévenin equivalent circuit, (*b*) measurement of the Thévenin voltage, and (*c*) measurement of the Thévenin resistance.

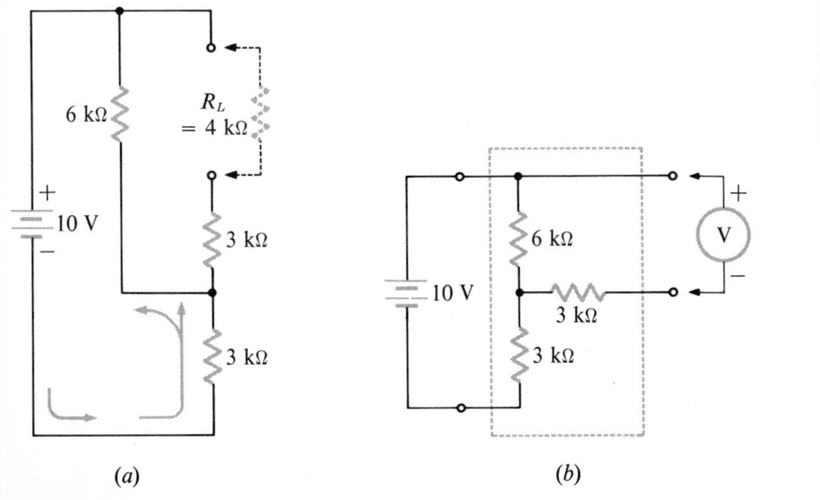

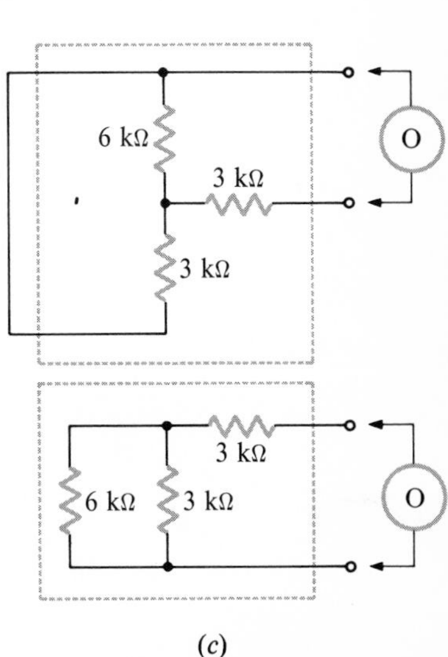

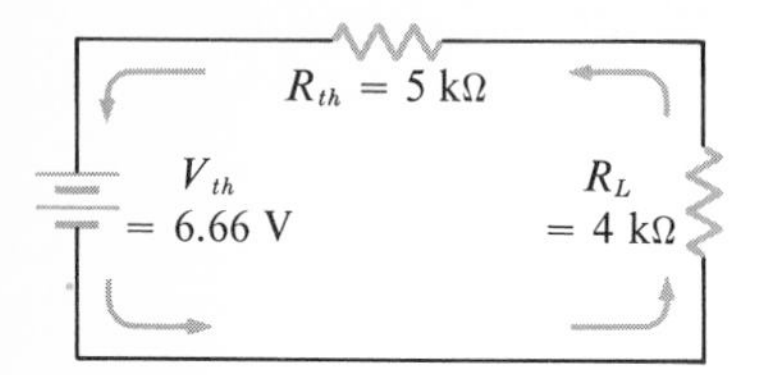

FIG. 2-9 The Thévenin equivalent circuit.

as follows. Since the actual voltage supply has essentially zero resistance, remove it from the circuit and short out the two terminals of the circuit where the supply was removed, as indicated in Fig. 2-8*c*.

Next attach the ohmmeter where the load resistance would normally be connected. Measure and record the resistance of the circuit from the load terminals. The resistance just measured is the Thévenin resistance of the circuit. Remove the ohmmeter from the circuit and remove the short circuit where the voltage supply must be connected.

Measurement of the Thévenin voltage Connect the 8- or 10-V supply to the circuit as indicated in Fig. 2-8*b*. The actual voltage used is unimportant, but an assumed value of 10 V makes the computation very simple. However, be sure to measure accurately, at the terminals of the supply, the supply voltage that is actually used.

Connect the voltmeter to the output load terminals from which the load resistance has been removed. Accurately measure the voltage at this point and record it as the *Thévenin voltage* (V_{th}). Notice that this Thévenin voltage is not the same as but is *lower* than the supply voltage. This is the *apparent* voltage that the circuit uses to deliver current to the load resistance through the Thévenin resistance.

The load voltage and current Once the Thévenin voltage and the Thévenin resistance has been determined, reconnect the load resistance where it belongs. The load resistance "sees" the circuit as it appears in Fig. 2-9. With the load resistance in place, current can again pass through the entire circuit, including the load resistance. The load current also passes through the Thévenin resistance and *seems to be the result of the Thévenin voltage*. When current passes through the Thévenin resistance, however, a voltage appears across it, since series resistances make up a voltage divider. The remaining voltage appears across the load resistance.

In order to compute the Thévenin voltage, if it cannot be measured, observe the redrawn circuit of Fig. 2-8*b*. Since the load resistor is not connected, no current can pass through the 3-kΩ resistor. The 10-V supply is divided across the series resistances, 3 and 6 kΩ. There is therefore no apparent voltage across the 3-kΩ resistance connected at the bottom end of the load resistance; the Thévenin voltage must be the same as the voltage across the 6-kΩ resistance. It may be computed as the ratio of two resistances and the supply voltage,

$$V_{th} = V_S \frac{R_1}{R_1 + R_2} \qquad (2\text{-}32)$$

Thus

$$V_{th} = 10 \text{ V} \times \frac{6 \text{ k}\Omega}{3 \text{ k}\Omega + 6 \text{ k}\Omega} = 10 \times 0.666 = 6.666 \text{ V}$$

The Thévenin resistance may be computed by means of the redrawn circuits of Fig. 2-8*c*. The shorted-out power supply effectively returns the top end of the 6-kΩ resistor to the bottom of the 3-kΩ resistor. The two resistances appear in parallel, with a parallel sum of 2 kΩ (3 × 6/3 + 6). When this 2-kΩ resistance is added to the 3-kΩ resistance with which it is in series, the total Thévenin resistance becomes 2 + 3 = 5 kΩ.

The voltage across the load depends on the size of the load resistance used, since a different current will pass for each value of load resistance used. The same current passes through the internal or Thévenin resistance as passes through the load (see Fig. 2-9). The load voltage is proportional to the ratio of the load resistance to the total circuit resistance,

$$V_L = V_{th} \frac{R_L}{R_L + R_{th}} \qquad (2\text{-}33)$$

Hence

$$V_L = 6.66 \text{ V} \times \frac{4 \text{ k}\Omega}{4 + 5 \text{ k}\Omega}$$

$$= 6.66 \times 0.445 = 2.96 \text{ V}$$

When the Thévenin voltage and resistance have been computed or measured, the voltage across the load for *any* load resistance can be found without recalculating the whole circuit from the beginning. This, in most cases, is not true for other systems.

An electronic generator, a battery, or a car generator has internal resistance and a Thévenin voltage. For this reason, when lots of current is delivered, the voltage at the load decreases proportionally. Witness what happens to the battery in a car when it has to deliver power to a starter on a cold morning. From a 12-V battery only about 7 or 8 V may get to the load that is the starting motor. The internal resistance of the voltage supply must *always* be considered.

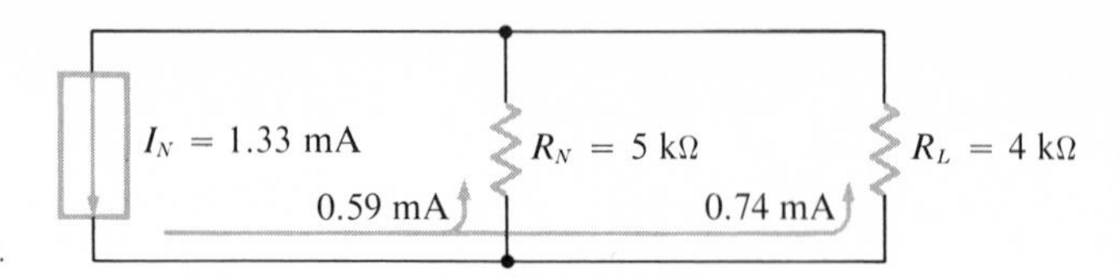

FIG. 2-10 The Norton equivalent circuit.

THE NORTON EQUIVALENT CIRCUIT

There are occasions when it is not the voltage at the load that is of concern, but the current. The *Norton equivalent circuit* was developed in a fashion similar to the Thévenin concept. If the same circuit is used, the relationship should become clear.

Instead of considering the internal voltage source, let us think of it as a constant-current source. The internal resistance is the same as in the Thévenin circuit, but in the equivalent circuit it appears to be in parallel with the current source (see Fig. 2-10).

Since the circuit is the same as the one used to explain the Thévenin concept, certain equivalents should be pointed out. The Thévenin resistance equals the Norton resistance. The Norton current I_N, found by Ohm's law, is

$$I_N = \frac{V_{th}}{R_N} \qquad R_N = R_{th} \tag{2-34}$$

so that

$$I_N = \frac{6.66\text{ V}}{5\text{ k}\Omega} = 1.33\text{ mA}$$

The Norton current so computed represents the short circuit, or *maximum current*, that can be delivered if the load resistance is zero Ohms. The current not drawn by a load resistance must always then pass through the Norton resistance. In this way the current-divider action of the parallel circuit is maintained.

Equation (2-34) may be used to compute the load current when the Thévenin voltage is not known. That is,

$$I_L = \frac{I_N R_N}{R_N + R_L} \tag{2-35}$$

Thus

$$I_L = \frac{1.33 \text{ mA} \times 5 \text{ k}\Omega}{5 \text{ k}\Omega + 4 \text{ k}\Omega} = \frac{6.66 \text{ V}}{9 \text{ k}\Omega} = 0.74 \text{ mA}$$

As with the Thévenin circuit equivalent, when the load resistance is changed, the circuit values need not be computed again. For example, if a load resistance of 10 kΩ is substituted for the 4-kΩ load, the new load current becomes

$$I_L = \frac{1.33 \text{ mA} \times 5 \text{ k}\Omega}{5 + 10 \text{ k}\Omega} = \frac{6.66 \text{ V}}{15 \text{ k}\Omega} = 0.44 \text{ mA}$$

When a series-parallel circuit is being used in its usual circumstances, it is quite complex. It may be necessary to resort to equivalent circuits such as the Thévenin circuit to determine the load voltage or the Norton circuit to determine the load current. The two theorems are frequently used together in the solution of a circuit. While the applications we have discussed were presented in terms of direct-current (dc) circuits, the theorems hold just as true with alternating-current (ac) circuits.

SUMMARY OF CONCEPTS

1. Force, expressed in Newtons, is not the same as voltage, which is expressed in Newton meters per Coulomb of charge.
2. Current is expressed in Amperes. An Ampere is one Coulomb of charge (6.25×10^{18}) moving past a point in one second.
3. Power is the rate of doing work. Expressed electrically in Watts, it is the product of the voltage and current used, or voltage squared divided by resistance.
4. Ohm's law establishes the relationship of voltage V, current I, and resistance R (or conductance G), $V = IR$.
5. Conductance is the reciprocal of resistance ($I = VG$) and is expressed in Siemens (formerly mhos).
6. The basic unit of capacitance C is the Farad, which is the capacitance needed to store one Coulomb of charge with one Volt of potential difference.
7. The Henry, the basic unit of inductance L, is the property of a coil that induces a voltage of one Volt when the current changes at a rate of one Ampere per second.
8. A series circuit acts as a voltage divider, since the same current passes through each resistance.
9. Kirchhoff's voltage law illustrates that the sum of the voltages in a series circuit equals zero. Stated another way, the sum of the voltages in a series circuit, not including the source, is equal in magnitude to the source voltage.
10. The total resistance in a series circuit is the sum of the individual resistances.
11. Power, dissipated as heat from most resistances, increases as the square of the current ($P = I^2R$).
12. The parallel circuit acts as a current divider, since the voltage is the same across all resistances in the network.
13. The total current in a parallel circuit is the sum of the individual branches of current.
14. The total resistance of a parallel circuit is smaller than the resistance of the smallest branch, since current in addition to this current passes through the other branches.

15. The total resistance of a parallel circuit may be determined from the resistance or the conductance values [$R_I = 1/(1/R_1 + 1/R_2 + 1/R_3)$ or $R_T = 1/(G_1 + G_2 + G_3)$].

16. Series-parallel circuits are handled as special series circuits with each set of parallel branches reduced to its single equivalent value.

17. When a series-parallel circuit is very complex and the voltage applied to the load must be calculated, the circuit may be reduced to its Thévenin equivalent. From this form the load voltage may be computed for any size of load resistance.

18. The Norton equivalent circuit may be used to compute the load current in complex series-parallel circuits. The Norton resistance appears across the current source that represents the possible short-circuit Norton current. This circuit is often used in conjunction with the Thévenin circuit.

GLOSSARY

Ampere The unit of current, one Coulomb of charge moving past a point in one second.

Capacitance The charge that can be stored in a capacitor, expressed in Farads.

Conductance The reciprocal of resistance; the property of a circuit that allows current to pass when a voltage is applied.

Coulomb The basic unit of electric charge. One Coulomb is the charge on 6.25×10^{18} electrons.

Current The motion of charges in one general direction through a conductor; generally electron current, but sometimes hole current.

Drift The means by which electron current moves in a zigzag path through a conductor.

Electron The elementary negatively charged particle in the atom, generally in orbit around the nucleus. A little larger than the proton, its mass is approximately 1/1,850 that of the proton.

Energy The ability to do work, either stored in position (potential energy) or due to motion (kinetic energy).

Farad The basic unit of capacitance; one Coulomb of charge stored under one Volt of potential difference.

Force A factor that produces a change in momentum, as when a mass in motion is accelerated or retarded. A force applied to a body at rest will produce motion even though it is only a slight surface distortion.

Henry The basic unit of inductance. One Henry of inductance will cause an induced voltage of one Volt if the current through the coil is changed by one Ampere per second.

Inductance That property of a coiled conductor which causes an induced voltage in the coil when it is cut by a moving magnetic field.

Joule The basic mks unit of energy or work. One Joule is equal to one Newton meter.

Kinetic energy The energy due to motion.

Mass That property of matter which, when acted upon by gravity, is called weight. For example, one kilogram of lead weighs one kilogram on earth but will still have one kilogram of mass on the moon, where it weighs only 0.66 lb.

Meter The unit of length in the metric system, about 39.37 in.

Momentum The property of a mass in motion at some constant velocity, i.e., mass times velocity. Force is required to change momentum, i.e., to change the velocity of the mass.

Newton The basic unit of force. One Newton equals one kilogram meter per second^2.

Ohm The unit of resistance or opposition to current. That amount of resistance needed to limit current to one Ampere when one Volt is applied to the circuit.

Parallel circuit A multipath circuit through which current may pass.

Potential difference The difference in potential energy between two points or charges, a voltage.

Potential energy Stored ability to do work, capable of being released as kinetic energy.

Power The rate of using energy, or the number of Joules used per second. One Joule per second equals one Watt.

Resistance Opposition to current, usually expressed in Ohms.

Series circuit A circuit in which all elements follow one after another, so that the same total current must pass through each resistance of the series.

Siemens (mho) The unit of conductance, which is the inverse of resistance ($G = 1/R$).

Thévenin voltage The apparent supply voltage as "seen" by the load resistance of any circuit.

Volt The unit of potential difference. One Volt is the amount of voltage needed to cause one Ampere of current to pass through one Ohm of resistance.

Voltage divider A series of resistances in which the sum of the individual resistor voltages equals the supply voltage.

Watt The unit of power, equal to one Joule per second. One Watt is the amount of work done in moving one Ampere of current through one Ohm of resistance with an applied potential difference of one Volt.

REFERENCES

Boylestad, Robert L.: *Introductory Circuit Analysis*, Charles E. Merrill Books, Inc., Columbus, Ohio, 1968.

Cooke, Nelson M., and Herbert F. R. Adams: *Basic Mathematics for Electronics*, 3d ed., McGraw-Hill Book Company, New York, 1970.

Gillie, Angelo C.: *Electrical Principles of Electronics*, 2d ed., McGraw-Hill Book Company, 1969.

Grob, Bernard: *Basic Electronics*, 2d ed., McGraw-Hill Book Company, New York, 1965.

Lippin, Gerard: *Circuit Problems and Solutions*, vol. 1, Hayden Book Company, Inc., New York, 1967.

Middleton, Robert G., and Milton Goldstein: *Basic Electricity for Electronics*, Holt, Rinehart and Winston, Inc., New York, 1966.

REVIEW QUESTIONS

1. What is force? What units are used in the definition of force?

2. How is force related to work?

3. In what way is potential energy concerned with work?

4. How does power differ from energy? What units are used in defining power?

5. What is the basic unit of electric charge?

6. How many electrons must pass a point in 1 s for 1 A of current?

7. Define current.

8. What is a Volt? How is it related to force?

9. Define the Watt in terms of energy, charge, and time.

10. What is the meaning of Ohm's law?

11. What is resistance? What unit is used?

12. How is conductance related to resistance?

13. Describe the Farad in terms of charge and potential difference.

14. What is inductance? How is it related to a changing magnetic field?

15. Explain what is meant by Kirchhoff's current law.

16. What is the difference between a series and a parallel circuit?

17. State Kirchhoff's voltage law for a series circuit.

18. How is power usually dissipated in an active circuit?

19. Why is a series circuit described as a voltage divider?

20. Why is a parallel circuit a current divider?

21. What is the advantage of using a Thévenin equivalent circuit?

22. State the Thévenin relationships.

23. Why should a Norton equivalent circuit be used sometimes instead of the Thévenin equivalent?

24. How are the Norton equation constants related to the Thévenin equation constants?

25. In solving a series-parallel circuit, why is it necessary to solve the parallel segment first?

PROBLEMS

1. How many Newtons of force are required to accelerate a mass of 2 kg at 5 m per second per second?

2. How much work is done when an application of 15 N of force moves an object 5 m?

3. If 100 J of energy are used in 10 s to perform some work, how many Watts of power are consumed?

4. Since 6.25×10^{18} electrons will yield 1 C of charge, how many Coulombs are there in 18.75×10^{18} electrons?

5. If 18.75×10^{18} electrons pass through a circuit in 1 s, how many Amperes of current are involved?

6. If 3 C of electrons move through a circuit using 6 J of energy, what is the applied voltage?

7. When 0.5 A of current pass through a resistance due to the application of 10 V of potential difference, how much power must the resistance dissipate or release?

8. How much voltage is needed to move 3 A through 50 Ω of resistance?

9. What is the conductance offered by 20 Ω of resistance?

10. How many Coulombs of charge may be stored in a 50-μF capacitor with 100 V applied?

11. If a capacitor will store 3.12×10^{18} electrons with 100 V applied, what is the capacitance in Farads? In microfarads?

12. If an inductance will cause a voltage of 75 V to be induced when the current changes from 1 to 4 A in 1 s, what is the inductance of the coil in Henries?

13. If a circuit uses a 5-V power supply and has 3,000-Ω resistance, how much current passes through the circuit?

14. What is the total resistance of the circuit shown?

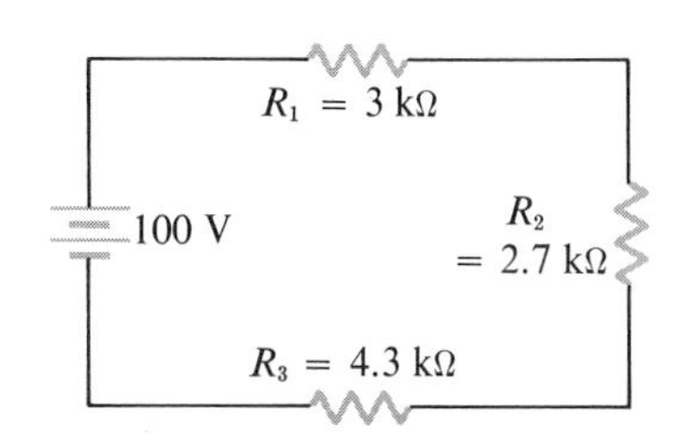

15. How much current moves through the circuit of Prob. 14?

16. What is the voltage across each resistance of the circuit in Prob. 14?

17. What is the total resistance of the circuit shown?

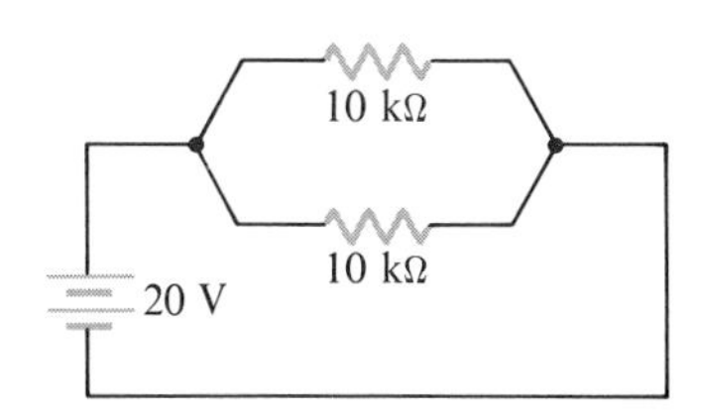

18. How much current passes through each resistance in Prob. 17?

19. What is the voltage across each resistance in Prob. 17?

20. How much power is consumed by the circuit in Prob. 17?

21. What is the total resistance of the circuit shown?

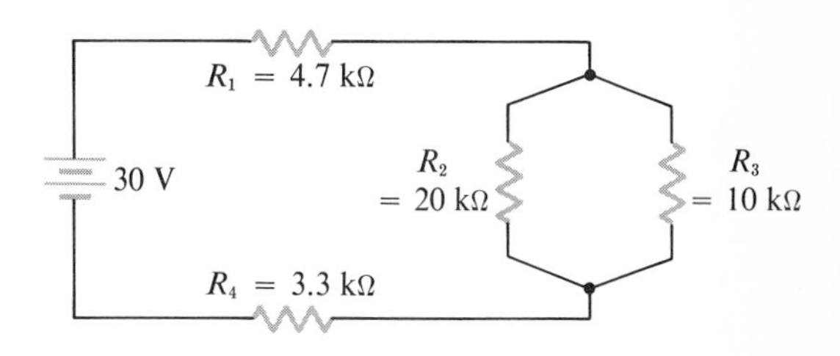

22. What is the total current of the circuit in Prob. 21? What is the current through R_2? Through R_3?

23. What is the voltage across R_2 and R_3 in the circuit of Prob. 21?

24. In a complex circuit the Thévenin voltage is 4.5 V and the Thévenin resistance is 9 kΩ. What is the voltage across the 5-kΩ load resistance? What is the load current?

25. Use Eq. (2-37) to find the Norton current of Prob. 24.

CHAPTER 3

OBJECTIVES

1. To relate the vocabulary of ac theory to that of dc theory
2. To define and relate terms such as cycle, Hertz, peak-to-peak, peak, average, root mean square, and period
3. To point out similarities and differences between alternating current and pulsating direct current
4. To develop reactance concepts from their relationship to the universal time-constant curves for *RC* and *RL* circuits
5. To demonstrate the significance of current and voltage phase relationships
6. To explore the vector development of series and parallel *RLC* circuits
7. To relate impedance and resistance
8. To develop the voltage and current concepts related to resonant conditions

CONCEPTS IN ALTERNATING-CURRENT THEORY

In the real world of electronics, a condition of pure direct current rarely exists for any length of time. At any instant of time, however, current and voltage do obey the laws set up for direct current. If the current or voltage changes value quickly, the circuit will respond during the change period in a manner that is difficult to describe from the data presented in Chap. 2. For this reason it is essential to consider the fundamental reactions of a circuit under dynamic, or changing, conditions.

MEASUREMENT OF AC VOLTAGE

You have probably noticed that during a thunderstorm there is a great deal of static and interference on the radio. The tremendous bursts of energy released during electrical storms generate "sparks" that are really high-frequency waves of energy. The question immediately comes to mind, "Does lightning discharge from the ground to the cloud, or vice versa?" The fact is that it moves in both directions at a very high rate of speed. The discharge thus alternates in direction.

Although the lightning bolt is rather difficult to handle, it does obey the same electrical and physical laws that cover the rest of the physical world. Much better known is the low-frequency alternating current that is used to light most homes. It does the same job that was done by direct current many years ago, but it is easier to handle.

Because power was consumed in the resistance of long wires when direct current was used, the homes at long distances from the generating plants found that the voltage left was too low to be useful. Most of it had been lost by the voltage-divider action of the series resistance of the wires. When alternating current was put into use, however, the voltage could be boosted by a transformer, so that the same amount of power could be transmitted to the homes at a lower current. At the end of the line another transformer was used to drop the voltage to the needed low value while boosting the current to the value needed. The procedure is no mystery, but it does require some explanation.

When current makes one complete alternation of direction, it is said to complete one *cycle*. That is to say, the voltage and current start out at zero, increase positively to some maximum value, then decrease back to zero, increase in a negative direction to some maximum value, and then return to zero. Figure 3-1 shows how this cycle of alternation is related to one turn of an ac generator.

Notice that when the coil or armature of the generator is not cutting the lines of magnetic force, no voltage or current is generated. However, when the coil is cutting at the maximum number of lines of force per second, at 90° to the field, the voltage generated is maximum; hence the current is maximum.

As the direction in which the generator armature cuts the lines of force reverses (down to up), so does the direction of the voltage that is generated. The change in direction occurs for each alternation, or

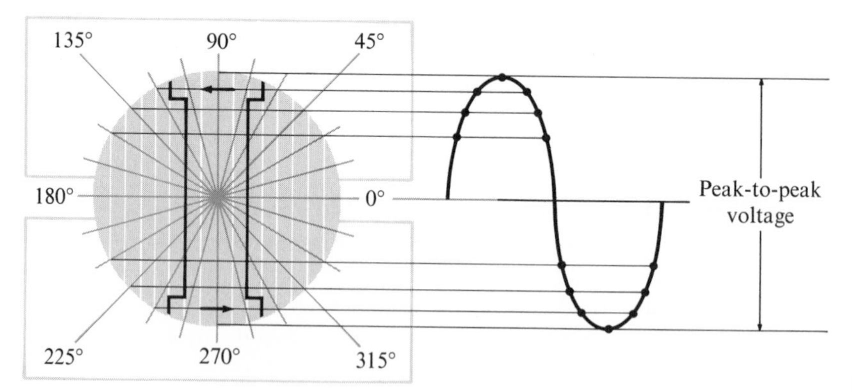

FIG. 3-1 One cycle of alternating current as it relates to the single turn of an ac generator.

cycle of operation. The cycle, like the circle, is made up of 360°. Note that at 0, 180, and 360° the current and voltage are zero. At 90 and 270° the voltage and current are at maximum values.

PEAK-TO-PEAK VOLTAGE

It is frequently necessary to measure the voltage of alternating current by means of an oscilloscope or a meter. On an oscilloscope the voltage waveform appears in a form much like that of the cycle in Fig. 3-1. Since both peaks of the waveform are easy to see, the voltage is measured from one peak to the next. This form of measurement gives the *peak-to-peak voltage*.

In the study of amplifiers and other instruments in electronics peak-to-peak measurement of voltage is employed, but it should not be confused with the measurements obtained with meters.

PEAK VOLTAGE

When an alternating current has been *rectified* or changed into a pulsating direct current with all pulses going in one direction, the oscilloscope will display a waveform pattern like that in Fig. 3-2. Since only one direction is used, with alternating current rectified to direct current, the measurement is called *peak* rather than peak to peak.

The peak voltage measurement is important, especially in dealing with power supplies, since the capacitors used to filter the pulsations and produce smooth dc charge up to this value. Again, peak voltage cannot be measured directly with a meter under normal circumstances, but it may be observed on an oscilloscope.

ROOT-MEAN-SQUARE VOLTAGE

Perhaps the most commonly used measurement of alternating voltage and current is *root-mean-square* (rms) measurement. This is because it is the dc equivalent of the ac waveform.

Once again, note that each cycle of alternating current is made up of 360° of rotation, and that in the first 90° the voltage rises to maximum value. This waveform is called the *sine wave* because it increases and decreases according to a mathematical expression called the *sine function*. In case you are not familiar with sines, every right-angle triangle has two angles that are less than 90°. When the angle opposite the right angle is 45°, the other acute angle is also 45°

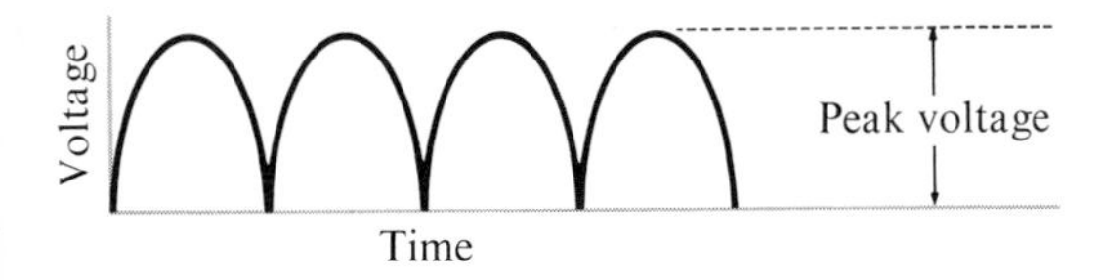

FIG. 3-2 Full-wave rectified pulsating direct current and peak-voltage measurement.

(one-half of 90°). When the length of the side opposite either 45° angle is divided by the side opposite the right angle, the hypotenuse, the resulting number is always 0.707. This number (or ratio) is the *sine of 45°*.

Since the 45° is just half of 90°, which is a point of maximum voltage or current, it just happens that the power at this point in the rotation of the generator is also one-half the maximum value. Thus the voltage at which the power is one-half the maximum value may be determined by multiplying the maximum, or peak, voltage by the sine of 45°, or 0.707. If the peak voltage is 10 V, the rms voltage must be $0.707 \times 10 = 7.07$ V.

Since the power consumed by a resistance is the same with 7.07 V rms as it is with 7.07 V dc, the rms voltage has come to be the most common measurement for describing alternating current and voltage. It is difficult to describe rms voltage accurately, however, unless the voltage happens to be sine wave in origin. Meters can be calibrated to read rms voltage on their ac scales, but most of them actually measure the *average* voltage.

AVERAGE VOLTAGE

The average voltage of an alternating sine wave is about 63.6 percent of the peak voltage, as indicated in Fig. 3-3. This value may be determined in several ways. One method is to take the value of each point on the sine wave, add the values, and take an average. The algebraic method is somewhat simpler.

Recall that the circumference of a circle is $2\pi r$, or 2×3.14 times the radius. Thus one cycle is made up of two half-cycles, each of which is equal to 3.14 times the radius. The angular equivalent of this is 2π (rad).

The average voltage in one half-cycle is found by taking the average of the sum of the voltages at each degree of rotation of the cycle. This is the same as dividing the peak voltage by π, or 3.14. For the average voltage of the entire cycle this value is multiplied by 2.

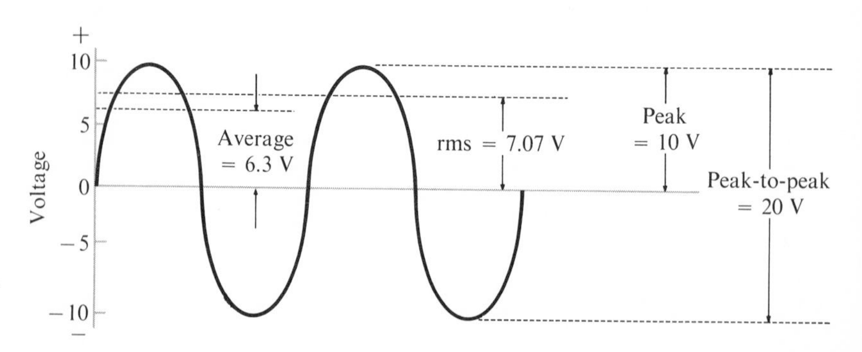

FIG. 3-3 A comparison of the peak-to-peak voltage to peak, rms, and average voltage.

The average voltage for one half-cycle is 0.318 times the peak voltage, and for the *whole* cycle it is 0.318 times the peak-to-peak voltage or 0.636 times the peak voltage. That is,

$$V_{rms} = \sin 45° \times V_p = 0.707 \times V_p \qquad (3\text{-}1)$$

$$V_{av} = \frac{V_{p\text{-}p}}{\pi} = \frac{V_{p\text{-}p}}{3.14} = V_{p\text{-}p} \times 0.318$$

$$= V_p \times 2 \times 0.318 = V_p \times 0.636 \qquad (3\text{-}2)$$

PULSATING DIRECT CURRENT

Alternating current periodically reverses direction and polarity, but the average voltage is zero, since it goes just as high in the negative direction as it does in the positive direction. Frequently the change in current will *look* the same when viewed on an oscilloscope, but the average current or voltage will not be zero. When the direction of current does not reverse, even though the waveform still appears as a sine wave, it is called *pulsating direct current*, and not alternating current. Figure 3-4 shows the difference between the two. Note that a 20-V peak-to-peak ac voltage is actually 0 V ± 10 V, whereas the pulsating dc voltage of 20 Volts peak to peak is actually 10 V ± 10 V. At no time does the pulsating dc voltage drop below zero, or become a negative value.

The effect of alternating current and pulsating direct current may be very similar; however, whenever a dc level is present in electronic circuits it must be isolated from the dc level of the next circuit by means of a capacitor unless the two levels are the same. Amplifiers operate to amplify ac voltages by using their pulsating-dc counterparts, as we shall see in Chap. 8.

THE MAGNETIC FIELD

When a conductor is carrying current, there is always a *magnetic field* around that conductor (see Fig. 3-5). The direction of the magnetic

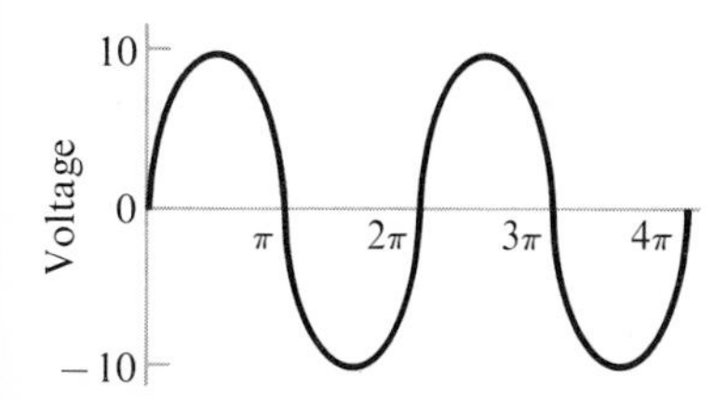

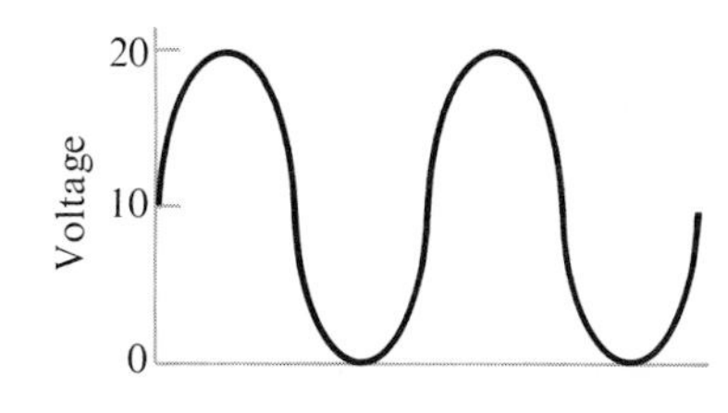

FIG. 3-4 A comparison of alternating current and pulsating direct current.

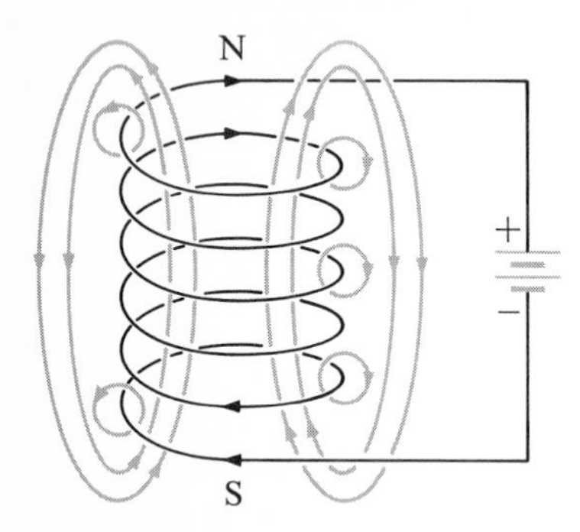

FIG. 3-5 The magnetic field associated with an inductor carrying current.

field follows the *left-hand rule*; that is, if the conductor is grasped so that the thumb of the left hand points in the direction of the electron movement (negative to positive), the direction of the fingers around the conductor indicate the direction of the magnetic field.

As the current increases, more lines of force are generated and the magnetic field expands. The lines develop from the center of the conductor and are forced outward, because like polarities oppose. The magnetic field strength is a function of the current, not of the voltage. *The greater the current* (electrons per second), *the stronger the magnetic field.* Thus a few electrons traveling at high speed may cause the same magnetic field as lots of electrons traveling at low speeds. As the current decreases, the magnetic field decreases accordingly. If the current changes direction, the magnetic field will also change direction.

When a conductor is wound into a coil, the lines of force in the magnetic field tend to reinforce each other, that is, to strengthen the field. If the coil is held in the left hand with the fingers going in the direction of the current movement in each turn, the thumb will point to the direction in which the reinforced magnetic field emerges from the end of the coil. This is known as the *north pole.* Because of their repulsion property, the lines of force that emerge from the north pole do not cross other lines of force, but instead return to the coil at the opposite end, or south pole.

When a magnetic field is expanding or contracting, the lines of force cut across the turns of the coil. The interaction of the lines of force with the electrons of the conductor is such that it moves them in a direction at right angles to the direction of cutting—in other words, through the conductor. The current that moves is the result of the *induced voltage.* The faster the magnetic field cuts the conductor, the higher the induced voltage.

THE UNIVERSAL TIME CONSTANT

In order to understand the operation of a circuit using alternating current we must now consider the sudden-change effects of direct current. They are generally called *transient effects,* since they are only passing through. Transient effects in a resistive circuit are similar to normal dc effects. When the circuit contains capacitance or inductance, however, there is a short period of time during which the capacitor must charge or the inductive magnetic field must expand. This time lag follows the same curve for both resistance-capacitance and resistance-inductance circuits; hence it is known as the *universal time constant.*

THE RESISTANCE-CAPACITANCE (*RC*) CIRCUIT

When a circuit contains some capacitance in series with the resistance, as soon as voltage is applied to the circuit, current will rush to charge up the capacitor. The only factor limiting the current flow at the instant the switch is closed is the resistance of the circuit. However, as shown in Fig. 3-6, as current moves into the capacitor, the collection of charges soon develops a potential difference equal to that of the voltage supply. The resistance simply limits the charge development to a longer time because it limits the current flow.

It can be shown mathematically from Ohm's law that the product of the resistance and capacitance of the circuit has the dimensions of time, expressed in seconds. When the resistance, expressed in Ohms, is multiplied by the capacitance, in Farads, the product is *one time constant,* defined as *the time it takes for the capacitor to charge to 63.2 percent of full charge, the charge or voltage of the supply.*

During the second time-constant period the capacitor will charge 63.2 percent of the remaining 36.8 percent, or up to 86.4 percent of full charge. During the third time-constant period the charge increases 63.2 percent of the remaining voltage, or up to 95 percent. At the end of four time constants the voltage is at 98.2 percent, and after five time constants, at 99.4 percent, which for all practical purposes is full charge. Thus

$$\text{Time constant} = \text{resistance} \times \text{capacitance} = RC \qquad (3\text{-}3)$$

$$\text{Charge time} = 5 \text{ time constants} = 5\,RC \qquad (3\text{-}4)$$

For the circuit of Fig. 3-6 the length of one time constant is 1 s, and so the charge time is 5 s,

$$\text{Time constant} = 50\ \text{k}\Omega \times 20\ \mu\text{F} = 5 \times 10^4 \times 2 \times 10^{-5}$$

$$= 10 \times 10^{-1} = 1\ \text{s}$$

$$\text{Charge time} = 5\,RC = 5 \times 1\ \text{s} = 5\ \text{s}$$

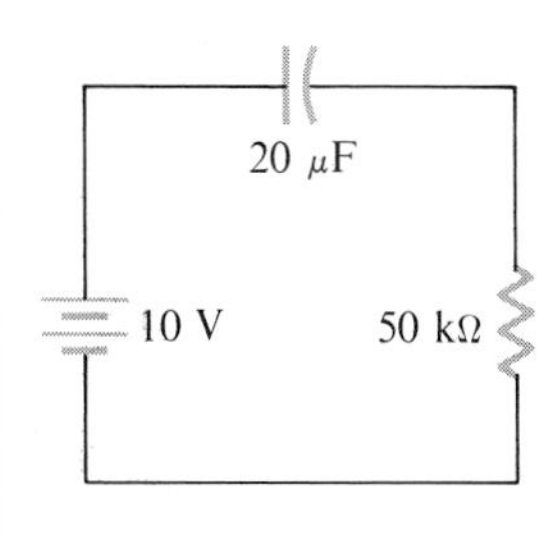

(*a*)

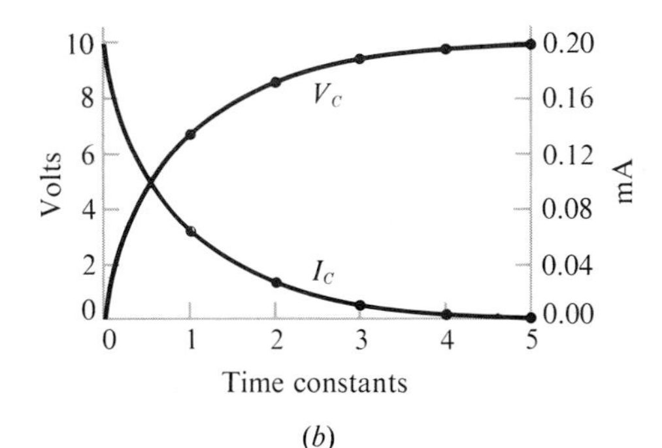

(*b*)

FIG. 3-6 (*a*) The *RC* circuit and (*b*) its charge-time-constant curves.

Generally the capacitances and resistances used are much smaller than these. When the resistance is 100 Ω and the capacitance is 0.002 μF, one time constant is only

$$\text{Time constant} = 1 \times 10^2 \times 2 \times 10^{-9} = 2 \times 10^{-7}\ \text{s}$$
$$= 0.2 \times 10^{-6} = 0.2\ \mu\text{s}$$

This is 0.2 millionths of a second, $2 \times 10^{-7} = 200 \times 10^{-9}$, or 200 *nanoseconds* (ns). The charge time is $5 \times 0.2 \times 10^{-6}$ s, or 1 μs.

It should be emphasized that as the voltage on the capacitor approaches full supply voltage, the current drops to zero. The current that flows in the circuit at any instant depends on the potential difference between the supply voltage and the voltage on the capacitor. The voltage on the capacitor is thus given by

$$V = \frac{Q}{C} \qquad (3\text{-}5)$$

where Q = charge stored, Coulombs
C = capacitance, Farads

In Fig. 3-6 the 20-μF capacitor at full charge after 5 s has about 10 V across it. The number of Coulombs of charge needed to produce this voltage may be determined by rearranging Eq. (3-5),

$$Q = CV$$
$$= 20 \times 10^{-6}\ \text{F} \times 10\ \text{V} = 200 \times 10^{-6}$$
$$= 2.0 \times 10^{-4}\ \text{C} \qquad (3\text{-}6)$$

Since 1 C of electrons is 6.25×10^{18}, the number of electrons stored in the capacitor must be about

$$Q = (6.25 \times 10^{18}) \times 2 \times 10^{-4}$$
$$= 12.5 \times 10^{14}\ \text{C}$$

This enormous number of electrons is stored in the capacitor in only 5 s. The current is limited to a maximum of 0.20 mA (0.0002 C/s) at the beginning and drops to zero at the rate indicated by the curve of Fig. 3-6.

In review, at the beginning of the first time constant the capacitor voltage is zero and the current is maximum. After five time constants the voltage is maximum and the current is zero. The shape of the curve is universal in terms of time constants, but the time of each constant and the voltage change.

THE RESISTANCE-INDUCTANCE (*RL*) CIRCUIT

When an inductance is in a circuit, it takes some time for the circuit to stabilize, just as in the *RC* circuit. However, the transient effect during this short time is the opposite of that in the *RC* circuit. The voltage developed across the coil is *maximum at time zero,* so that the *potential difference* between the source and the coil is zero. The current, as a result, is zero at time zero.

In the dc circuit of Fig. 3-7, as soon as the switch is closed the rate of expansion of the magnetic field is the greatest and the induced voltage is highest, equal to the supply voltage. The current at this instant is zero. The rate of expansion decreases, however, as the final dc position of the magnetic field is approached. When the magnetic field is no longer cutting through the conductor, the induced voltage is zero. At this point the potential difference between the supply voltage and the induced voltage is again equal to the supply voltage, and the current is at maximum.

During one time constant, the period of time needed for the current to increase to 63.2 percent of its maximum value, the induced voltage is decreasing to 36.8 percent of its maximum value. As with the *RC* circuit, the charge time for the *RL* circuit is five time constants. However, the manner of change is not the same.

Two factors control the length of time of the time constant: the amount of inductance and the resistance of the circuit, which limits the current flow to some maximum value. Since the current is limited by

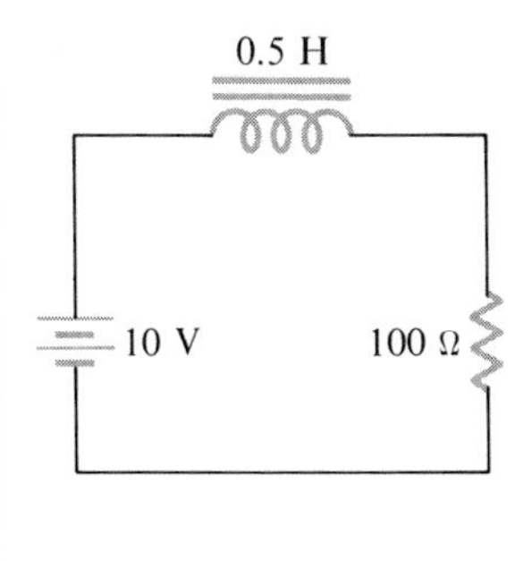

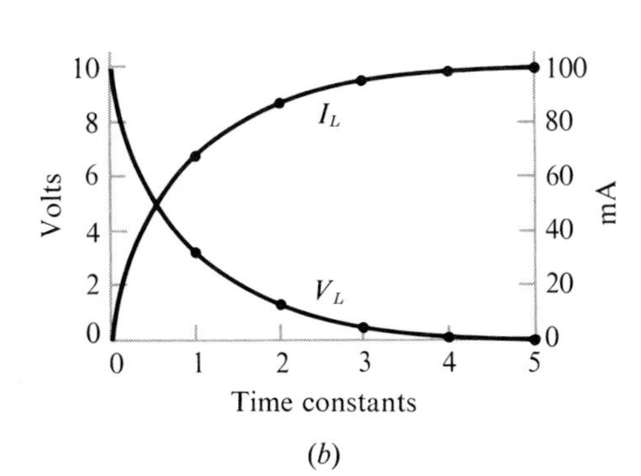

FIG. 3-7 (*a*) The *RL* circuit and (*b*) the universal time-constant curves.

the resistance of the coil, as well as by the resistance of the rest of the circuit, the rate of expansion of the magnetic field through the inductance is correspondingly faster. For larger values of inductance the magnetic field is also increased. Therefore the length of time of one time constant depends directly on the size of the inductance and inversely on the size of the resistance. Since the resistance limits the current flow, the larger the resistance, the faster the rate of change and the sooner the current reaches its maximum, but lower, value. Stated in simple terms, current can reach a low value faster than it can reach a high value.

The length of one time constant in an *RL* circuit, then, is

$$\text{Time constant} = \frac{L}{R} = \frac{0.50\ \text{H}}{100\ \Omega} = 0.005\ \text{s} \qquad (3\text{-}7)$$

$$\text{Charge time} = 5\frac{L}{R} = 5 \times 0.005 = 0.025\ \text{s}$$

AC CIRCUIT PHASE RELATIONSHIPS

If, instead of a dc supply, an ac generator is used, the difference in time at which the induced voltages and currents in inductive and capacitive circuits reach maximum becomes important. Since the induced voltage and the current cannot both be at maximum at the same time, they are said to be *out of phase*.

In the *RC* circuit of Fig. 3-6 we saw that when the current is maximum the capacitor voltage is zero, and conversely, at time zero it is the current that is maximum. When the points are plotted in terms of the ac waveform for 0, 90, 180, 270, and 360°, for both current and voltage, the result is as shown in Fig. 3-8.

When the current curve is extended backward to the point at which it crosses the zero axis, it can be seen that, in terms of time, the current is ahead of the capacitor voltage that opposes it by a quarter-

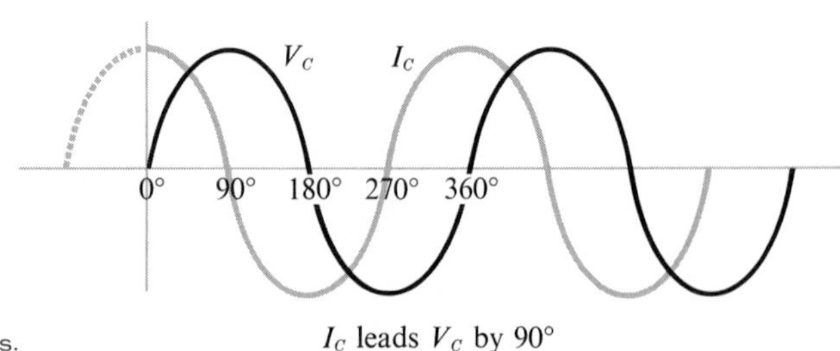

FIG. 3-8 Current-voltage *RC* phase relationships.

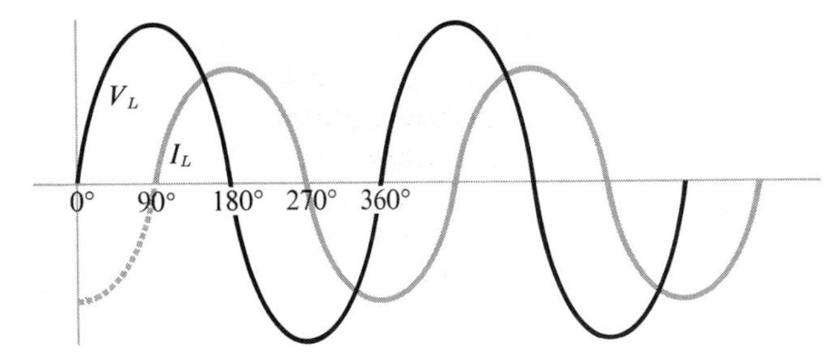

FIG. 3-9 Voltage-current *RL* phase relationships.

cycle, or 90°. That is, the current is *leading* the voltage by 90°. Note that we are discussing the *countervoltage*, or *back voltage* developed on the capacitor, and not the supply voltage. This phenomenon is found in every ac circuit containing any capacitance. Since two conductors separated by some insulating dielectric make up a capacitor, some tiny amount of capacitance exists in the best circuits.

In the *RL* circuit of Fig. 3-7 we saw that at time zero the induced voltage is maximum and the forward current is zero. Conversely, at the time of full charge the induced voltage that opposes the forward current is zero, and so the current is maximum. When these points are plotted for the ac situation, as indicated in Fig. 3-9, the resulting curves are just the opposite of the *RC* curves. The induced voltage leads the forward current by 90°, or one quarter-cycle.

Note that this is the voltage-current relationship across only the coil—more specifically, across the inductance of the coil. For a pure inductance the phase difference would be 90°. However, since the coil has resistance, and for a resistance the voltage and current are *in phase*, or occur at the same time, the phase shift will not be exactly 90°. It is this resistance problem that makes experiments with time constants difficult, especially when the techniques are unfamiliar.

ANGULAR VELOCITY

In our discussion so far we have dealt with both dc and ac voltages. It is possible, however, to treat any current or voltage as dc for some specific instant in time. It is not difficult to calculate the voltage of an ac waveform at some point of generator rotation, say, 40° after the waveform passes through zero. However, if we employ the unit of time, the second, instead of degrees of rotation, the computation becomes somewhat more complex. It is then necessary to know the speed with which the alternator or ac generator is turning.

Since the alternator is rotating, the speed of the armature may be measured in two ways: in revolutions per minute or in degrees per second. For most computations the best unit of measurement is developed from *angular velocity, the number of degrees of rotation per second.* For a complete picture we must delve further into the concept of the *radian.*

The *radian* is based on the radius of a circle, which is, of course, half the diameter. If we were to take the exact length of some radius and lay it out around the perimeter of the circle, we would find that the circumference is 6.28 times the radius. Since 6.28/2 is 3.14, or π, this is the familiar expression $2\pi r$, which gives the distance around the circumference in linear units. We would also find that the radius length covered 57.3° of arc. A radian, then, is the *number of degrees the generator rotates as the outermost point of its armature moves a distance equal to one radius, or 57.3°.*

If one radian is 57.3°, then the circumference of the circle is 6.28 rad ($6.28 \times 57.3 = 360°$). Mathematically this expression is generally stated as 2π rad, which is the equivalent in *angular* units of $2\pi r$.

Once we know that one alternation, or one cycle, is 2π rad, then we can determine the angular velocity. The dimensions of angular velocity are degrees and time, and translated into units this becomes radians per second. If there were only one cycle per second, the angular velocity would be simply 2π rad/s, or 2π rad per cycle times *frequency*, the number of cycles per second. This *angular velocity* ω is usually expressed mathematically as

$$\omega = 2\pi f \qquad (3\text{-}8)$$

where f = frequency

Frequency is expressed in *Hertz* (Hz). For a frequency of 400 Hz, the angular velocity ω would be $6.28 \times 400 = 1{,}256$ rad/s.

The significance of the angular velocity lies in the fact that the voltage developed at any instant on either a capacitor or an inductance depends on how fast the supply voltage and current are changing.

REACTANCE

Ordinary resistance should be distinguished at this point from the reactance of a circuit. Resistance is a property of a conductor by which the positive core of the atoms of the material strongly attract the outermost electrons. The more strongly the electrons are attracted or bound into the atom or molecule, the greater the material's resistance. Thus resistance limits current flow, or opposes the motion of the electrons from the negative potential to the positive potential.

However, other forms of opposition may also be present in any circuit. For example, if the supply voltage is 10 V and the circuit has a capacitor with an instantaneous voltage of 5 V in such a direction as to oppose the supply voltage, the net current at that moment behaves as if it were limited by some resistance. In reality the limiting factor is the potential difference for the total circuit, 10 V − 5 V = 5 V. This limiting factor, which is actually a countervoltage, is called *reactance*, usually denoted by X; like resistance, it is expressed in Ohms.

Reactance is commonly considered in two forms: the *capacitive reactance* just mentioned and its counterpart, *inductive reactance*. We shall explore these two reactances in terms of the time-constant concept. This in turn will clarify the need to observe phase relations.

CAPACITIVE REACTANCE

In a circuit such as that of Fig. 3-10, where the supply voltage is from an ac generator, the capacitor has an opportunity to absorb a charge only during the first quarter-cycle, or 90°. During the next 90° the supply voltage drops to zero, and during the third 90° it reverses directions, so that the capacitor must attempt to charge in the opposite direction.

Since the supply voltage is constantly changing, the capacitor is never able to charge completely. However, if the capacitance is very small and the frequency of alternation is very slow, the capacitor will develop a fair amount of countervoltage. As the frequency is increased, the time per quarter-cycle decreases, so that the amount of charge also decreases. If the size of the capacitor is increased, it also takes longer to accumulate full charge. Thus there are two variables: the angular velocity of the generator and the size of the capacitor. When either of these variables increases, the amount of countervoltage decreases.

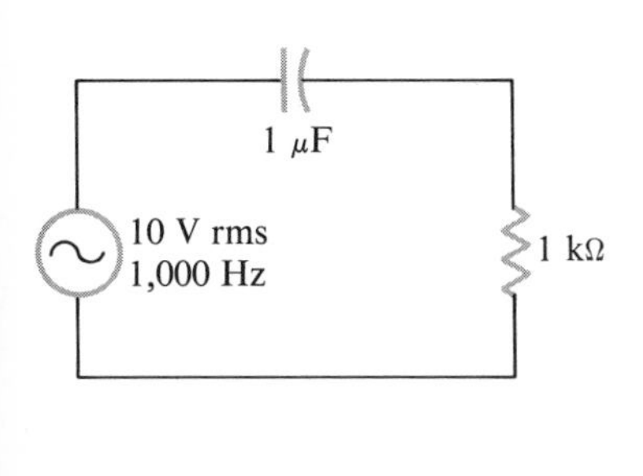

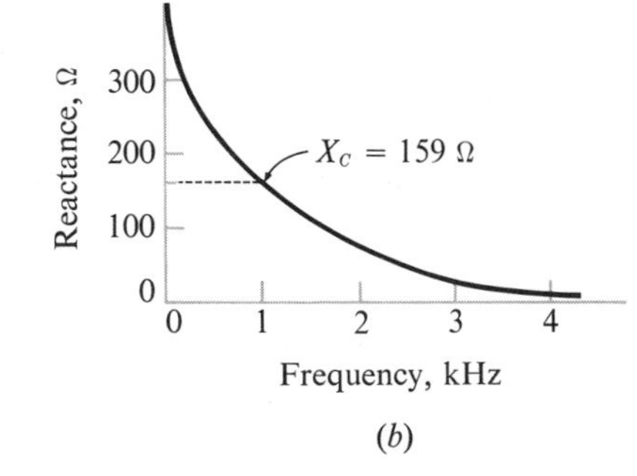

FIG. 3-10 (*a*) The *RC* circuit using alternating current, and (*b*) the amount of reactance compared to frequency of operation.

This inverse relationship indicates the form of the mathematical expression for *capacitive reactance* X_C,

$$X_C = \frac{1}{2\pi fC} = \frac{1}{\omega C} \tag{3-9}$$

where C = Farads
f = Hertz
$\omega = 2\pi f$

Capacitive reactance, then, is the effective opposition voltage offered by a capacitor in an ac circuit, but over the entire cycle, not just at some point in time. It increases as the frequency and capacity decrease, which allows greater accumulation of charge per unit time.

In the circuit of Fig. 3-10 the reactance to the supply voltage is 159 Ω at 1,000 Hz,

$$\begin{aligned} X_C &= \frac{1}{2\pi fC} \\ &= \frac{1}{6.28 \times 1{,}000 \times 1 \times 10^{-6}} = \frac{0.159}{1 \times 10^{-3}} \\ &= 0.159 \times 10^{3} = 159\ \Omega \end{aligned} \tag{3-10}$$

Since 1/6.28 is a constant and has a value of 0.159, Eq. (3-9) reduces

$$X_C = \frac{0.159}{fC} \tag{3-11}$$

In review, capacitive reactance is caused by the effect of the countervoltage developed on a capacitor during that part of the cycle when the supply voltage is increasing. As such, the reactance is accompanied by a phase shift between the circuit current and this countervoltage.

CAPACITIVE PHASE SHIFT

According to the information plotted in Figs. 3-8 and 3-6, the voltage developed on a capacitor during charge does not increase and decrease in step with the current in the circuit. As a matter of fact, in a purely

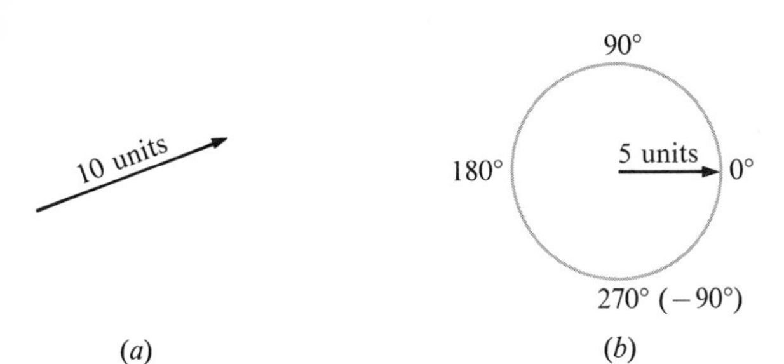

FIG. 3-11 (a) A vector of 10 V, (b) a vector of 5.0 mA at 0° phase shift.

capacitive circuit with an ac source the voltage and current are out of phase by 90°, with the current leading the voltage.

The capacitive phase shift is a function of the capacitive reactance, and not of the resistance. For this reason, since all circuits contain some amount of resistance, the total phase shift of the circuit is always less than the 90° expected.

The simplest way to illustrate phase shift is in terms of vectors. A vector is symbolized by an arrow of some specific length, pointing in a specific direction. A quantity such as 10 V, for example, would be represented by an arrow whose length is 10 units. A current of 5 mA with 0° of phase shift would be represented by an arrow five units long pointing in the direction of 0°. Since vector arrows can indicate both quantity and direction, they afford a simple means of comparing variables that may differ in two dimensions. For example, in the circuit of Fig. 3-12*a*, suppose the voltage is measured as 10 V in phase with the current in the resistance and 5 V when it is 90° out of phase with the current in the capacitance. The two voltages can be compared vectorally by constructing two arrows, one 10 units long and the other 5 units long, pointing in a 90° angle from each other as shown in Fig. 3-12*b*.

As Fig. 3-12 indicates, when two voltages are in different directions, or out of phase with each other, there is a net voltage at some

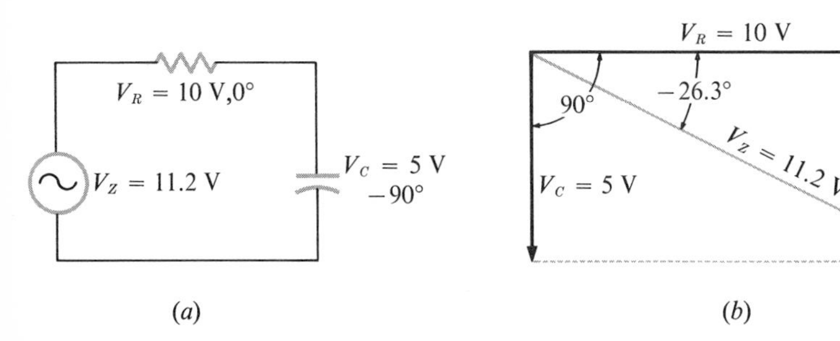

FIG. 3-12 (a) An *RC* ac circuit, and (b) vector comparison of its component voltages.

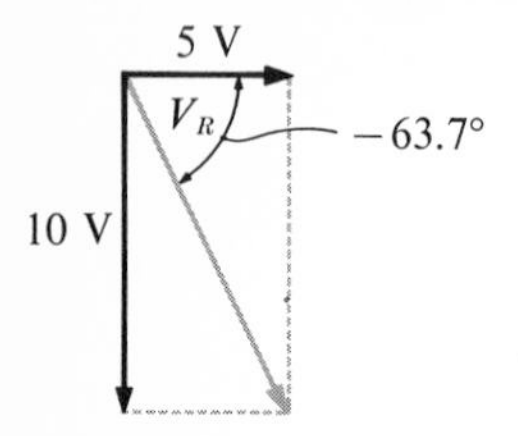

FIG. 3-13 Vector diagram with the phase angle approaching −90°.

angle other than 90°, but less than 90°. If a parallelogram is constructed, as in Fig. 3-12*b*, it can be seen that the resulting voltage is greater than 10 units but less than 15 units. By application of the Pythagorean theorem, this voltage can be computed as about 11.2 V.

If the voltages in Fig. 3-12 were reversed, with the 10 V across the capacitor and 5 V across the resistance, the vector diagram would be as shown in Fig. 3-13. It can be seen that as the ac voltage across the capacitor increases as a result of the increased reactance, the resulting total out-of-phase voltage shifts to approach −90°. Although there is resistance in the circuit, the resultant can never reach the absolute −90° limit. It will, however, always lag behind the current of the circuit.

INDUCTIVE REACTANCE

The inductive circuit of Fig. 3-14 may be analyzed in much the same way as the capacitive circuit of Fig. 3-10. As current starts to move through the inductance, the magnetic field begins to expand. However, expansion in one direction can occur only while the current is increasing—that is, during the first quarter-cycle. During the second quarter-cycle, when the current is decreasing, the magnetic field is collapsing. When the alternating current starts to move in the opposite direction the magnetic field again expands, but with the opposite polarity. The net result is that the magnetic field is able to expand only during the first and third quarter-cycles.

Since the voltage that is induced into a coil of wire depends on the rate at which the lines of force cut across the wire, the *frequency* of the alternating current is important. The magnetic field takes longer to expand to its limit at a low frequency than at a high frequency. It therefore follows that the voltage induced into a coil operating at a low frequency, such as 60 Hz, would be much less than the induced voltage at say, 1,000 Hz.

The induced voltage is always in such a direction as to oppose the forward movement of current through the coil of wire and therefore resembles resistance. However, it is not resistance, but a counter-voltage, in this case called *inductive reactance*. The fact that this

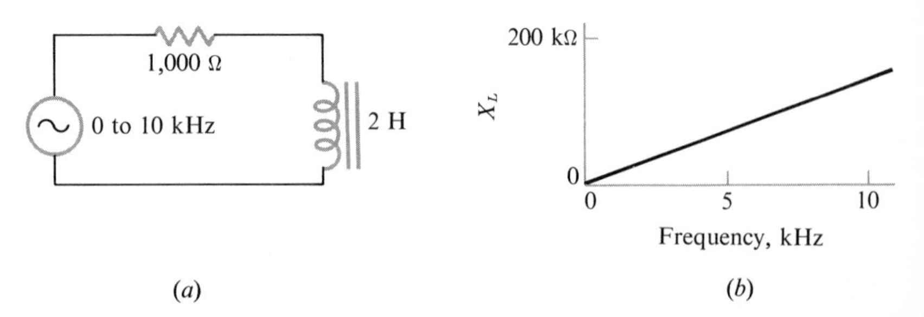

FIG. 3-14 (*a*) The *RL* circuit using alternating current, and (*b*) the amount of reactance in comparison to the frequency of operation.

countervoltage increases with higher frequencies makes it easy to develop the mathematical expression for reactance.

As for capacitive reactance, the angular velocity of the generator must be known in order to determine the rate of cutting of the coil by the moving lines of force. This rate is, of course, related to the frequency. Since one complete rotation is 2π rad, and frequency is in rotations per second, the angular velocity $\omega = 2\pi f$ is expressed in radians per second.

The actual number of lines of force associated with a given current in a coil depends on the inductance of the coil itself, expressed in Henrys. Therefore the expression for inductive reactance X_L becomes

$$X_L = 2\pi f L = \omega L \qquad (3\text{-}12)$$

where $f =$ Hertz
$L =$ Henry
$\omega = 2\pi f$

Inductive reactance, then, is the effective opposition voltage offered by an inductor in an ac circuit, but over the entire cycle, not at just one point in time. It increases linearly with an increase in frequency, as indicated in Fig. 3-14. The equation shows that as the inductance is increased, the reactance also increases.

If a simple circuit such as that of Fig. 3-14 were analyzed to determine the inductive reactance at 1,000 Hz, the resistance of the coil would be disregarded, and only the inductance would be considered. That is,

$$\begin{aligned} X_L &= 2\pi f L \\ &= 2 \times 3.14 \times 1{,}000 \text{ Hz} \times 2 \text{ H} \\ &= 6.28 \times 2 \times 10^3 = 12.56 \text{ k}\Omega \end{aligned} \qquad (3\text{-}13)$$

This represents the countervoltage induced in the coil, and it leads the current by 90°. As with capacitive reactance, this countervoltage opposing the forward flow of current is expressed in Ohms.

INDUCTIVE PHASE SHIFT

As we have seen, in a reactive circuit the countervoltage, or reactance, is out of phase with the current in the circuit. It leads the current by a phase angle of up to 90°, or a quarter-cycle. It is not possible to isolate an inductance from the resistance of the wire with which it is

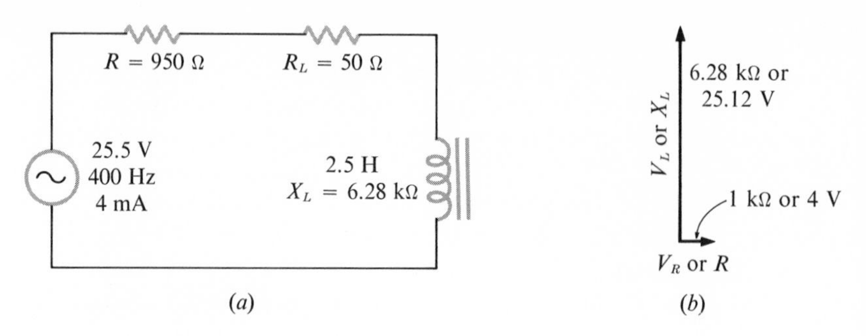

FIG. 3-15 (*a*) The true *RL* circuit and (*b*) its phase relationships.

wound, and so the phase angle can never reach 90° for the entire circuit. However, if the inductance and resistance are considered separately, as in Fig. 3-15, the inductive reactance is found to be 90° out of phase with the total resistance.

The inductive reactance in this case is

$$\begin{aligned} X_L &= 2\pi f L \\ &= 6.28 \times 400 \text{ Hz} \times 2.5 \text{ H} \\ &= 6.28 \text{ k}\Omega \end{aligned} \tag{3-14}$$

This reactance represents a voltage that leads the current by 90°, so that a vector diagram can be plotted showing both the resistance and the reactance. From Ohm's law, voltage is proportional to resistance; hence a vector comparison of the voltage across the resistance and the countervoltage is the same as comparing the size of the resistance to the size of the inductive reactance.

IMPEDANCE

The amount of current that passes through a complete circuit powered by alternating-current sources depends not only on the resistance of the circuit, but also on the capacitive and inductive reactances. This *total* opposition to current is called *impedance Z*. In order to determine the impedance of a circuit it is necessary to know the amount of resistance, capacitive reactance, and inductive reactance present.

Figure 3-16 shows the total resultant opposition to current flow, Z, as the vector sum of X_L and R. The amount of capacity in this circuit is so low as to be ignored. However, it should be remembered that all circuits have some inductance and some capacitance, and that at some frequency these amounts may need to be accounted for.

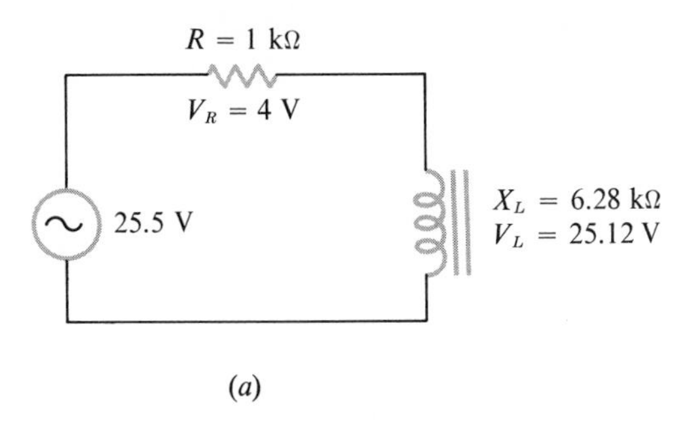

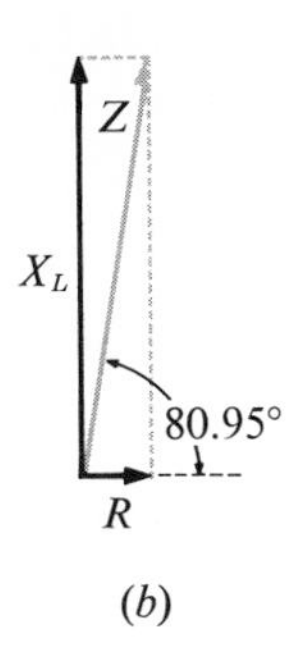

FIG. 3-16 (*a*) A series *RL* circuit and (*b*) its phase relationships of X_L, R, and Z, the total resultant opposition to current.

IMPEDANCE AND THE PHASE ANGLE

The actual phase angle between the supply voltage and the voltage across the resistance or the inductance may be computed from the trigonometric functions in Table 3-1. It should be emphasized that the resultant voltage in the circuit is the supply voltage, and so the sum of the inductive and resistive voltages must be a vector sum. Consider Fig. 3-16. Note that the voltage across the inductance is 25.12 V and the voltage across the total resistance is 4 V. If these two values were added, the total voltage would appear to be 29.12 V. The supply voltage, however, is only 25.5 V.

If this problem is plotted on a large graph, the values can be measured directly from the graph. It is much easier, however, to apply simple trigonometry, since the sides of any 90° triangle have a fixed mathematical relationship regardless of the length of any of the sides. In this case the sine of the phase angle is the inductive voltage divided by the resultant or supply voltage V_Z, and the tangent of the phase angle is the inductive voltage divided by the resistive voltage. Thus, if α = phase angle, then

$$\sin \alpha = \frac{V_L}{V_Z} = \frac{X_L}{Z} \qquad (3\text{-}15)$$

$$\tan \alpha = \frac{V_L}{V_R} = \frac{X_L}{R} \qquad (3\text{-}16)$$

First let us determine what the actual phase angle is for Figs. 3-15 and 3-16. Either the voltages or the reactance and resistance may be used, since they are proportional. From Eq. (3-16),

$$\tan \alpha = \frac{X_L}{R} = \frac{6.28\ \text{k}\Omega}{1.00\ \text{k}\Omega} = 6.28 \qquad (3\text{-}17)$$

TABLE 3-1 Trigonometric functions

A°	RADIANS	SIN A	COS A	TAN A	CTN A	SEC A	CSC A	RADIANS	A°
0	.0000	.0000	1.0000	1.0000	....	1.000	....	1.5708	90
1	.0175	.0175	.9998	.0175	57.29	1.000	57.30	1.5533	89
2	.0349	.0349	.9994	.0349	28.64	1.001	28.65	1.5359	88
3	.0524	.0523	.9986	.0524	19.08	1.001	19.11	1.5184	87
4	.0698	.0698	.9976	.0699	14.30	1.002	14.34	1.5010	86
5	.0873	.0872	.9962	.0875	11.43	1.004	11.47	1.4835	85
6	.1047	.1045	.9945	.1051	9.514	1.006	9.567	1.4661	84
7	.1222	.1219	.9925	.1228	8.144	1.008	8.206	1.4486	83
8	.1396	.1392	.9903	.1405	7.115	1.010	7.185	1.4312	82
9	.1571	.1564	.9877	.1584	6.314	1.012	6.392	1.4137	81
10	.1745	.1736	.9848	.1763	5.671	1.015	5.759	1.3963	80
11	.1920	.1908	.9816	.1944	5.145	1.019	5.241	1.3788	79
12	.2094	.2079	.9781	.2126	4.705	1.022	4.810	1.3614	78
13	.2269	.2250	.9744	.2309	4.331	1.026	4.445	1.3439	77
14	.2443	.2419	.9703	.2493	4.011	1.031	4.134	1.3265	76
15	.2618	.2588	.9659	.2679	3.732	1.035	3.864	1.3090	75
16	.2793	.2756	.9613	.2867	3.487	1.040	3.628	1.2915	74
17	.2967	.2924	.9563	.3057	3.271	1.046	3.420	1.2741	73
18	.3142	.3090	.9511	.3249	3.078	1.051	3.236	1.2566	72
19	.3316	.3256	.9455	.3443	2.904	1.058	3.072	1.2392	71
20	.3491	.3420	.9397	.3640	2.747	1.064	2.924	1.2217	70
21	.3665	.3584	.9336	.3839	2.605	1.071	2.790	1.2043	69
22	.3840	.3746	.9272	.4040	2.475	1.079	2.669	1.1868	68
23	.4014	.3907	.9205	.4245	2.356	1.086	2.559	1.1694	67
24	.4189	.4067	.9135	.4452	2.246	1.095	2.459	1.1519	66
25	.4363	.4226	.9063	.4663	2.145	1.103	2.366	1.1345	65
26	.4538	.4384	.8988	.4877	2.050	1.113	2.281	1.1170	64
27	.4712	.4540	.8910	.5095	1.963	1.122	2.203	1.0996	63
28	.4887	.4695	.8829	.5315	1.881	1.133	2.130	1.0821	62
29	.5061	.4848	.8746	.5543	1.804	1.143	2.063	1.0647	61
30	.5236	.5000	.8660	.5774	1.732	1.155	2.000	1.0472	60
31	.5411	.5150	.8572	.6009	1.664	1.167	1.942	1.0297	59
32	.5585	.5299	.8480	.6249	1.600	1.179	1.887	1.0123	58
33	.5760	.5446	.8387	.6494	1.540	1.192	1.836	.9948	57
34	.5934	.5592	.8290	.6745	1.483	1.206	1.788	.9774	56
35	.6109	.5736	.8192	.7002	1.428	1.221	1.743	.9599	55
36	.6283	.5878	.8090	.7265	1.376	1.236	1.701	.9425	54
37	.6458	.6018	.7986	.7536	1.327	1.252	1.662	.9250	53
38	.6632	.6157	.7880	.7813	1.280	1.269	1.624	.9076	52
39	.6807	.6293	.7771	.8098	1.235	1.287	1.589	.8901	51
40	.6981	.6428	.7660	.8391	1.192	1.305	1.556	.8727	50
41	.7156	.6561	.7547	.8693	1.150	1.325	1.524	.8552	49
42	.7330	.6691	.7431	.9004	1.111	1.346	1.494	.8378	48
43	.7505	.6820	.7314	.9325	1.072	1.367	1.466	.8203	47
44	.7679	.6947	.7193	.9657	1.036	1.390	1.440	.8029	46
45	.7854	.7071	.7071	1.000	1.000	1.414	1.414	.7854	45
A°	RADIANS	COS A	SIN A	CTN A	TAN A	CSC A	SEC A	RADIANS	A°

From the trigonometric tables, the angle whose tangent is 6.28 is $\alpha = 80.95°$, and

$$\sin 80.95° = 0.9876 \qquad (3\text{-}18)$$

Since $\sin \alpha = V_L/V_Z$, the equation can be rearranged to the form that will yield the resultant or supply voltage V_Z,

$$V_Z = \frac{V_L}{\sin \alpha} = \frac{25.12 \text{ V}}{0.9876} = 25.5 \text{ V} \qquad (3\text{-}19)$$

There are other methods of solving phase-shift problems, but this method is the shortest one when all the factors are considered. From these data it is also possible to show by Ohm's law that the current in this circuit is 4 mA.

For a circuit such as Fig. 3-17 there are two possible approaches to determining the total impedance. If the supply voltage and total current are known, the impedance may be computed from Ohm's law. If only the component values and the frequency of the generator are known, a vector diagram may be used.

Recall that in the capacitor the current leads the voltage by 90°—that is, the phase angle is −90°. Across the inductance the voltage leads the current by 90°. The phase angle is zero in the resistance.

Figure 3-17 shows a series *RLC* circuit with a 10-V rms supply voltage which delivers a current of 3.66 mA to the circuit. From Ohm's law, the impedance Z, the total opposition to current flow, is

FIG. 3-17 (*a*) An *RLC* circuit with (*b*) a vector-diagram solution and a Pythagorean solution for impedance.

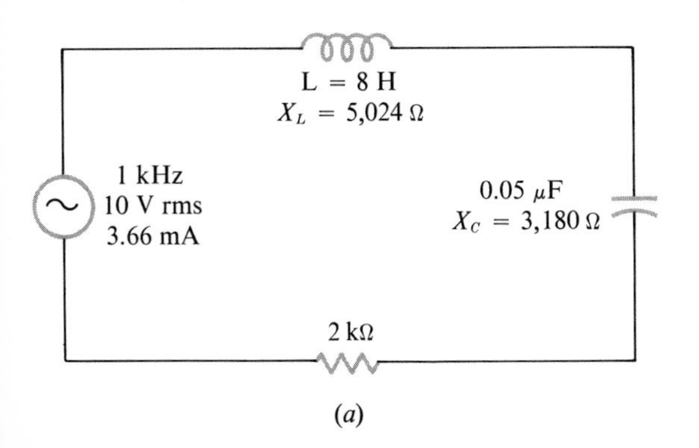

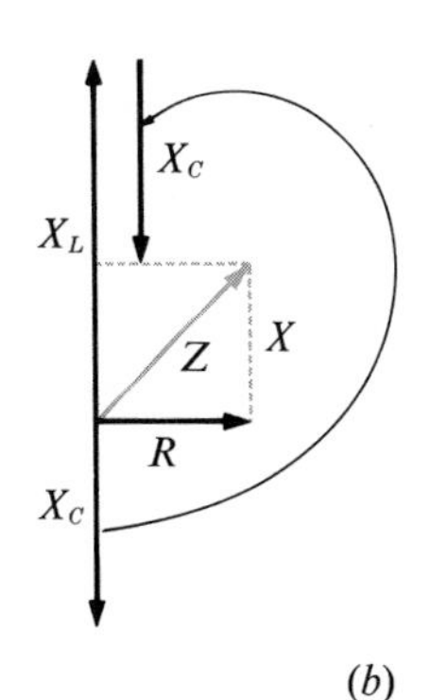

$$X = X_L - X_C = 5{,}024 - 3{,}180 = 1{,}844\ \Omega$$

$$Z = \sqrt{R^2 + X^2} = \sqrt{2{,}000^2 + 1{,}844^2} = \sqrt{7.43 \times 10^6} = 2{,}732\ \Omega$$

$$Z = \frac{v}{i}$$
$$= \frac{10 \text{ V rms}}{3.66 \text{ mA rms}} = 2{,}732\ \Omega \qquad (3\text{-}20)$$

where v is the ac voltage and i is the alternating current. Note that capital letters generally indicate dc or instantaneous values of voltage or current.

From the component values and frequency given, the inductive reactance X_L is 5,024 Ω and the capacitive reactance X_C is 3,180 Ω. Since these two values are 180° out of phase, or in opposite directions, the total reactance X of the circuit is the difference between the two values; that is,

$$X = X_L - X_C$$
$$= 5{,}024 - 3{,}180 = 1{,}844\ \Omega \qquad (3\text{-}21)$$

This indicates that there is an excess inductive reactance of 1,844 Ω. If the capacitive reactance were larger, the difference would be a *negative* number, and the reactance vector would have to be drawn pointing downward.

When the value of reactance X has been determined, the vector diagram may be completed as in Fig. 3-17. The total impedance of the circuit, then, is due primarily to the inductive influence in the circuit, since the impedance or resultant vector is in the upward direction.

The value of impedance may be determined by the Pythagorean theorem, as indicated in Fig. 3-17, but this still does not give us the phase angle between the resistance and the impedance. What we really need is the number of degrees by which the inductive voltage leads the total current in the circuit, which is the same phase angle. This may be determined by the same process used for Fig. 3-16. From Eq. (3-16),

$$\tan \alpha = \frac{X}{R} = \frac{1{,}844\ \Omega}{2{,}000\ \Omega} = 0.922 \qquad (3\text{-}22)$$

From trigonometric tables, the angle whose tangent is 0.922 is 40.67°, and

$$\sin 40.67° = 0.678 = \frac{X}{Z} \qquad (3\text{-}23)$$

Hence

$$Z = \frac{X}{\sin \alpha}$$

$$= \frac{1{,}844}{0.678} = 2{,}732\ \Omega \qquad (3\text{-}24)$$

The value of Z determined by both methods checks out to be the same, but the latter method is preferred since it also yields the phase angle between the supply voltage and the circuit current with no further computation.

Although we have little use for phase angle α at this point, it will be of considerable importance in our study of amplifiers, especially transistor amplifiers. It is also needed to understand the concept of resonance.

RESONANCE

It is not within the scope of this short review of the concepts of alternating current to investigate the subject of resonance in depth. However, let us consider some of the conditions under which resonance occurs and make some simple observations.

From the circuit and the vector diagram of Fig. 3-17 it is apparent that the series circuit allows the same current to pass through all components in the circuit. Notice also that the voltage across the inductance leads this current by 90°, while the voltage across the capacitor lags the current by 90°. Only the voltage across the resistance is in phase with the current.

If, instead of the frequency of 1,000 Hz, some lower frequency were used, the inductive reactance would decrease and the capacitive reactance would increase. At some frequency the two reactances would be of the same magnitude, or size; this is the frequency known as *resonance*. One definition of resonance, then, is *that circuit condition that exists when the value* (magnitude) *of inductive reactance is equal to the value of the capacitive reactance*. In a series circuit this condition can exist at only one frequency, the resonance frequency. It can easily be seen that since the inductive reactance and capacitive reactance are out of phase with each other by 180°, the net result in the circuit is zero reactance at resonance.

The circuit and diagram of Fig. 3-18 illustrate the nature of resonance. At one frequency, f_r, both X_L and X_C equal 1,000 Ω, with the net result that they cancel, leaving only the resistance in the circuit to limit the ac current. With a 10-V ac source, the measured voltage across the resistance in the circuit would be 10 V. Only the resistance would be dissipating any power, however, since what the capacitor stores during one half-cycle it discharges during the other half-cycle. The magnetic field of the inductor expands when the capacitor discharges, but the inductor supplies energy to charge the capacitor as the magnetic field collapses.

Figure 3-18 also illustrates that above resonance the circuit reacts like an inductive circuit, but below resonance it reacts capacitively. Whether the circuit is series or parallel, the equation for determining the resonance frequency is the same. The total inductance and the total capacitance of the circuit must be known. This will be clearer if we derive the equation from the assumption that $X_L = X_C$ at resonance. Thus

$$2\pi f L = \frac{1}{2\pi f C} \qquad X_L = X_C \tag{3-25}$$

Multiplying both sides of the equation by f yields

$$f\,2\pi f L = \frac{1}{2\pi C} \tag{3-26}$$

and dividing both sides by $2\pi L$ and combining terms yields

$$f^2 = \frac{1}{2\pi\ 2\pi L\ C} = \frac{1}{4\pi^2 LC} \tag{3-27}$$

FIG. 3-18 (*a*) A graphical plot of the change in reactance with frequency, with reference to the point of resonance in (*b*) the series *RLC* circuit.

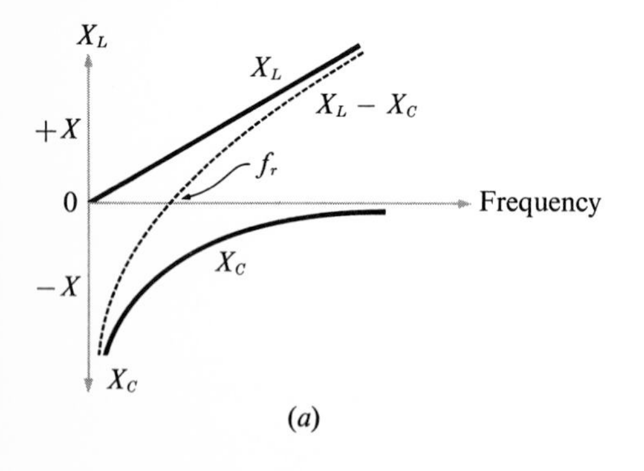

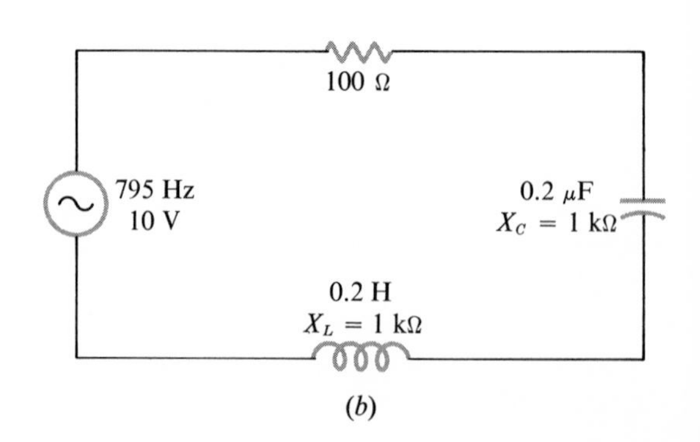

Then, taking the square root of both sides, we have

$$f = \frac{1}{2\pi\sqrt{LC}} = \frac{0.159}{\sqrt{LC}} \qquad (3\text{-}28)$$

If we substitute the values in the circuit of Fig. 3-18, we find that the resonance frequency in this case is

$$f = \frac{0.159}{\sqrt{0.2\ \text{H} \times 0.2\ \mu\text{F}}} = \frac{0.159}{\sqrt{0.2 \times 0.2 \times 10^{-6}}} = \frac{0.159}{\sqrt{0.04 \times 10^{-6}}}$$

$$= \frac{0.159}{\sqrt{4 \times 10^{-8}}} = \frac{0.159}{2 \times 10^{-4}} = \frac{0.159 \times 10^{4}}{2} = \frac{1{,}590}{2}$$

$$= 795\ \text{Hz} \qquad (3\text{-}29)$$

The reactances of both the inductance and capacitance may be computed at the resonant frequency to check that their values are indeed identical at 1,000 Ω each.

The circuit of Fig. 3-18 is a simple one, but its behavior at frequencies near resonance is of interest. We have already seen that only resistance limits the current flow at resonance, so that this is the condition of maximum current. If the frequency of the generator is either increased or decreased, however, the current will decrease. This should be easily seen, since below resonance capacitive reactance adds to the resistance vectorally, and above resonance the inductive reactance adds vectorally. When the current is plotted against frequency for the circuit of Fig. 3-18, the results are as shown in Fig. 3-19.

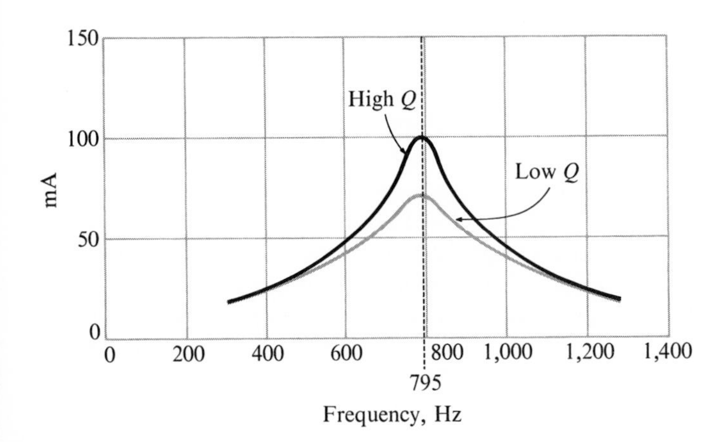

FIG. 3-19 Current versus frequency in an *RLC* circuit for two different values of resistance that result in two values for *Q*.

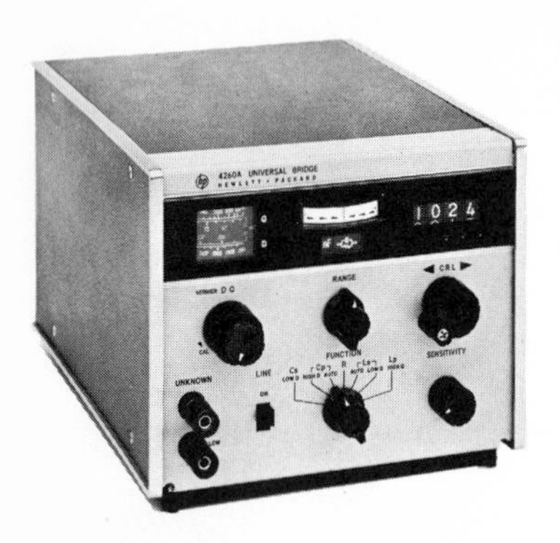

FIG. 3-20 The Hewlett-Packard Model 4260A Universal Bridge for quick accurate measurement of inductance, capacitance, resistance, *Q* factor, and *D* factor (*courtesy Hewlett-Packard*).

These findings may be checked by a simple experiment if a signal generator is available. Since it is not easy to measure ac current, an indirect method must be used. Current passing through a resistance makes itself known by the size of the voltage measured across the resistance. Thus, if the voltage across the 100-Ω resistance is measured for a whole range of frequencies from 100 to 1,500 Hz, the effect may be observed. Remember that capacitors and inductors rarely have exactly the value indicated on them. Thus the results may vary somewhat from those predicted unless the exact values are determined. The highest voltage should be recorded when the circuit is in resonance.

If, instead of the 100-Ω resistance, a larger value is chosen, the resonance frequency will not be affected. However, the amount of current is dependent on the resistance according to Ohm's law ($I = V/R$). Less current will pass at resonance, and the curve will be somewhat flatter in appearance.

It is difficult to obtain a coil without a fairly large resistance, but one with the smallest resistance available should always be used in experiments of this type. The series resistance should be made rather large in relation to the resistance of the coil to eliminate as much experimental error as possible.

FACTOR OF MERIT

The quality of an *RLC* circuit is sometimes referred to in terms of its *factor of merit Q*. The *Q* factor of a circuit is the *ratio of the inductive reactance to the resistance*,

$$Q = \frac{X_L}{R} \tag{3-30}$$

The larger the resistance of the circuit, the smaller the current flow and the smaller the *Q*. The circuit of Fig. 3-18 and the resonance curve of Fig. 3-19 indicate some unusual circumstances related to this *Q* factor.

Note that at the resonant frequency a maximum current of 100 mA passes through the circuit and that the entire supply voltage is measured across the resistance. The two reactance voltages are 180° out of phase and therefore cancel each other. When Ohm's law is used to compute the voltage across the two reactances, however, it is found that each is 100 V instead of 10 V. This phenomenon is known as the *resonant rise in voltage* and is observable in resonant circuits. The amount of rise in voltage depends on the value of the circuit *Q*. Since the Q of the circuit of Fig. 3-18 is $X_L/R = 1{,}000/100 = 10$, we would expect to find a rise in voltage across the inductance at resonance equal to $QV_s = 10 \times 10\ \text{V} = 100\ \text{V}$. Stated in other terms,

$$V_L = QV_s \qquad (3\text{-}31)$$

where V_s = ac supply voltage

From the point of reference of current flow and Ohm's law, the voltage at resonance across both the inductor and the capacitor may be computed and checked against Eq. (3-31). Since the current in the series circuit passes through both components, which have identical values of reactance but are opposite in phase, we should find voltages of $V = 0.1\ \text{A} \times 1\ \text{k}\Omega$, or 100 V. Again it should be emphasized that the only power dissipated in a reactive circuit is in the *resistance*, because both inductances and capacitances return the power to the circuit.

Whether the circuit is series resonant or parallel resonant, the Q factor is useful in determining how well voltages at the resonant frequency may be passed or rejected. For more detailed coverage of the resonance properties of series and parallel circuits see the references at the end of the chapter.

SUMMARY OF CONCEPTS

1. Alternating current at any instant in time is direct current.
2. Alternating current may be measured in several ways: peak to peak, peak, root mean square, or average.
3. Alternating current follows Ohm's law.
4. The reaction of a capacitor or inductor to alternating current may be explained in terms of the universal time constant.
5. A capacitor will charge in five time constants, where one time constant equals the series resistance times the capacitance, expressed in seconds.
6. The magnetic field associated with an inductor will expand or collapse in five time constants. One time constant equals the ratio of inductance to series resistance.
7. The voltage on a charging capacitor depends on the current that has passed into the capacitor ($V = Q/C$).
8. In an ac circuit the voltage on the capacitor lags the circuit current by 90°, or one quarter-cycle.
9. The capacitive opposition to current, called capacitive reactance, is the countervoltage built up on the capacitor through partial charge ($X_C = 1/2\pi fC = 1/\omega C$).
10. The countervoltage developed on an inductance passing alternating current, called inductive reactance, depends on the rate of change of the current ($X_L = 2\pi fL = \omega L$).
11. In an ac circuit the inductor voltage leads the circuit current by 90°.
12. Inductive reactance and capacitive reactance are out of phase with the circuit current by 90°, so that there is partial or total cancellation of reactance in a circuit containing both inductance and capacitance.
13. The resonant frequency is the frequency at which both inductive and capacitive reactances are equal.
14. At resonance only the resistance of a circuit limits the circuit current passage.
15. The phase relationships of voltage, current, and resistance in an *RLC* circuit may be represented by vectors whose lengths represent the values and whose direction from the origin represent the phase angle between the voltage and current or reactance and resistance.

16. Impedance, the total opposition to current in an *RLC* circuit, may be determined by vector analysis, by the Pythagorean theorem, or by simple trigonometry.

17. The factor of merit of a circuit, the Q factor, is the ratio of the inductive reactance to the resistance ($Q = X_L/R$).

18. The voltage across a capacitor or inductor at resonance will be larger than the supply voltage by a factor of Q, known as the resonant rise in voltage ($V_L = QV_S$).

19. A high-Q circuit will pass or reject a resonance-frequency signal better than a low-Q circuit.

GLOSSARY

Alternation A complete reversal of direction, such as a change from a positive direction to negative and then back again.

Angular velocity The number of degrees of rotation, per second, where one cycle per second is a rotation of 360° per second (2π rad/s).

Average voltage A means of describing the value of alternating voltage with a single number, about 63.6 percent of the peak voltage, the arithmetic average of all voltages of the complete cycle of voltage.

Cosine A trigonometric function of some angle of rotation. In describing a right-angle triangle, the cosine of the angle in question is the number that is the ratio of the length of the side next to the right angle to the length of the hypotenuse.

Cycle One complete alternation of direction, that is, 360° of rotation, as in alternating current.

Induced voltage The voltage present across the terminals of a coil of wire when it passes through a magnetic field and cuts lines of force, always in a direction opposing the voltage source that caused it.

Impedance The total opposition to current in a circuit, composed of resistance and inductive and capacitive reactance.

Lines of force The invisible elastic lines that make up the force field associated with every magnetic source. They originate at the north pole of a magnet and return to the south pole.

Peak-to-peak voltage The voltage, or potential difference, between the positive and negative peaks of the ac sine wave.

Peak voltage The voltage, or potential difference, between the zero voltage reference and the peak of the sine-wave shape of an alternating voltage.

Phase angle The number of degrees of rotation between two voltages or currents, or between voltage and current.

Phase shift The amount by which two voltages, two currents, or a voltage and a current have shifted out of phase from the condition where each starts from zero at the same time.

Polarity The positive or negative direction of voltage or current; the north and south poles of magnetism.

Pulsating direct current A voltage which may increase and decrease either periodically or randomly, but without a change in polarity.

Radian The length of the radius laid out along the circumference of the circle. One radian takes in 57.3° of arc or rotation.

Reactance The countervoltage developed on a capacitor or inductor in an ac circuit, always in a direction opposed to the source voltage.

Resonance A condition in an ac circuit where the frequency is such that the inductive and capacitive reactances cancel because they are equal in magnitude but opposite in polarity.

Root-mean-square voltage The measure of the effective value of a sine-wave voltage, current, or power. Its actual value is equal to 0.707 times the peak voltage.

Sine A trigonometric function of some angle, e.g., the ratio of the side opposite the angle in question to the hypotenuse of the right triangle.

Sine wave The graphical plot of voltage against time of the normal ac waveform, the most common signal form. It is actually a plot of the sine of the angle of rotation of the ac generator multiplied by the peak voltage for the entire cycle of operation.

Tangent A trigonometric function of some angle, e.g., the ratio of the length of the side opposite the angle in question to the length of the side adjacent the angle in a right triangle.

Time constant A universal value, usually expressed as a percentage, corresponding to the time it takes for a capacitor to charge to 63.2 percent of full charge or for a magnetic field to expand to 63.2 percent of its limit.

Transient effect A sudden occurrence of voltage or current that pulses through a circuit and may or may not be repeated.

Vector A graphical means of representing both the magnitude and direction of a force, voltage, current, or resistance, where the magnitude is represented by the length of the vector arrow and the direction is indicated by its direction.

REFERENCES

Gillie, Angelo C.: *Electrical Principles of Electronics*, 2d ed., McGraw-Hill Book Company, New York, 1969, chaps. 8–21.

Grob, Bernard: *Basic Electronics*, 2d ed., McGraw-Hill Book Company, New York, 1965, chaps. 11–23.

Lippin, Gerard: *Circuit Problems and Solutions*, vol. 1, Hayden Book Company, Inc., New York, 1967, chaps. 6–14.

Middleton, Robert G., and Milton Goldstein: *Basic Electricity for Electronics*, Holt, Rinehart and Winston, Inc., New York, 1966, chaps. 7–16.

Shrader, Robert L.: *Electrical Fundamentals for Technicians*, McGraw-Hill Book Company, New York, 1969, chaps. 8–37.

Weick, Carl B.: *Principles of Electronic Technology*, McGraw-Hill Book Company, New York, 1969, chaps. 6–14.

REVIEW QUESTIONS

1. What is the difference between alternating current and direct current?

2. Describe the ac cycle in terms of the generator motion.

3. At what points in the cycle is the generator cutting the greatest number of lines of force per second?

4. What is the difference between peak voltage and peak-to-peak voltage?

5. How is rms voltage related to peak voltage?

6. Why is the normal ac-voltage waveform called a sine wave?

7. In what way could a sine wave not be an alternating-current wave?

8. Is it possible to have a 10-V *p-p* waveform whose average over the entire cycle is 0 V? Explain.

9. Describe what is meant by time constant in an *RC* circuit.

10. How many time constants are required to essentially charge up a capacitor?

11. Why is the time-constant curve called a universal time-constant curve?

12. Why does it take time to charge a capacitor?

13. At what time in the expansion of the magnetic field around a coil of wire carrying current is the induced voltage greatest?

14. What is the polarity of the induced voltage in a coil compared to the forward voltage?

15. What is meant by the left-hand rule?

16. In an *RC* circuit using alternating current, what is the phase relationship between the voltage and the current?

17. In an *RL* circuit does the current lead the induced voltage, or vice versa? Explain.

18. What is an advantage of using angular velocity in radians per second for ac computations?

19. What is the relationship between inductive reactance and induced voltage?

20. What is the phase relationship between inductive reactance and capacitive reactance?

21. How does inductive reactance change with an increase in frequency of the current?

22. What are some advantages in using vector analysis in describing ac circuit actions and phase relations?

23. What is impedance?

24. What is the difference between impedance and inductive reactance?

25. Can the phase shift in a coil ever reach 90°? Why?

26. When the total phase shift in an ac circuit is 0° and the circuit contains resistance, capacitance, and inductance, what is the condition called? What circuit component is limiting the current?

27. What are some of the effects of resonance?

PROBLEMS

1. Convert 50 V rms into peak and peak-to-peak units.

2. What is the average voltage of a 75-V peak sine wave?

3. If a circuit contains a capacitor and a resistor and the dc supply voltage is 15 V, what will be the voltage across the capacitor after one time constant?

4. How much time is needed to charge a 30-μF capacitor through a 25-kΩ resistance?

5. If the supply voltage in Prob. 4 is 15 V, what is the charging current at time zero? At time of full charge?

6. When the 30-μF capacitor is charged under 15 V of potential difference, how many Coulombs of charge are stored in the capacitor?

7. If a circuit contains a 0.25-H inductor and a 200-Ω resistor and has a supply voltage of 10 V, how much is the charging current after one time constant? After five time constants?

8. In the circuit of Prob. 7, what is the induced voltage in the coil at time zero? After one time constant?

9. What is the capacitive reactance of a 0.05-μF capacitor in a 60-Hz ac circuit?

10. What is the inductive reactance of a 5-H coil in a 60-Hz circuit? A 400-Hz circuit?

11. In a series circuit containing a 0.02-μF capacitor and a 10-kΩ resistance, powered by a 60-Hz 10-V rms supply, what is the capacitive reactance?

12. In Prob. 11, how much out of phase are the current and voltage in the capacitance?

13. In Prob. 11, how much is the impedance of the circuit?

14. In Prob. 11, how much is the total current?

15. What is the angular velocity, in radians per second, of a generator turning at 1 kHz?

16. Plot a graph of the capacitive reactance of a 0.05-μF capacitor for frequencies from 0 Hz to 4 kHz.

17. Plot a graph of the inductive reactance of a 2-H inductor for frequencies from 0 Hz to 4 kHz.

18. Draw a vector diagram for the reactances and resistance in a 400-Hz ac circuit containing a 0.4-μF capacitor in series with a 500-Ω resistor and a 0.1-H inductor.

19. In Prob. 18, what is the total reactance?

20. Use a vector diagram to compute the impedance of the circuit in Prob. 18.

21. Use trigonometric functions to compute the impedance and phase angle of Prob. 18.

22. If the capacitive reactance in a circuit is 4,700 Ω and the inductive reactance is 4,200 Ω, what is the total reactance?

23. What is the resonant frequency of a series circuit with a 0.004-μF capacitor and a 50-mH coil?

24. If the inductive reactance of a coil is 25 kΩ and its resistance is 500 Ω, what is its Q value?

25. If the coil in Prob. 24 is in a circuit powered by a 10-V ac source, what is the voltage across the coil at resonance?

CHAPTER 4

OBJECTIVES

1. To relate the basic concepts and theory to experimental evidence and verification
2. To consider the effect of impedances on circuit design and measurements
3. To develop the ability to handle test equipment in a proper manner
4. To relate the functions of certain test equipment to solutions of circuit measurement problems

BASIC INSTRUMENTATION AND THE ELECTRONIC SYSTEM

Basic to the study of electronics concepts is learning how to make measurements and how to make them accurately. Most first attempts at making measurements end in frustration. The student may not know whether he is measuring the equipment or the circuit, what is going on in the circuit, or whether anything is working. In short, he does not know what to look for, and so he has trouble finding it.

It is always a good idea to know what you are looking for before you proceed with any measurement. However, it is equally important to know how to use the equipment, so that when you find what you are looking for you will recognize it. Further, it is a good idea to know how much the test equipment you are using affects the accuracy of your measurements.

EXPERIMENTAL MEASUREMENTS

In almost every case theory becomes practical only when it can be experimentally verified. There are some theories that have been very difficult to verify by conventional means, such as Einstein's theories of relativity. Nevertheless, although these theories cannot be verified in their entirety, sufficient data can be collected about parts of them to suggest that the balance is equally valid.

A theory is not a fact. It is only a possible explanation of some fact, to be used until a better or more complete explanation is developed. However, most of solid-state electronics is based on rather sound theories of atomic structure and electrical concepts, verified by means of very complex test equipment. Much of the basis for solid-state electronics measurement would in fact have been completely overlooked or unobservable 30 years ago. Theory states that certain effects should occur in less than a billionth of a second and be less than a nanoampere or nanovolt in magnitude. Without man's ability to rise to the challenge with the necessary measuring equipment, such minute amounts could never have been detected. We assume here that the theories and concepts we are discussing are valid, but each one must hold up under experimental investigation.

Perhaps one of the most frustrating problems in electronics measurements is that of circuit loading. A loaded circuit or system will perform differently from a circuit with no load—just as a loaded truck going up a steep hill performs differently from an empty one. The problem of the designer is to develop a system that will operate properly with the load it is designed to carry.

In electronics the *load* usually refers to the amount of current or power that will be taken from the system. As we saw in Chaps. 2 and 3, the amount of current depends on what limits it. Current is limited by resistance or, more specifically, by impedance. The amount of power consumed by the load depends on the impedance of the load and how much current is delivered. Consequently, anything that changes the impedance of the system also alters the power requirement.

For maximum transfer of power from one subsystem to the next the output impedance of one subsystem must match the input impedance of the next. To provide for maximum voltage transfer, the output impedance of one subsystem must be very small compared to the input impedance of the next.

THE VOLT-OHM-MILLIAMMETER

An instrument that is common in most electronics laboratories and workrooms is the *Volt-Ohm-Milliammeter* (VOM). This handy self-contained measuring instrument has a variety of uses. However, it should not be relied upon for making accurate measurements. The circuitry of the VOM is such that it draws a significant amount of current from a circuit when it is used to measure voltages. Thus the load added to the circuit by the measuring process makes accurate measurement impossible. Of course, this is also true to some extent of every other measuring instrument. For some circuits, however, the VOM should not be used at all.

Figure 4-1 indicates how a VOM would be connected into a circuit to measure current. Note that the current must pass through the meter and that it is in series with the rest of the circuit. By its very nature, it must also have some resistance. This resistance will limit the current to some value lower than it was before measurement. This slight error is one that is not easy to avoid without expensive equipment. The amount of error introduced depends on which scale is being used.

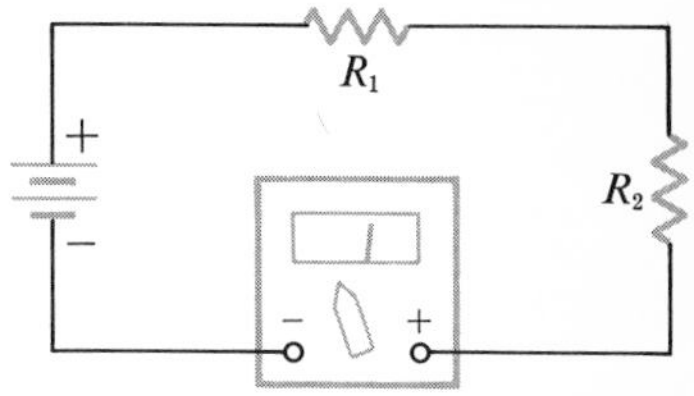

FIG. 4-1 Proper connection of the VOM in a circuit for measuring current.

Another error introduced in the current readings on a VOM is the calibration error. Because of the manner by which meter faces are made and the construction of the needle mechanism, most inexpensive VOMs have an error of about ±3 percent. Some of this error is offset if the meter is set on a range that allows the current reading to be made in the middle third of the scale. The meter should be read from directly in front of the scale, and not at an angle. The spacing between lines is usually such that viewing at an angle yields an error known as *parallax*. This is the same parallax problem that results when a camera viewfinder and the lens are not at the same point of observation, so that the camera does not take a picture of exactly what is viewed in the viewfinder.

To use the current ranges of the VOM circuit polarity *must* be observed. The black, or negative, terminal of the VOM must be connected in the circuit so that it is closest to the negative terminal of the supply. The meter must be placed in *series* with the current flow; putting a VOM in parallel with a resistance to measure current is likely to destroy the meter. Full-scale deflection requires only about 50 mV. When the meter is on a current range and is placed across a resistance with much current flow, there is generally far more than 50 mV across it. Figure 4-2 shows the proper way of connecting the VOM into a transistor circuit to measure collector current.

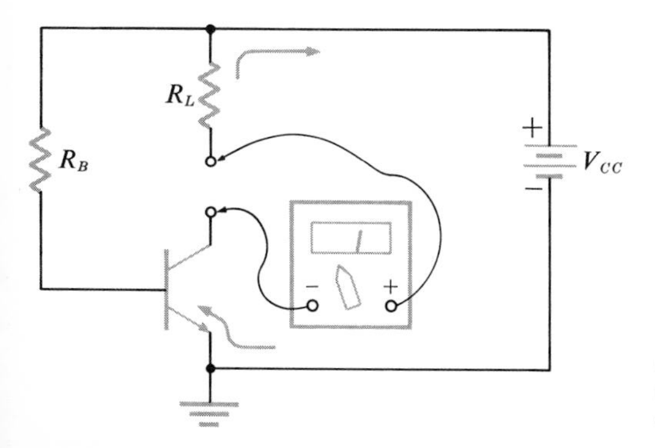

FIG. 4-2 Proper connection of the VOM in a transistor circuit for measuring collector current. The range switch must be set on the correct range.

VOLTAGE MEASUREMENT

The VOM is constructed in such a way as to allow a wide range of voltages to be measured. As indicated in Fig. 4-3, it actually uses the same milliammeter movement, but connected in series with a precision resistor and calibrated to read voltage. Each range simply has a different precision resistance in series with the meter to limit the current to that needed to produce full-scale deflection.

The typical quality VOM has a meter that will give a full-scale deflection with 50 μA of current. The meter is generally referred to as a 20,000-Ω/V meter. From Ohm's law, it can be shown that if full-scale deflection is for 1 V, the 20,000-Ω/V meter requires 50 μA of current. That is,

$$\begin{aligned} V &= IR \\ &= 50\ \mu\text{A} \times 20{,}000\ \Omega = 50 \times 10^{-6} \times 20 \times 10^{3} \\ &= 1{,}000 \times 10^{-3} = 1\ \text{V} \end{aligned} \tag{4-1}$$

The Ohms-per-Volt rating is generally referred to as the *sensitivity* of the meter. It is actually the reciprocal of the full-scale current (I_{fs}) of the meter movement,

$$\text{Sensitivity} = \frac{1}{I_{fs}} = \frac{1}{\text{Amperes}} = \frac{1}{\text{Volts/Ohms}} = \frac{\text{Ohms}}{\text{Volt}} \tag{4-2}$$

FIG. 4-3 (*a*) A microammeter connected to a series circuit to measure voltage; (*b*) a diagram indicating how the microammeter may be internally changed to measure voltage with the addition of series resistance and a voltage scale to the meter face.

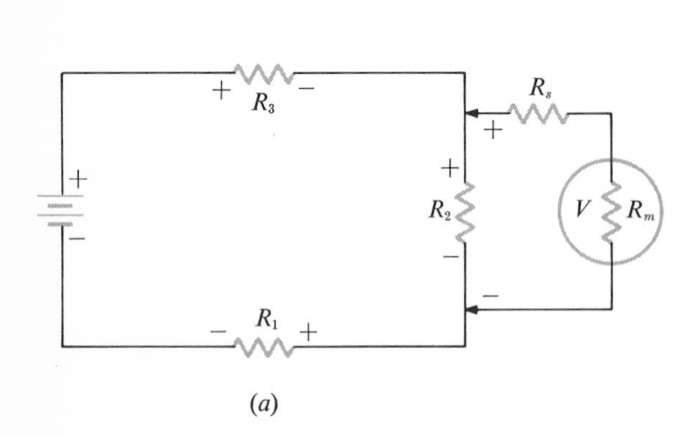

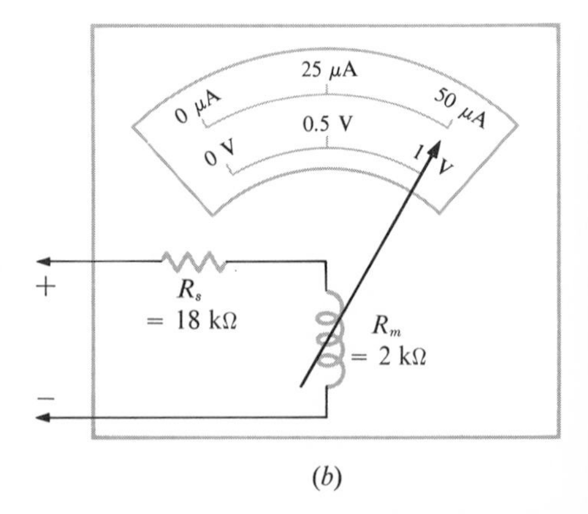

Another common and inexpensive VOM uses a 1,000-Ω/V meter movement. This would be one that has a 1.0-mA full-scale current reading.

If a VOM with the 20,000-Ω/V meter is examined, it will be found that the series resistance is not actually 20,000 Ω. The reason is quite simple. The sum of the series resistance and the meter resistance must be 20,000 Ω if 1 V is to give a full-scale reading when 50 μA of current is passing through the meter. Therefore, if the meter resistance is 2,000 Ω, the series resistance must be 18,000 Ω. In other words,

$$V = (R_{\text{meter}} + R_{\text{series}})I \tag{4-3}$$

$$R_{\text{in}} = R_{\text{meter}} + R_{\text{series}} \tag{4-4}$$

If it is desired to measure voltages greater than 1 V, and it usually is, the range of the meter may be extended by substituting a larger-value series resistance. The current will still be limited to 50 μA and the voltage across the meter will not exceed its safe value. The larger voltage will be found across the series resistor. Figure 4-4 shows how a simple switching of input resistors may be used to extend the range of the simple voltmeter.

The sensitivity of the VOM remains a function of the meter movement, but the input resistance does not remain the same for each range. Equation (4-4) shows that as the series resistance is changed to change the full-scale range of the meter, the input resistance changes accordingly. Figure 4-4 points out that for the 0- to 10-V range the meter would have a 200-kΩ input impedance. For the 100-V range the input impedance would be about 2 MΩ.

The significance of the input impedance for the voltmeter becomes apparent when we try to measure voltage. Refer again to the circuit of Fig. 4-3. The voltmeter is in parallel with resistance R_2, and

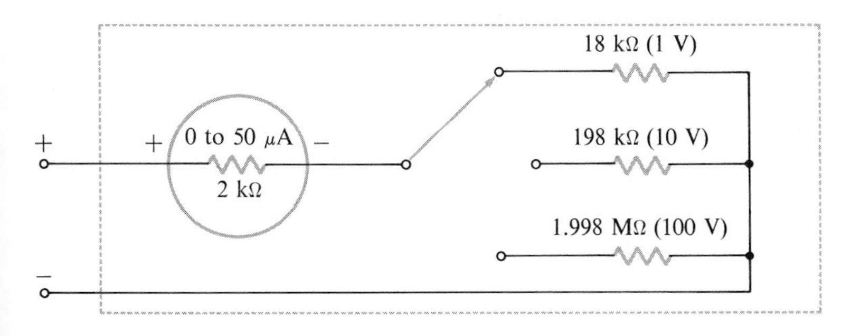

FIG. 4-4 Connecting the 0 to 50 microammeter as a multirange voltmeter.

so part of the current that would normally pass through R_2 will be diverted to the voltmeter. The voltage across the resistance R_2 will thus be somewhat lower than it was before the voltmeter was connected. Since the voltmeter and the resistance R_2 are in parallel, the same voltage will be found across both of them.

The amount of current taken from the circuit of Fig. 4-3 to yield a voltage of 1.0 V for the 20,000-Ω/V meter is only 50 μA, but for the 1,000-Ω/V meter the current would be 1.0 mA. If the total current in the circuit were only 10.0 mA, the error in measurement of voltage with the 1,000-Ω/V meter would be significant. The circuit would be so loaded that approximately 1 mA of current would have been drawn away, about 10 percent of the total. In contrast, 50 μA would be only about 0.5 percent.

For accuracy in work with solid-state circuitry the VOM should not be used if anything in the line of electronic voltmeters is available. Care should always be taken when measuring the voltage between the base and the emitter of a transistor with a VOM. The nature of the low-impedance circuit will almost always yield an incorrect answer.

Calibration error for the voltmeter section of the VOM occurs on the same basis as for the milliammeter. The parallax error is also the same. The most serious source of error other than calibration is, of course, that due to the loading factor. If the loading factor is to be kept low (less than 1.0 percent), the input impedance to the voltmeter should be greater than 100 times the resistance across which the voltage is being measured. If this precaution is observed, no more than 1 percent of the current of the circuit will ever be drawn by the meter. Of course, this can become a real problem in many circuits. Fortunately there are many excellent electronic voltmeters that can perform this task at a reasonable price.

RESISTANCE MEASUREMENT

Just as Ohm's law involves current, voltage, and resistance, a milliammeter can be made to measure resistance. The reasoning is much the same as that discussed for the voltmeter. The main difference is that to measure resistance a calibrated voltage source might be maintained within the VOM to produce the current needed by the meter. Figure 4-5 gives a typical ohmmeter circuit. Note that the only current used by the meter comes from this internal source.

The resistance R_{adj} in the ohmmeter provides a means by which the meter may be zeroed as the internal battery begins to age. If the two leads of the ohmmeter are shorted, the amount of current flow can be adjusted to the full-scale reading of the meter by adjusting R_{adj}. This condition corresponds to zero resistance in the external circuit. It is always necessary to zero an ohmmeter before attempting to make resistance measurement.

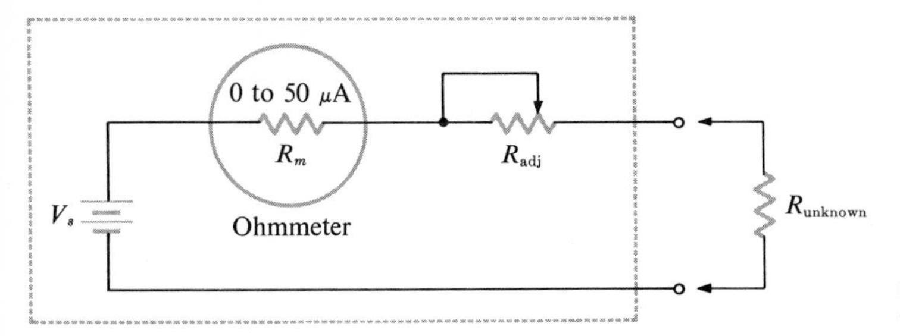

FIG. 4-5 Connecting the 0 to 50 microammeter as an ohmmeter.

With the meter zeroed, an unknown resistance may be connected between the leads of the ohmmeter. The amount of current passing through the resistance is a function of the size of the resistance and the open-circuit voltage. The resistance R_{adj} corresponds to the Thévenin resistance, or internal resistance, of the ohmmeter and must be considered in the determination of the open-circuit voltage, that is, the Thévenin voltage. A review of Chap. 2, especially Eqs. (2-35) and (2-36), should help at this point. The value of the unknown resistance may be computed from the current indicated by the meter. A new meter face may be made, or resistance values may be printed on the same face as the voltmeter and milliammeter.

For some unknown resistance greater than zero a current less than maximum full-scale current will be registered on the meter (provided it has first been zeroed). The amount of meter deflection should be the same as the ratio of R_{adj} to the sum of R_{adj} and R_{unknown}. Thus the actual current through R_{unknown} is

$$I = \frac{V_S}{R_{\text{adj}} + R_{\text{unknown}}} \qquad (4\text{-}5)$$

and the meter deflection is

$$\frac{I}{I_{fs}} = \frac{R_{\text{adj}}}{R_{\text{adj}} + R_{\text{unknown}}} \qquad (4\text{-}6)$$

If the resistance needed to zero a 50-μA meter with a 1.0-V supply is 20,000 Ω (including the meter resistance), an unknown resistance giving a half-scale deflection of 25 μA would also be 20,000 Ω; according to Eq. (4-6),

$$0.5 = \frac{20\ \text{k}\Omega}{20\ \text{k}\Omega + R_{\text{unknown}}}$$

and so

$$R_{\text{unknown}} = \frac{20\ \text{k}\Omega}{0.5\ \text{k}\Omega} - 20 = 20\ \text{k}\Omega \tag{4-7}$$

For a quarter-scale deflection, 12.5 μA, R_{unknown} would have to be 60 kΩ, and so on. Calibration of the meter continues in this fashion and approaches infinity as the meter approaches zero deflection. Note that zeroing the meter is actually allowing full-scale current to pass, while zero current indicates infinite resistance. The ohmmeter thus operates in the opposite direction from the voltmeter and the milliammeter.

The ohmmeter, like the voltmeter and the milliammeter, may be used for a variety of ranges by proper choice of current-limiting resistances. Ranges typically run from $R \times 1$ to $R \times 1$ MΩ; that is, the meter reading should be multiplied by the setting of the range switch.

A large number of VOM-type meters are available on the market. Some are of the type just described. Other models are in reality a form of electronic multimeter and have a very high input impedance and much better accuracy than the simple resistance types. The electronic varieties frequently use two FETs in a balanced circuit that give input impedances up to 10 or 20 MΩ or more. These will be discussed later.

EXPERIMENTAL USE

It is somewhat difficult to experiment accurately with all unknown quantities, but a lot can be learned by experimenting with a quality voltmeter. If a 0- to 1-mA or 0- to 50-μA meter is available, connect it into the circuit of Fig. 4-6*a*. Be certain that the potentiometer is set for maximum resistance. Slowly decrease the resistance of the potentiometer until the current indicated on the meter corresponds exactly to full-scale deflection. Set the 2.5-kΩ potentiometer to maximum resistance and connect it in parallel with the meter. Slowly reduce its

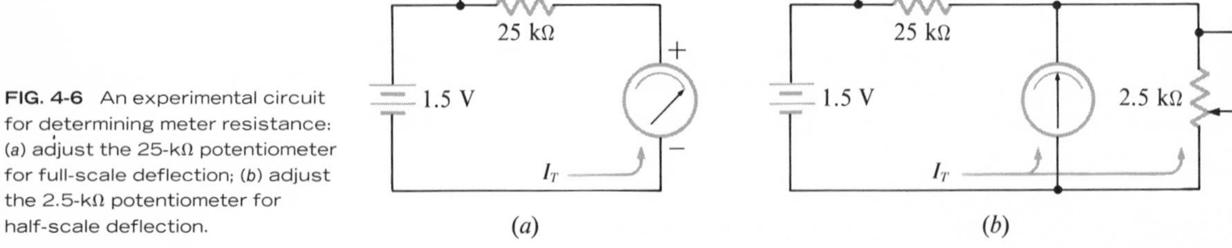

FIG. 4-6 An experimental circuit for determining meter resistance: (*a*) adjust the 25-kΩ potentiometer for full-scale deflection; (*b*) adjust the 2.5-kΩ potentiometer for half-scale deflection.

resistance until the meter reads one-half its full-scale value. Disconnect the power and the 2.5-kΩ potentiometer without changing its setting. With an accurate ohmmeter, measure the value of the 2.5-kΩ potentiometer. This resistance is equal to the meter resistance. Note the full-scale current and record it.

The parallel method of determining meter resistance follows Ohm's law and Norton's theorem. The voltage across a parallel circuit is the same across each branch. If the current through the meter drops to one-half its value when alone in the circuit, the other resistance in parallel must be of equal value for the same amount of current to pass through both.

Using Ohm's law, compute the voltage that is needed, with full-scale current and the meter resistance, to give full-scale deflection. For example, if the meter resistance is 2,000 Ω and the current is 50 μA, the voltage required to produce full-scale deflection would be only 100 mV, or 0.100 V.

Meter shunts Once the resistance of a milliammeter has been determined, a wide variety of applications are open. If the meter has a maximum current rating of 50 μA, it is of little use for measuring larger values of current without modification. This modification, called a *shunt*, is merely a resistance placed in parallel with the meter to shunt part of the current around the meter.

If the meter in the previous experiment was found to have an internal resistance of 2,000 Ω, the shunt resistance needed to change it to read 500 μA rather than 50 μA must be 222 Ω. Let us see why. The new current range is 10 times greater than that of the meter; therefore the resistance of the meter and its shunt should be 10 times smaller, or 200 Ω. Stated another way, 50 μA must go through the meter and 450 μA must go through the shunt. That is, 0.9 of the total current must go through the shunt:

$$R_{\text{shunt}} = \frac{I_{fs}}{I_T - I_{fs}} R_{\text{meter}}$$

$$= \frac{50\ \mu\text{A}}{500\ \mu\text{A} - 50\ \mu\text{A}} \times 2{,}000\ \Omega = 222\ \Omega \qquad (4\text{-}8)$$

For the sake of accuracy, the shunt is always a precision resistor, often made of resistance wire wound on a small spool. Here we shall use a potentiometer instead of a precision resistor.

Connect the circuit of Fig. 4-7, being careful to set the potentiometer in series with the meter to its maximum resistance. Apply power

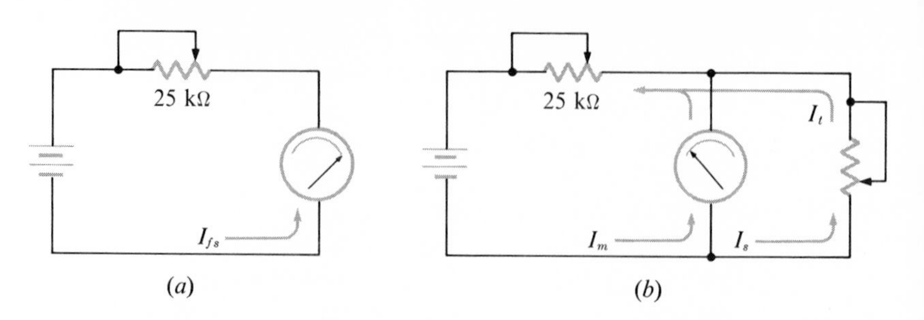

FIG. 4-7 An experimental circuit for determining the value of shunt resistance needed to extend the range of the meter by a factor of 10: (*a*) adjust the current to full scale; (*b*) add the shunt resistance and adjust for one-tenth of full-scale current.

to the circuit and adjust the 25-kΩ potentiometer to give a full-scale reading on the meter. Now insert the 500-Ω potentiometer, set at maximum resistance, in parallel with the meter as indicated in Fig. 4-7*b*. Adjust the 500-Ω potentiometer to give a 1/10-scale reading on the meter. Disconnect the power and remove the 500-Ω potentiometer without changing its setting. With an accurate ohmmeter, measure its resistance. It should measure about one-ninth the value of the meter resistance. That is, R_{meter} in parallel with $R_{\text{shunt}} = 1/10\, R_{\text{meter}}$, or 9/9 in parallel with $1/9 = 1/10$.

If the meter range needs to be extended to read 5 mA instead of 500 μA, simply insert the proper values into Eq. (4-8) to determine what value of shunt to use.

The milliammeter as a voltmeter Connect the circuit of Fig. 4-8*a*. If a flashlight cell is used for power, be sure it is a fresh one for accurate results. Otherwise it is sufficient to measure the terminal voltage under load and record this value. Adjust the 25-kΩ potentiometer to give full-scale deflection of the meter. Disconnect the power supply and measure the resistance of the potentiometer.

Now connect the circuit of Fig. 4-8*b*. If an 8- or 9-V power supply is not available, use an ordinary transistor radio battery. The

FIG. 4-8 (*a*) Adjustment of a milliammeter or microammeter to read 1.5 V full scale, and (*b*) the resulting voltmeter used to measure voltage in a circuit while the voltage is monitored with a standard voltmeter.

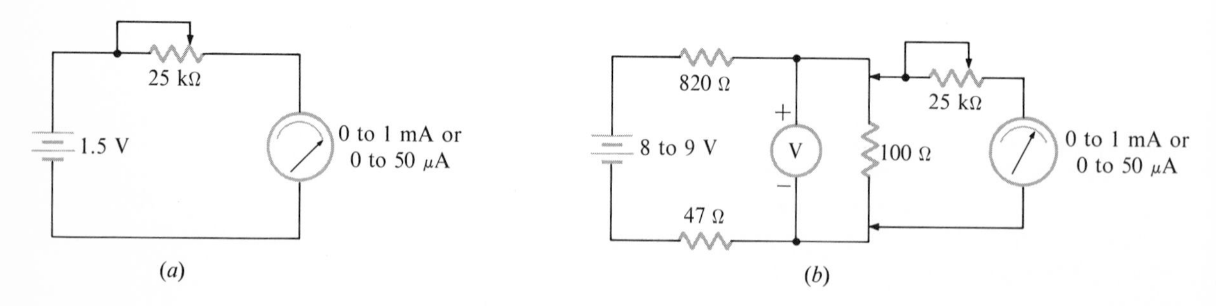

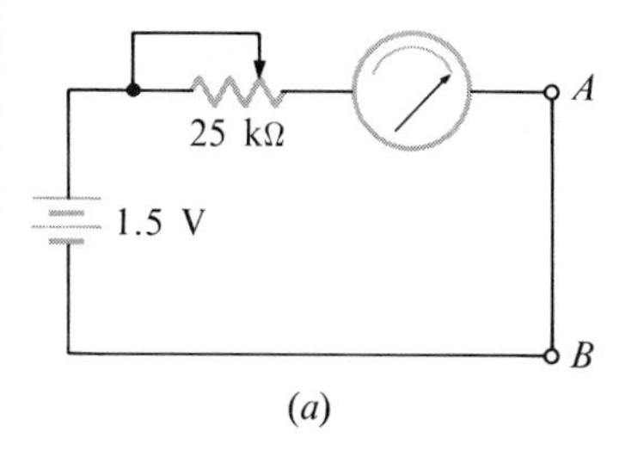

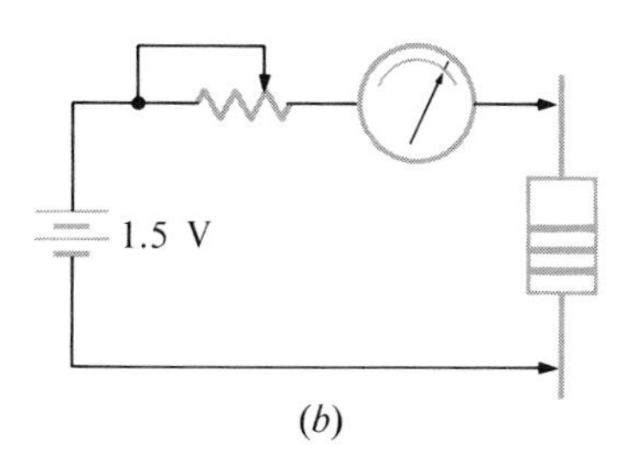

FIG. 4-9 Calibration of the nonlinear scale of an ohmmeter: (*a*) adjust the zero (full-scale) position of the needle; (*b*) mark deflection for known values of resistance.

circuit allows about 10 mA through the circuit, which will provide about 1 V across the 100-Ω resistance and about 0.5 V across the 47-Ω resistance. Measure these two voltages with accurate voltmeters if possible, and record the readings.

Connect the voltmeter across the 100-Ω resistance. Since the meter is calibrated to read 1.5 V full scale, it should now read about 75 percent of full scale. When the voltmeter was attached to the circuit to monitor the voltage was there a change in reading? Any change in reading is due to the loading effect of the voltmeter as it draws current from the circuit.

If a 0- to 1-mA full-scale milliammeter is available, repeat the entire experiment with this meter. Note that the loading effect for the 0- to 1-mA meter is greater than for the 50-μA meter.

The milliammeter as an ohmmeter Using the same 1.5-V cell, 50-μA or 1-mA meter, and 25-kΩ potentiometer as in the previous experiment, connect the circuit of Fig. 4-9*a*. Adjust the meter to full scale as before; this will be used as the zero resistance point on the scale.

Disconnect the short circuit between points *A* and *B*. Connect the 100-Ω resistance between points *A* and *B*, and with a marking pen mark on the glass face of the meter the point at which the needle comes to rest. Remove the 100-Ω resistance and replace it with a 1,000-Ω resistance. Again mark the reading on the meter face. Repeat this for 10 kΩ, 100 kΩ, and 1 MΩ. Note that the lower values of resistance can be read more accurately than larger values. Rezero the meter to some value other than zero and remeasure the 10- and 100-kΩ resistances. Do you see how much error can be involved in resistance measurements if the meter is not checked for its zero position before a reading is made?

THE ELECTRONIC VOLTMETER

Electronic voltmeters are generally far more accurate and less likely to load a circuit than the VOM. Although they generally do not have a direct means for measuring current, some of the newer digital

FIG. 4-10 An electronic multipurpose voltmeter for measuring dc and ac voltages and resistance (*courtesy Hewlett-Packard*).

multimeters do the same task as the VOM and give a direct readout to much greater accuracy at the same time. Their cost, of course, is also several times that of the VOM. Mass production and the use of large-scale integration has made certain models of digital multimeters available for less than $300, about the cost of many high-quality electronic voltmeters of the analog or meter-reading variety.

The procedure for using the electronic voltmeter is the same as for the VOM. While the VOM is also capable of measuring alternating voltages, the electronic voltmeter does the task more accurately with less loading.

Figure 4-11 shows one circuit for an electronic voltmeter using two field-effect transistors (FETs). Various schemes have been worked out for only one FET; others use a variety of switches and resistances to provide for several ranges and functions. The circuit of Fig. 4-11 can easily be built and checked out as a voltmeter project in most laboratories. The 5-kΩ potentiometer is used to balance the current flow between the two FETs. They are balanced when zero current flows through the meter, that is, when the voltages at both source terminals are equal. The 20-kΩ potentiometer limits the maximum meter current.

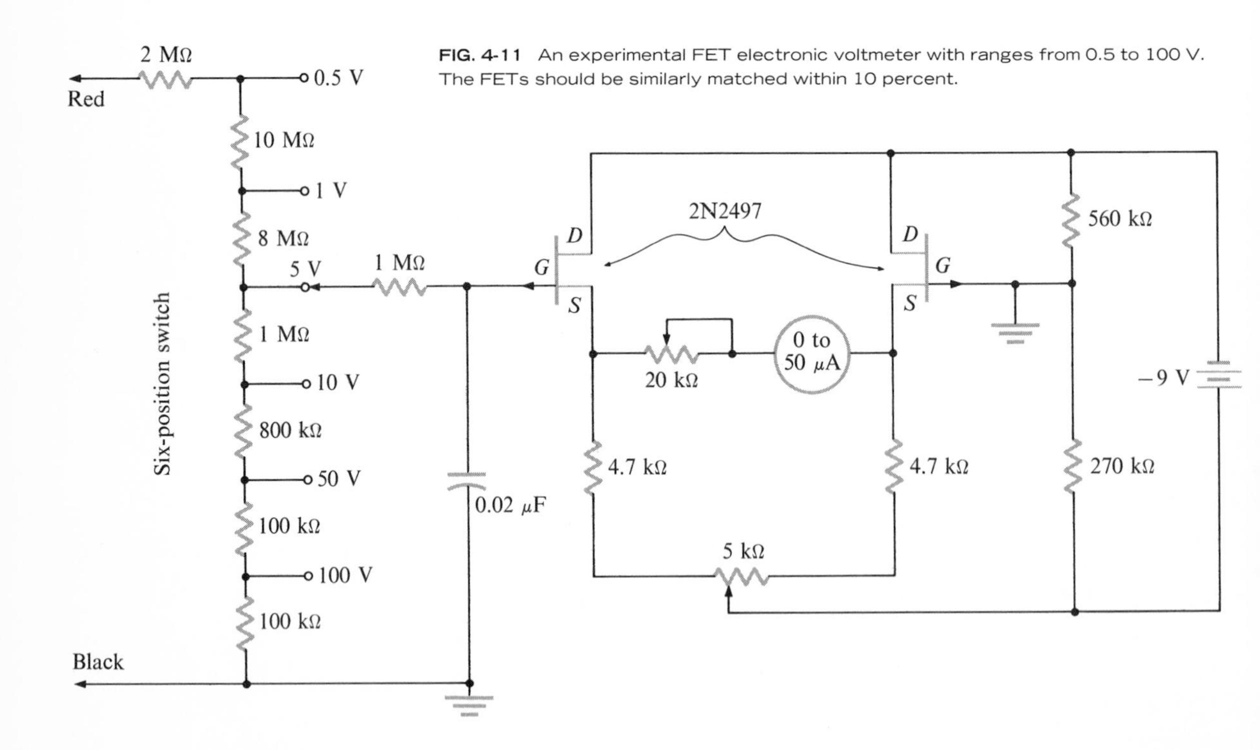

FIG. 4-11 An experimental FET electronic voltmeter with ranges from 0.5 to 100 V. The FETs should be similarly matched within 10 percent.

Most electronic voltmeters are set up so that the ac voltages can be measured on the dc scale. The circuit uses a full-wave bridge rectifier and is corrected to read rms voltage only. If a peak-to-peak measurement is needed, the scale for Volts peak to peak must be used. Even then the peak-to-peak voltage must be that from a sine wave, or it will be rather inaccurate. The range that gives a reading in the middle of the scale should always be used.

FIG. 4-12 A dual-trace triggered-sweep oscilloscope for laboratory or industrial use (*courtesy Tektronix, Inc.*).

THE OSCILLOSCOPE

Although the electronic voltmeter gives much more accurate measurements than the VOM, it tells nothing about the actual shape of the waveform. For this reason an oscilloscope should be used whenever possible for determination of alternating voltages. The oscilloscope is an instrument that provides a graphic display of changing voltage with respect to time. This is accomplished simply by the motion of a tiny beam of electrons from left to right across the face of a *cathode-ray tube* (CRT). As the beam moves across the face of the tube, it is deflected upward for a positive input voltage and downward for a negative input voltage. In other words, the oscilloscope has a vertical voltage axis and a horizontal time-base axis.

Let us first consider how the CRT works. Observe Fig. 4-13. An electron current through the filament of the CRT heats it to the point where electrons are boiled out. These electrons are accelerated toward the screen of the CRT by a high positive voltage on the accelerating anode. The beam is focused so that when it strikes the phosphors on the screen, a tiny spot on the surface glows. The beam normally strikes the surface of the screen at the center.

If a positive voltage is applied to the top vertical deflection plate and a negative voltage applied to the lower plate, the beam will be pushed and pulled upward. The amount of deflection depends on the

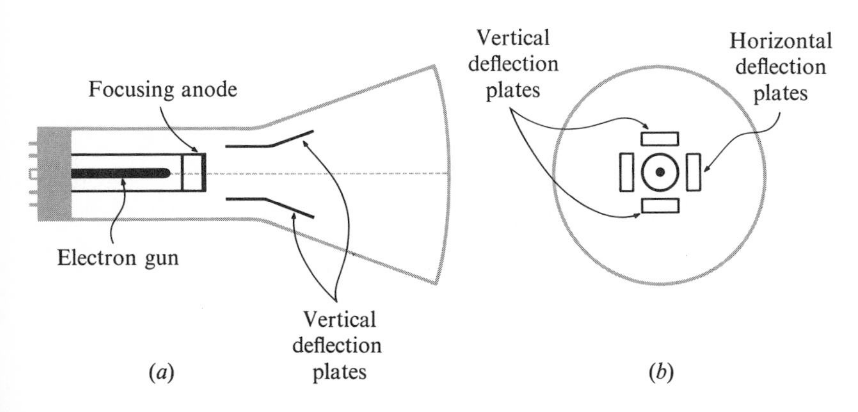

FIG. 4-13 Cross section of a cathode-ray tube: (*a*) side view of the electron gun, focusing anode, and vertical deflection plates; (*b*) face view showing the deflection plates, electron gun, and focusing anode.

amount of voltage applied. Similarly, a positive voltage applied to the right-hand plate and a negative voltage applied to the left-hand plate will pull and push the beam to the right. Reversing the polarity of the voltages will move the beam to the left.

If a sawtooth voltage such as that indicated in Fig. 4-14 is applied to the horizontal deflection plates, the beam would start at the left of the tube face and move to the right. The distance it moves—that is, the length of the trace—depends on the amplitude of the sawtooth waveform. The speed with which it moves from left to right depends on the frequency of the sawtooth wave. The speed with which the trace returns to the left of the screen depends on how fast the voltage drops from its highest positive voltage to its lowest negative voltage.

Figure 4-15 shows the controls of a typical laboratory oscilloscope. There are two beam positioning controls, one for vertical and one for horizontal. Usually these are set so that the beam is centered. There is also a control to vary the vertical gain and a control to vary the horizontal gain. To allow for a wide range of input voltages a vertical attenuator is used. There is also a sawtooth frequency control to lock the waveform in one place.

The first step in using an oscilloscope is to familiarize yourself with the controls. Locate the on-off switch, turn it on, and wait for the internal components to warm up to a stable temperature. This may take 5 min or so. Next advance the intensity control until a spot, or *trace*, appears on the screen. The trace should not be any brighter than needed, since excessive brightness may permanently burn the phosphors on the screen.

When the beam has been located, it should be centered by means of the horizontal and vertical centering controls. It may be necessary to turn up the horizontal gain control to produce a line. Any length of line may be used. Once the line is obtained, it should be focused by means of the focus control until it is as sharp as possible.

FIG. 4-14 (*a*) The sawtooth-shaped voltage applied to the horizontal deflection plates of the CRT, and (*b*) the CRT with deflection plates connected to show functions of vertical and horizontal amplifiers.

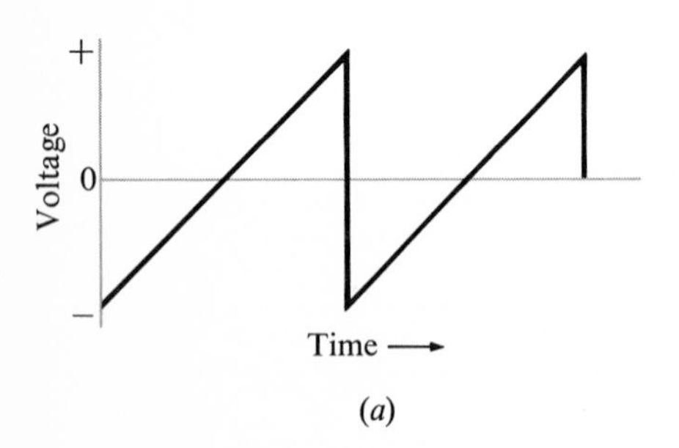

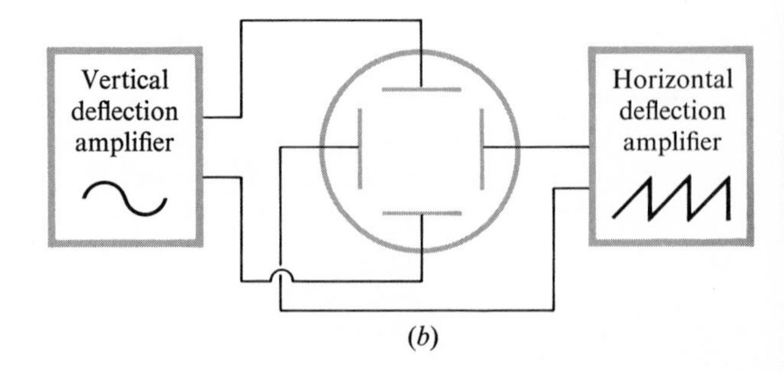

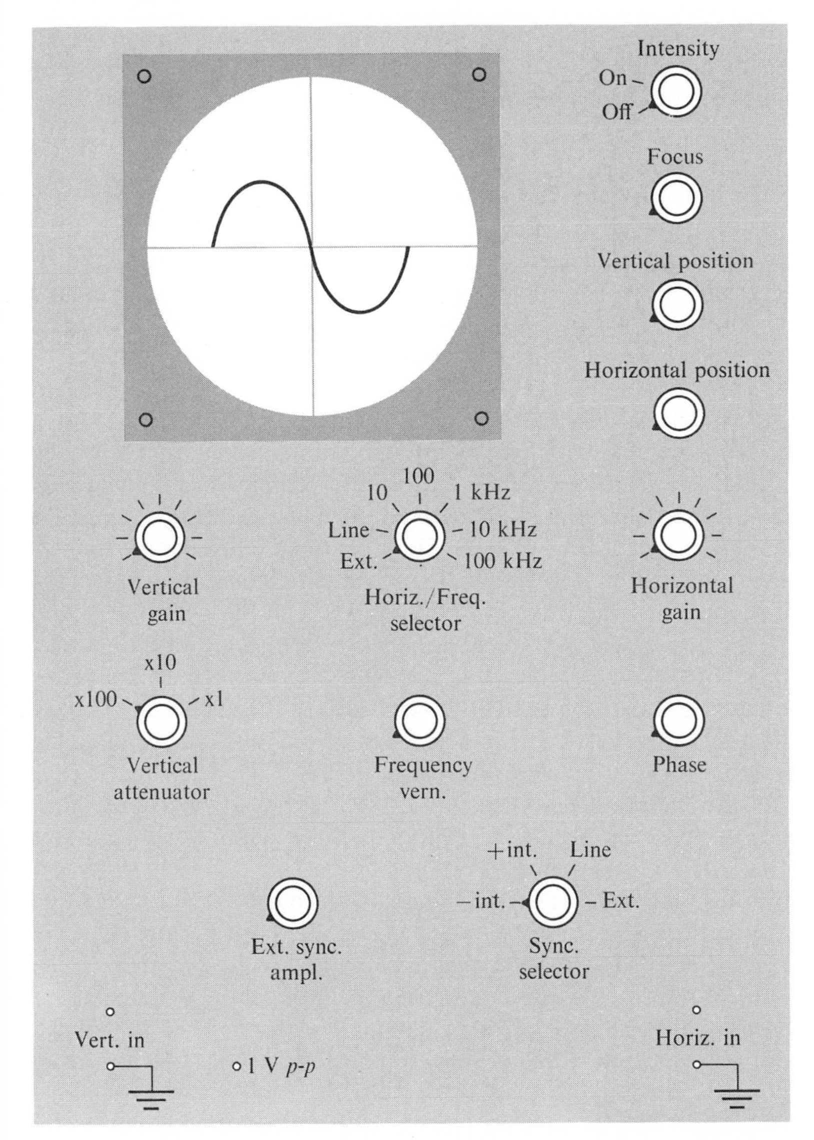

FIG. 4-15 Diagram of a typical laboratory oscilloscope and its controls.

If a sine-wave generator is available, connect its output terminals to the vertical input terminals after some suitable warmup time. Adjust the output of the generator until a signal appears on the screen that is at least two spaces high. Probably the waveform will seem to be moving. Find the "sync" selector switch. It usually has several ranges: plus, minus, line, and external sync. Switch the selector to the plus position. Check the output frequency of the generator and set the coarse frequency switch to the range closest to that of the generator. Adjust the fine frequency control until the waveform comes to a stop. There may be one or more complete cycles showing on the screen. If there is only one, the sawtooth sweep frequency is the same as the input frequency of the generator. If two cycles are showing, the sawtooth generator is operating at just one-half the sine-wave generator frequency.

The vertical gain control and vertical attenuator switch are very important. If the signal voltage at the scope input is too large for the screen when the attenuator switch is set on the X 1 position, reduce the vertical gain. If the signal is still too large, turn the attenuator switch to the X10 or X100 position. The attenuator thus reduces the signal voltage by a factor of 10 or 100, depending on the setting.

VOLTAGE MEASUREMENT

To measure voltage with an oscilloscope the instrument must first be calibrated, and for this a source of accurate voltage must be obtained. Generally the calibrating voltage used on more expensive scopes is a square wave of 1.0 or 10.0 V. On most simple laboratory scopes, however, the source is generally a sine wave from the power line. On some scopes this sine wave is 1.0 V *p-p*, and on others it is 0.4 V *p-p*. In any case the calibrating voltage must be connected to the vertical input terminals, and the vertical gain and attenuator must be set to some convenient size. Usually the amplitude of the calibrating voltage is adjusted by the vertical gain control to give a full-scale reading. On many scopes full scale is 10 divisions.

If the calibrating voltage of 1.0 V covers 10 divisions, each division or space equals 0.1 V. Of course, if the calibration takes place with the attenuator set on the X10 position, it will still yield 0.1 V per division on that scale. If the attenuator is changed to the X1 position and the vertical gain control is not changed, the scope is more sensitive by a factor of 10. Each space now is calibrated to read 0.01 V.

Once the oscilloscope has been calibrated, it is possible to measure a wide range of ac or pulsating-dc voltages. The ac voltage should be connected to the vertical input of the scope, with care that the ground or common point in the circuit being measured is also connected to the ground of the scope. If this precaution is not taken, particularly with circuits connected directly into the ac power line, such as the ac-dc radio, a short circuit may occur between the circuit and the scope.

Do not change the setting of the vertical gain once the calibration is complete. Set the attenuator switch to a range that will allow a waveform larger than one space in height, but less than full scale. Compute the value of voltage for each space for the attenuator position chosen. Count the number of spaces filled by the voltage waveform and multiply by the voltage per space to get the voltage of the waveform.

As an example, let us assume that the scope has been calibrated on the X10 range. If full scale is used for a 1.0-V signal, each space will equal 0.1 V *p-p*. When the attenuator is changed to the X1 position, each space will equal 0.01 V, or 10 mV *p-p*. If the ac signal being measured comes from an amplifier stage in an audio preamplifier, it may be so small that a meter will not respond accurately. Such a signal 2.4 spaces in height would have a value of 2.4 spaces times 0.01 V per space, or 0.024 V *p-p*. Figure 4-16 shows a typical test setup for measurement with the oscilloscope.

Sometimes the voltage waveform observed is from a standard ac line, which has a frequency of 60 Hz. In this case the waveform may be synchronized easily by switching the sync selector switch to *line*. This automatically sets the sawtooth sweep frequency at 60 Hz and locks it in phase with the ac line frequency. The line setting should never be used when the signal is other than 60 Hz, since the signal or waveform can be locked into position only with a sawtooth sweep of its frequency or its harmonic.

THE TRIGGERED OSCILLOSCOPE

One of the most exciting advances in the oscilloscope in recent years is the triggered sweep. This allows the sweep trace to be triggered at the beginning of each waveform and sweep at some particular predetermined and calibrated time base. This automatically locks or stops the waveform for easy accurate viewing. In addition, it allows the horizontal, or *X*, axis to be calibrated in seconds, milliseconds, and microseconds instead of Hertz (cycles per second). The time of each waveform can then be determined accurately rather than by guess.

The heart of the triggered scope is usually a Schmitt trigger circuit, which acts like a high-speed on-off switch. It can be adjusted to turn on when the voltage reaches some predetermined amount, say,

FIG. 4-16 A typical audio-signal generator and oscilloscope set up to monitor circuit action.

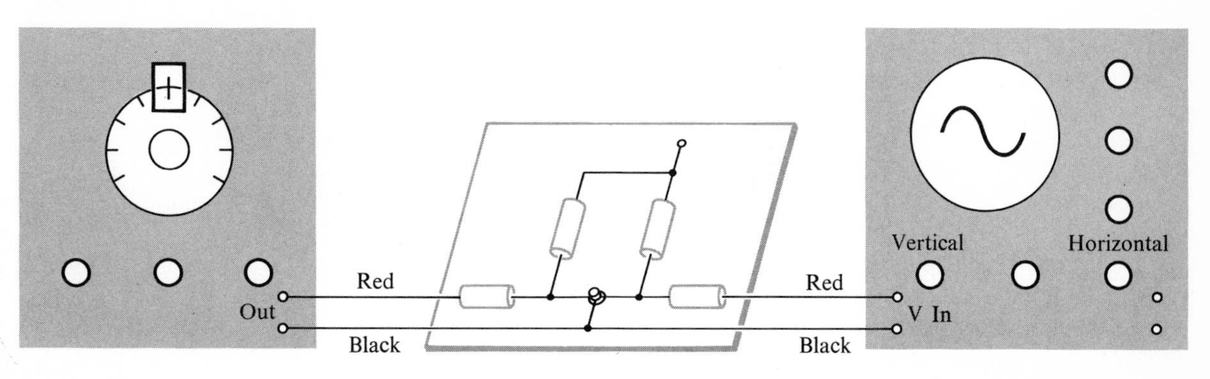

10 μV. It can also be adjusted to turn off at some particular voltage. This means that when no signal voltage is present at the input terminals, no waveform will be visible on the screen. When a waveform voltage exceeds the preset value, the Schmitt trigger circuit will turn on the sawtooth wave sweep and allow the waveform to be displayed.

Some triggered scopes trigger automatically, while others have trigger sensitivity and stability controls. These must usually be adjusted together, especially if the signal is of very low amplitude.

On most triggered scopes the vertical amplifier is also preset or calibrated as a switchable attenuator. The attenuator is calibrated or checked periodically as aging occurs. In other words, the scale is simply set at 10 mV per division, 100 mV per division, 1 V per division, and so on. Of course, if the experimenter wishes to change to the uncalibrated mode of operation for some measurement, this is generally possible. For accurate measurement, however, the control must be in the calibrated position.

DUAL-TRACE AND DUAL-BEAM OSCILLOSCOPES

Often in making amplifier voltage checks it is necessary to compare two voltages. For years this was done by using two oscilloscopes. An electronic switch permitted the scope input to be switched rapidly

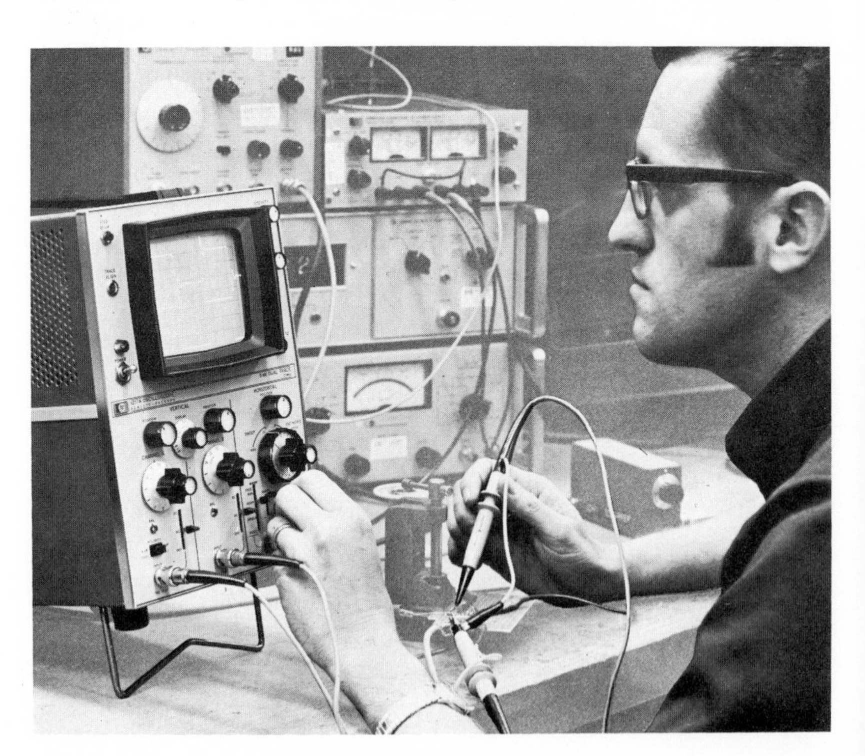

FIG. 4-17 A dual-trace triggered-sweep oscilloscope used to diagnose troubles in a fast-pulse circuit.

from one source to the other. If the switching rate was many times greater than the sweep speed, the two displays were visible at the same time. The same gain controls affected both signals, so that they could be compared easily.

Eventually a scope was developed that allowed a variety of vertical amplifiers to be used by a plug-in method. Two identical amplifiers are frequently made in the same case and are coupled to the scope input by the built-in electronic switch. They are called *dual-trace* vertical amplifiers. Further switching allows them to be used separately or together and added or subtracted algebraically. The two signals may be displayed on alternate traces or by the rapid switching usually called *chopping*.

The dual-trace scope has certain limitations in the frequencies that may be observed together. To overcome these limitations CRTs were developed that have two electron guns. The two signals then go to separate guns, and no switching of signals is necessary.

The advent of the dual-beam scope made possible further manufacturing economies. With two independent beams to work with, each can be fitted with an electronic dual-trace switch to yield a four-trace scope. Since transistors and integrated circuits are so tiny, all this can be put into a case much smaller than the single-trace scope of yesteryear.

FREQUENCY LIMITATIONS

An amplifier is limited in the range of frequencies it can amplify equally. Wideband amplifiers that can also amplify very small signals are very expensive. For this reason most oscilloscopes intended for laboratory use have a frequency limitation for flat response to about 500 kHz. This provides a usable display to about 5 MHz.

Many sophisticated oscilloscopes are capable of displaying signals in the microwave region by means of sampling techniques. Others can measure up to 100 MHz without the need to sample the signal train. However, the fact that a signal does not appear on the oscilloscope does not mean that no signal is present. It may merely be occurring at some frequency higher than that to which the scope will respond. In addition, fast rise pulses may not be displayed in their proper shape because the fast rise time corresponds to some frequency beyond the range of the scope. A good square pulse may thus appear rounded.

THE AUDIO-SIGNAL GENERATOR

Experimenting in electronics is very difficult without some form of signal generator. The usual choice is one that puts out a sine wave at frequencies of about 20 Hz to 20 kHz. However, this is by no means

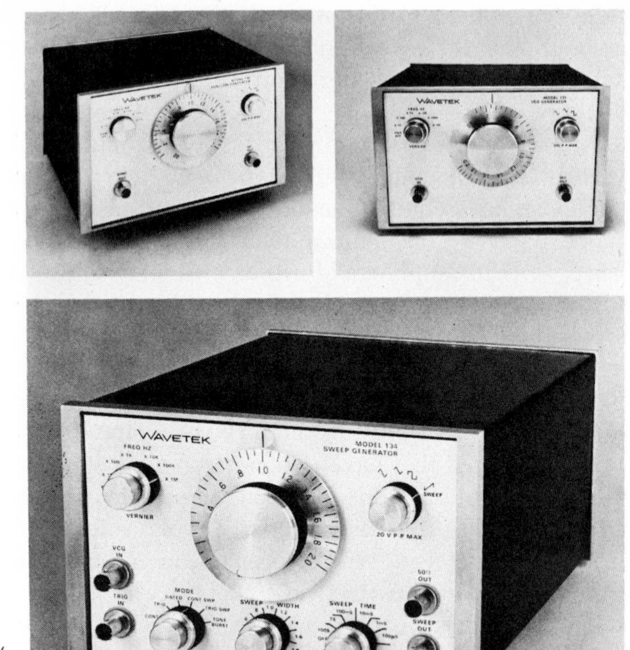

FIG. 4-18 A multifunction signal generator capable of producing various waveforms and of sweeping automatically from a low to a high frequency for fast circuit testing (*courtesy Wavetek*).

the only choice, or even the best one. Many of the newer signal generators also have a square-wave output, which is essential for certain amplifier measurements. The newest addition to the line of audio-signal generators is the *function generator*. It puts out sine waves, square waves, triangular waves, and sometimes pulses and sawtooth waves. The triangle wave offers advantages over the sine wave for amplifier testing. Frequencies available from some function generators extend from 0.001 Hz to over 5 MHz with little distortion.

Figure 4-19 shows the circuit of a simple solid-state sine-wave oscillator that can be built with readily available parts. It can be used to supply the signal for most of the amplifier experiments outlined in this text. The output frequency is between 1 and 2 kHz, the midfrequency area for most audio amplifiers. The output is variable to some degree, and the output impedance is low enough to operate most amplifiers.

For complete experimentation in solid-state electronics a good audio-signal generator is a priority item. An understanding of its use, functions, and limitations is also necessary. For example, its frequency range should be at least from 20 Hz to 20 kHz. This is adequate for audio work, although most transistors will function in amplifiers that

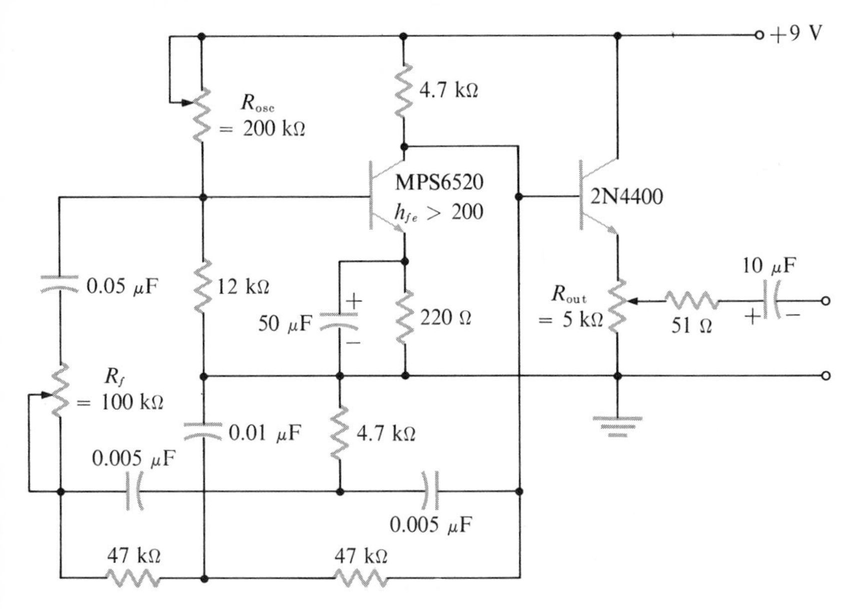

FIG. 4-19 An audio sine-wave oscillator using a twin-T feedback network and operating near 1 kHz. The oscillation is controlled by R_{osc} with waveform distortion controlled by R_f. The output impedance is less than 100 Ω and variable from about 10 mV to several Volts. The oscillator should operate with any silicon transistor having an h_{fe} greater than 200. Oscillation and distortion may be difficult to maintain at 9.0 V with a low h_{fe} transistor.

operate well beyond these limits. A range of 10 Hz to 100 kHz is highly desirable if frequency-response checks are to be made.

As mentioned above, some function generators have response beyond 1 MHz. This range permits both audio and AM radio alignment with a single instrument.

Two controls, and sometimes three, are found on all signal generators. These are the frequency control, the output or amplitude control, and the attenuator range control.

The frequency control usually consists of a range switch—10 to 100 Hz, 100 Hz to 1 kHz, 1 to 10 kHz, etc. Within each range the frequency may be varied from one limit to the next. The accuracy of the frequency dial is rarely better than 1 or 2 percent, but this is of no consequence in most conceptual problems. In most elementary course work only the 1- to 10-kHz range is used. The size and shape of the waveform are usually more important than the frequency.

The output controls warrant special attention. The output impedance of the generator is usually either 600 or 50 Ω, or nearly so. In order to preserve this output impedance, and at the same time preserve the fidelity of the waveform (its freedom from distortion), at low levels an attenuator is generally used.

The output, gain, or attenuation control potentiometer should not be used at its lower quarter-turn. At very low output the generator often displays some distortion. If the signal output is too large, the attenuator switch should be moved to the next position. That is, the

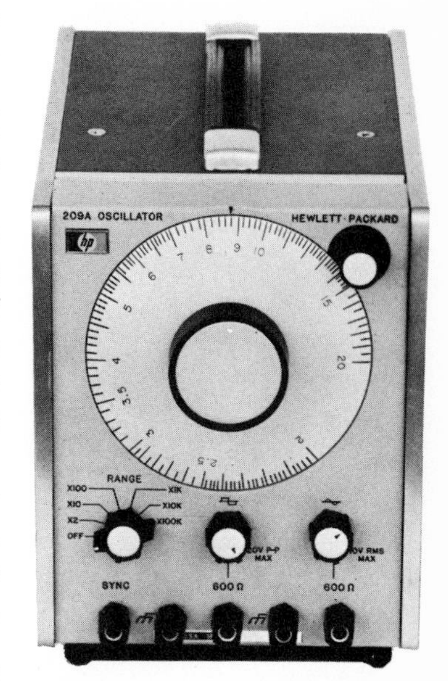

FIG. 4-20 A sine/square wave *RC* Wien bridge oscillator with highly stable wide-frequency-range output (*courtesy Hewlett-Packard*).

output should be reduced by a factor of 10, which will allow the output control to work in the middle of its range.

The attenuator switch may be marked off in decibels (dB). Attenuation by a factor of 10 equals 20 dB, a factor of 100 equals 40 dB, and so on.

Most signal generators are voltage-output instruments and should not be expected to deliver large amounts of current. For solid-state work in general the signal generator with a 50-Ω output impedance seems preferable to the 600-Ω instruments commonly used for vacuum-tube circuitry, although both will work. However, proper loading must always be provided for the generator as well as the circuit.

There are hundreds of different kinds of instruments in electronics, but the VOM, the electronic voltmeter, the oscilloscope, and the audio-signal generator are basic tools and must be mastered before experimentation in electronics can have much meaning.

FIG. 4-21 A low-cost integrated-circuit frequency counter (*courtesy Hewlett-Packard*).

SUMMARY OF CONCEPTS

1. The use of any measuring instrument disturbs an electronic circuit and introduces error.
2. Measurement of voltage requires a high-input-impedance voltmeter, or significant loading of the circuit will result.
3. The ammeter (milliammeter or microammeter) is the basic meter from which the voltmeter and ohmmeter are made.
4. The ammeter is always placed in series with the circuit current it is to measure.
5. The ammeter range may be extended by use of a shunt resistance.
6. The voltmeter is an ammeter with a high resistance to limit its current flow to full scale with a certain applied voltage.
7. The ohmmeter is an ammeter with a series current-limiting resistance and an internal current source. It measures the current flow through the unknown resistance for a known applied voltage.
8. The electronic voltmeter has a very high internal or input impedance. It tends to load a circuit less than the VOM, and its measurements are generally more accurate.
9. The oscilloscope provides a means of measuring both the size and shape of an alternating voltage.
10. The triggered scope displays waveforms which themselves trigger the trace. They are usually calibrated with a time base in seconds per division on the horizontal axis and Volts per division on the vertical axis.
11. Most audio-signal generators deliver a sine wave that is relatively pure and variable in amplitude.
12. The frequency range of most audio-signal generators is 20 Hz to 20 kHz but may go much higher.
13. Function generators are a source of several shapes of ac waveforms: sine waves, square waves, triangular waves, sawtooth waves, and sometimes pulses.
14. Common output impedances of signal generators are 600 Ω and 50 Ω; 50 Ω is better for solid-state experimenting.

GLOSSARY

Attenuator A device or circuit that lowers the voltage or power level of a signal.

Audio Generally the frequency range of 20 Hz to 20 kHz, but may extend lower or much higher.

Calibrate To adjust an instrument so that it will measure accurately on a particular scale.

Cathode-ray tube (CRT) The picture tube in an oscilloscope whose coated face will glow at the point where it is struck by cathode rays from the electron beam.

Deflection The amount of movement of an indicating device, such as a meter needle or scope trace, due to some change in voltage, current, or resistance.

Field-effect transistor (FET) A high-input-impedance transistor whose current is controlled by input voltage rather than current.

Full-wave bridge rectifier A device that can convert a whole sine wave into direct current.

Horizontal gain The oscilloscope control that controls the length of the horizontal trace.

Load The amount of current or power delivered by a circuit or system to a resistance or impedance of another circuit, often referred to as the resistance that limits the output current or power.

Loading error The error introduced by the impedance of the meter when voltage or current is measured.

Meter sensitivity The reciprocal of the full-scale current of an ammeter when it is used as a voltmeter.

Ohmmeter zeroing Adjustment of the meter's resistance to compensate for battery aging and to produce full-scale deflection (to 0 Ω) when leads are shorted.

Parallax error The error introduced in viewing two coinciding points from some angle, so that they appear not to coincide.

Schmitt trigger A circuit whose output voltage is high when the input is above a fixed voltage and low when it drops below a fixed point.

Vertical gain The control that determines the amount of vertical movement allowed in the oscilloscope trace.

REFERENCES

Gillie, Angelo C.: *Electrical Principles of Electronics*, 2d ed., McGraw-Hill Book Company, New York, 1969, pp. 220–253.

Grob, Bernard: *Basic Electronics*, 2d ed., McGraw-Hill Book Company, New York, 1965, pp. 116–141, 497.

Herrick, Clyde N.: *Electronic Circuits*, Charles E. Merrill Books, Inc., Columbus, Ohio, 1968, pp. 128–164.

Hickey, Henry V., and William M. Villines, Jr.: *Elements of Electronics*, 3d ed., McGraw-Hill Book Company, New York, 1970, pp. 160–206, 454–461.

Lippin, Gerard: *Circuit Problems and Solutions*, Hayden Book Company, Inc., New York, 1967, pp. 41–51.

Malvino, Albert Paul: *Electronic Instrumentation Fundamentals*, McGraw-Hill Book Company, New York, 1967.

Middleton, Robert G., and Milton Goldstein: *Basic Electricity for Electronics*, Holt, Rinehart and Winston, Inc., New York, 1966.

Partridge, G. R.: *Principles of Electronic Instruments*, Prentice-Hall, Inc., Englewood Cliffs, N.J., 1960.

Shrader, Robert L.: *Electrical Fundamentals for Technicians*, McGraw-Hill Book Company, New York, 1969, pp. 147–191.

Weick, Carl B.: *Principles of Electronic Technology*, McGraw-Hill Book Company, New York, 1969, pp. 137–161.

Zbar, Paul B.: *Electronic Instruments and Measurements*, McGraw-Hill Book Company, New York, 1965.

REVIEW QUESTIONS

1. What is meant by loading an electronic circuit?

2. If a VOM has ranges of 0 to 150 μA, 0 to 1.5 mA, 0 to 5 mA, 0 to 15 mA, and 0 to 150 mA, which range should be used to measure a current of 4.5 mA?

3. What is meant by calibration error?

4. Draw a circuit showing how to connect a milliammeter in a three-resistor series circuit.

5. Draw a circuit showing how to connect a voltmeter into the series circuit below to measure the voltage across R_1. Watch the polarity.

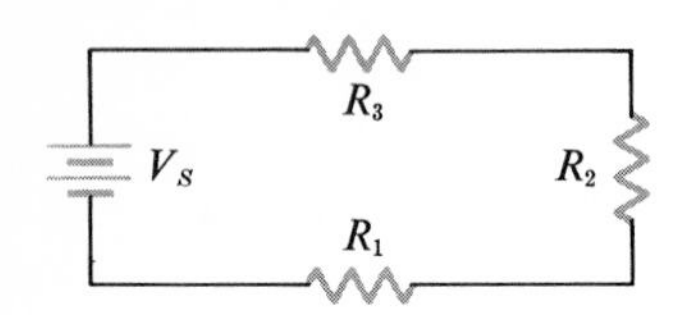

6. Why is a high input impedance needed for a voltmeter?

7. What is the function of the series resistance in a voltmeter?

8. What is meant by loading error?

9. Why is it necessary to zero an ohmmeter before making any measurement?

10. What are the differences between a VOM and an electronic voltmeter?

11. Explain the function of an ammeter shunt.

12. What are the functions of the deflection plates in the CRT of an oscilloscope?

13. Explain the reason for a sawtooth waveform for horizontal deflection.

14. What are the steps in measuring a 2-V *p-p* 1-kHz signal with a laboratory scope? A 0.5-V 60-Hz signal?

15. Why should the vertical gain control not be moved once the scope has been calibrated?

16. What precautions must be taken in measuring voltages in ac-dc equipment?

17. What is the difference between a laboratory scope and a triggered scope?

18. What is the difference between a dual-trace scope and a dual-beam scope?

19. What is the sequence of operations in setting an audio-signal generator to 400 Hz at 5 V *p-p*?

20. What range of frequencies is generally used in testing an audio amplifier?

21. What output impedances are commonly used for signal generators?

PROBLEMS

1. If a milliammeter has a full-scale deflection of 1.0 mA and a resistance of 100 Ω, what voltage would cause full-scale deflection of the meter?

2. If the calibration error of a milliammeter is ±3 percent and the meter reads 2.0 mA, what are the actual limits of current in the error of the reading?

3. Find the sensitivity of a voltmeter made from a 150-μA meter.

4. In the circuit below, what is the voltage across resistance R_2? What is the voltage when the indicated voltmeter is attached?

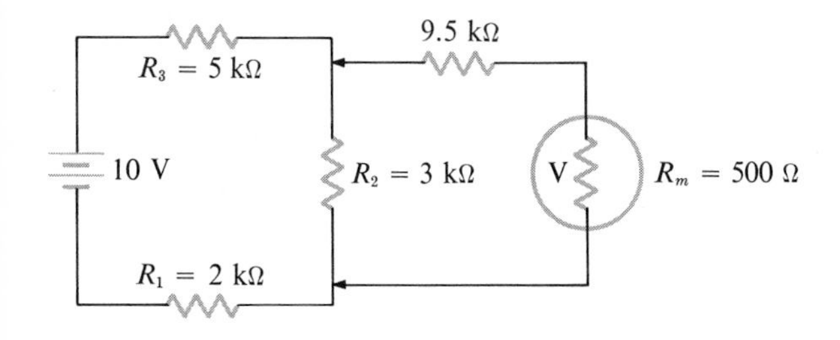

5. In Prob. 4, what is the input resistance of the voltmeter?

6. If a voltmeter resistance is 200 Ω and the series resistance for the 1.0-V range is 9.8 kΩ, what multiplier resistance would be needed to convert the meter to a 10-V range?

7. If a 100-μA full-scale meter needs to be converted to a 100-mA meter and the meter resistance is 1,000 Ω, what value of shunt resistance will be needed?

8. On a scope calibrated for 10 divisions equal to 1 V peak to peak on the X1 scale a signal is observed to fill four spaces. What is the voltage of the signal?

9. In Prob. 8, what is the value of each space on the X10 attenuator scale of the scope?

10. If a signal is found to be 4.7 divisions high on a triggered scope set at 10 mV per division, what is the voltage of the signal?

11. If the output frequency of a signal generator is set at 50 on the scale and the multiplier switch is set on the $R \times 100$ Hz range, what is the output frequency?

12. If the output of the signal generator in Prob. 11 is 3 V peak to peak on the X1 attenuator and the attenuator is reset to the X10 or −20-dB position, what is the new output voltage?

13. If a triggered scope is set on 10 μs per division and the period of one cycle displayed on the scope is 5.5 divisions, what is the time of one period? What is the frequency?

PART TWO

CHAPTER 5

OBJECTIVES

1. To introduce the conceptual basis of conduction in solid materials
2. To consider the concept of energy in an atomic structure as it relates to conduction
3. To investigate the nature of ionic and covalent bonding in the structure of crystalline substances
4. To compare the electron valence-band and conduction-band energy levels
5. To introduce the concepts of intrinsic and extrinsic current carriers
6. To compare random carrier diffusion to directional current-carrier drift
7. To consider the nature of the semiconductor energy levels, the valence band, and the conduction band
8. To consider the nature of the *NP* junction and its related depleted regions and potential barrier
9. To discuss hole current as it relates to the forward current across an *NP* junction
10. To consider the forward and reverse characteristics of the junction diode

CONDUCTION, THE SEMICONDUCTOR, AND THE *NP* JUNCTION

One of the major stumbling blocks in learning about diodes, transistors, and integrated circuits is the special meaning of apparently familiar words. Terms such as "hole," "carrier," "intrinsic," "extrinsic," "conventional current," "valence band," and "quantum" are often confusing even to the initiated, because their meanings seem to have been changed with new developments. In reality the meanings have not changed, but it is likely that the correct or complete meaning was not learned in the first place. The purpose of this chapter is to introduce the functional concepts needed to understand semiconductors regardless of the circumstances in which they are employed. Whether a semiconductor is used in a transistor radio or the latest space probe, the theory behind it is the same.

ENERGY

The key to an understanding of solid-state devices is the concept of energy. Energy may take such varied forms as heat, light, electrochemical, mechanical, potential, kinetic, and others perhaps less familiar. To start with, let us concern ourselves with kinetic and potential energy.

When a body is at rest, no matter how small the body happens to be, it may have energy due just to its position. That is, it may fall from this position and do some work in falling. Thus in its position prior to falling it has some amount of potential energy; that is, it is potentially able to do some work. Since work is defined as moving something through some distance, this potential, or stored, energy must be converted to energy motion, or kinetic energy, in order to do the work. The conversion of energy from one form to another must always be in balance. Thus the amount of potential energy lost by the particle or body must be equal to the other forms of energy gained by the particle or body. In this case the *total amount* of energy of the particle remains the same; it simply changes from potential energy to kinetic energy. In doing work, some of the kinetic energy is transferred, leaving the particle with less potential energy.

The picture becomes more complex when we turn to tiny particles such as atoms, electrons, and protons. An electron is never at rest. It always has kinetic energy. It has, or is said to have, a certain quantity of energy called a *quantum*. Specifically, the motion of the electron is described by its various *quanta* or quantum numbers. The electron spins on its axis, and it may spin in either direction (but not both). It may be in any one of a large number of possible orbits, each of which has a different energy level. No two orbits have exactly the same amount of energy, although many energy levels are very nearly the same. This electron may lose or gain energy only in quantum units; that is, it may gain sufficient energy to move to a higher-energy orbit, or it may lose energy and go to a lower available orbit. The energy lost may be emitted in the form of light or heat. By the same token, the energy gained may be by the absorption of light or heat energy.

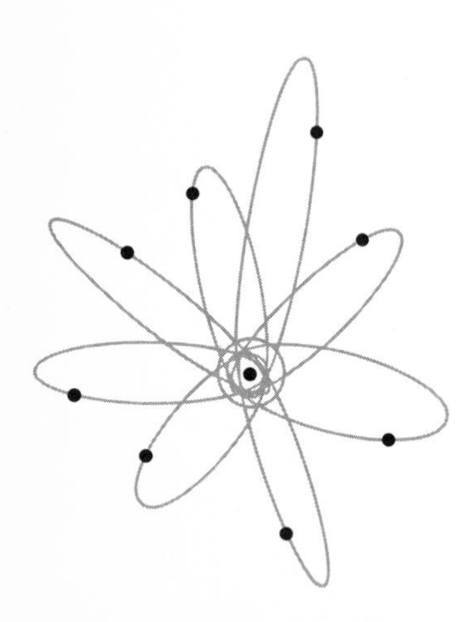

FIG. 5-1 Typical circular and elliptical orbits of electrons around the positive nucleus.

The fact that some orbits have energies that are very nearly the same does not mean that the orbits themselves are the same. An electron may travel in a circular orbit, a highly elliptical orbit, or one not so elliptical and still have almost the same energy. The transition from one type of orbit to the next may be very easy, then, if the material is simply heated or exposed to light.

Occasionally electrons may absorb so much energy that they are able to leave the atom altogether and become free. They may travel some distance through the material before they are captured by some other atom with the right energy level open. It was Einstein's famous *photoelectric* theory and equation that opened the door to many discoveries about which we are concerned. According to this equation, the energy acquired by an electron exposed to light is equal to *Planck's constant* (a mathematical constant used in quantum physics) times the frequency of the light striking the electron. That is,

$$E = hf \tag{5-1}$$

where E = energy
h = Planck's constant, 6.56×10^{-27} erg-s
f = frequency, Hz

The total amount of energy released by the material depends on the total number of electrons released, but the energy of each electron depends on the frequency of the light.

We know much more about the nature of the electron and its energy levels today, but the basic concept still holds. We know, for instance, that atoms are in a continual state of agitation, with electrons jumping from one state to another as they become available. This is most noticeable in the shiny metals. The shine of a metal is due to the absorption and release of energy as the electrons drop back into the atom. An electron may absorb enough heat energy to escape from the atom. This same electron may be captured by some other atom and fall into a different orbit, thus releasing a different amount of energy from that which it absorbed. The different energy is emitted in the form of visible light, rather than invisible light, or heat.

This freeness of electrons in metals is also what gives them their conductive properties. Since the normal position of the electron is in its orbit, it is said to be in its chemically ready or *valence state*. When it is released from the valence state, or band of energy, it is said to go into its *conduction state*, or the conduction band of energy. With conductor metals the two energy levels, the valence band and the conduction band, are so close in nature that in effect they overlap. In other materials, such as insulators, the electrons are tightly held in their orbits by the atom and thus find it difficult to escape. As a result they are practically nonconductors.

The nonconductor, or insulator, is interesting in that the orbits are almost all filled. The electrons must acquire a great amount of energy while they are in the valence band in order to move into the conduction band. They can do so, however—as you may know if you have tried to handle high-voltage wires with thin rubber gloves. In fact under certain conditions nonconductors such as wood, glass, rubber, and plastic may actually become good conductors.

BONDING

Metallic substances such as copper, silver, gold, and lead will unite with nonmetallic substances such as oxygen, chlorine, sulfur, and other materials by means of a *chemical bonding*. In this bonding the metallic substance having one or two relatively free electrons will give them up to nonmetallic substances. The two materials then undergo a chemical change known as *ionic bonding*. Copper and oxygen, for example, thus form copper oxide of one kind or another. In the reaction the copper loses its two outermost valence electrons to the oxygen, which needs two valence electrons. The resulting substance is quite stable and forms rapidly. To see this yourself, heat some shiny copper for a moment and watch the oxide form. In this reaction the copper atom became a copper *ion*, having a double positive charge, and the oxygen became an oxide *ion*, having a double negative charge.

The reason for this is that the most stable form for the ion to adopt is one in which its valence energy levels are complete. This generally requires eight electrons. Since copper has only two valence electrons in its neutral state, it is easier to give them up than to acquire an additional six. By the same token, oxygen needs only two valence electrons to be complete and so may obtain them easily from the "free" electrons of the copper. The compounds thus formed are said to be *ionic* because the atoms involved have given up or taken on valence electrons to become *charged*.

Covalent, or share, *bonding* may occur between elements of the same or other types without exchange of electrons. Some metal-like elements, as well as all crystals, form definite shapes as a result of this type of bonding. It is of interest here because semiconductor materials such as germanium and silicon crystalize by means of covalent bonding.

Elements in the group IV of the periodic table of elements (see Table 5-1) have four valence electrons. In order to become a stable grouping of atoms into molecules these elements must either lose four electrons, gain four electrons, or *share* four electrons. In the silicon or germanium crystal the structure is such that each atom is surrounded by four other atoms. Each of these four surrounding atoms shares one

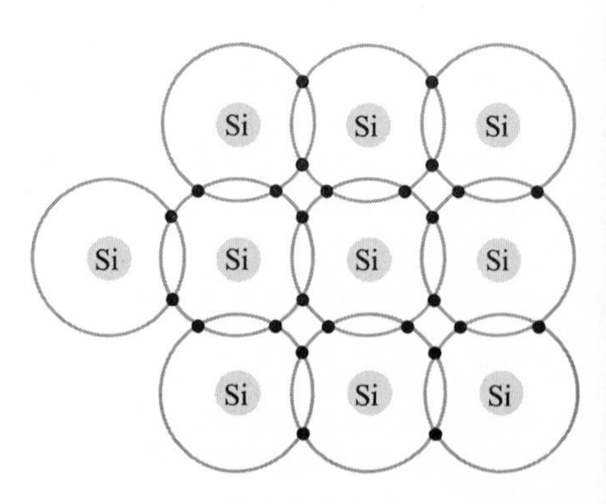

FIG. 5-2 A schematic representation of the sharing of electrons in a covalent bonding in the silicon crystal.

of its valence, or outer-shell, electrons with the central atom. With these four shared valence electrons, the valence *shell*, or energy level, of the central atom is in a stable configuration of eight electrons. These electrons are bound rather tightly and require significant amounts of energy to escape to the conduction band.

In the silicon crystal structure, or crystal lattice, each atom is surrounded by four other silicon atoms which share in the covalent bonding. Of course, the atoms on the surface have slightly different orientation, but the general concept still holds. The result is a strong crystalline structure, as indicated in Fig. 5-2.

VALENCE BANDS AND CONDUCTION

It can be shown mathematically that there are several possible energy levels or orbits that electrons in silicon, germanium, or copper may occupy which are not occupied. Instead, because of the bond energy, the temperature, and the closeness of other forces, the electrons generally find themselves tightly bound in the lower-energy orbits. In the case of copper, a good conductor, the distance between atoms is such that the attraction of the electron by the positive nucleus and the

TABLE 5-1 Periodic table of the elements

IA																	VIII
1																	2
H	IIA											IIIA	IVA	VA	VIA	VIIA	He
3	4											5	6	7	8	9	10
Li	Be											B	C	N	O	F	Ne
11	12											13	14	15	16	17	18
Na	Mg	IIIB	IVB	VB	VIB	VIIB		VIII		IB	IIB	Al	Si	P	S	Cl	Ar
19	20	21	22	23	24	25	26	27	28	29	30	31	32	33	34	35	36
K	Ca	Sc	Ti	V	Cr	Mn	Fe	Co	Ni	Cu	Zn	Ga	Ge	As	Se	Br	Kr
37	38	39	40	41	42	43	44	45	46	47	48	49	50	51	52	53	54
Rb	Sr	Y	Zr	Nb	Mo	Tc	Ru	Rh	Pd	Ag	Cd	In	Sn	Sb	Te	I	Xe
55	56	57	72	73	74	75	76	77	78	79	80	81	82	83	84	85	86
Cs	Ba	La	Hf	Ta	W	Re	Os	Ir	Pt	Au	Hg	Il	Pb	Bi	Po	At	Rn
87	88	89															
Fr	Ra	Ac															
				58	59	60	61	62	63	64	65	66	67	68	69	70	71
Lanthanides				Ce	Pr	Nd	Pm	Sm	Eu	Gd	Tb	Dy	Ho	Er	Tm	Yb	Lu
				90	91	92	93	94	95	96	97	98	99	100	101	102	103
Actinides				Th	Pa	U	Np	Pu	Am	Cm	Bk	Cf	Es	Fm	Md	No	Lw

repulsion by other electrons causes a shifting of the various possible energy levels. As a result, there exists a group of possible orbits virtually free of the influence of the crystal atoms which are in fact at lower energy than some orbits in the valence band. When an electron in one of these outer orbits makes the transition to one of these "free" orbits, it is virtually free to leave the atom under the influence of very little additional energy. Thus it is said to be in the conduction band or, more correctly, in one of the energy levels of the conduction band. In other words, this electron can now become part of a current flow. It is a *current carrier*. Note again that it is these outermost valence electrons which make up the electrical properties of a substance.

CURRENT CARRIERS

As indicated earlier, electrons are always in motion and are influenced by many energy sources. The absorption of heat or light energy, which causes an electron to move or make the transistion to the conduction band, gives only a random motion to the free electron. Remember that there are billions of such electrons in any conductor. This random, nondirectional motion does not constitute current flow. Since velocity means speed in a particular direction, each free electron travels at a different velocity. Moreover, it does not travel very far. The average distance traveled at room temperature is about 10^{-7} m, or about one-tenth of a millionth of a meter, before it is recaptured by another atom. Of course, in terms of the distance between atoms, this is a long trip.

DRIFT CURRENT

When a potential difference, or a voltage, is placed across a length of conductor, the electric field that is established tends to cause the free electrons to *drift* toward the positive terminal of the voltage source. This does not mean that they move in a straight line. Electrons in random motion are affected by the electric field, but they are also affected by atoms nearby. Thus they may change direction many times in their short life span. The net result, however, is that many short-lived carriers are transported first in one direction, and then in another, but their general drift is in the direction of the positive potential of the source voltage.

Since the free electrons, which are the current carriers, move in a net direction under the influence of an electric field, they become a *drift current*. It is impossible to know how far each individual electron, or current carrier, travels, or how many times it changes direction or even goes backward. It is possible to know how many go into one end of a conductor and how many come out the other end; the number is the same.

Conduction, then, is the result of many actions and energies. The electron that is shared in a covalent bond somehow breaks away by absorbing some energy from heat or light or electric field. It moves randomly unless an electric field is present. Under the influence of an electric field it will move in a zigzag path in the general direction of the positive potential source. During this trip it may be recaptured by some vacant covalent bond, but another electron will take its place. This *current flow in a general direction due to an electric field is known as a drift current.*

CURRENT FLOW IN SEMICONDUCTORS

Elements in group IV of the periodic table have four valence electrons. These elements, particularly germanium and silicon, display electrical properties that are neither those of conductors nor those of insulators, but are somewhere between these two extremes. Compounds of other elements evidence similar properties. Notable among these are galena lead, copper oxide, and gallium arsenide.

These properties may be better understood in terms of our previous discussion of energy levels. Recall that in conductors electrons in the covalent bonds can easily escape from the valence levels to the conduction levels, because the energies of these two bands are so nearly the same that in some cases they even overlap. In insulators there is a considerable difference in energy between the valence levels and the conduction levels—sufficient to constitute a *forbidden gap.* A great amount of energy would have to be added for an electron to cross this gap.

The semiconductor lies between these two extremes. Although there is a forbidden gap between the conduction and the valence bands, less energy is needed for electrons to make the transition than in insulator materials. In its pure state, a semiconductor is not a good conductor in any sense of the word. However, as we shall see, it can be made to conduct rather easily. At room temperature the forbidden energy gap for silicon amounts to about 1.12 electron Volts (eV); for germanium it is only 0.72 eV. The gap becomes greater as the temperature is decreased. However, as the temperature is increased and the electrons may absorb more energy, the gap is correspondingly reduced.

As the temperature rises, a greater number of available current carriers are generated in a semiconductor, since more can cross the forbidden gap. At room temperature, about 27°C (Celsius) or 300° K (Kelvin), the carrier concentration per cubic centimeter for silicon is about 2×10^{10}, or about 20 billion. For germanium the concentration is a thousand times greater, about 2.5×10^{13} carriers per cubic centimeter.

INTRINSIC CONDUCTIVITY

The covalent bonds of germanium and silicon are quite like the covalent bonds in carbon. The main difference, of course, is in the binding energy that holds the electrons in the bonds. When an electron does absorb enough energy to escape from the bond, it leaves behind a vacancy, or *hole*. Note that when the electron leaves the bond and goes into the conduction band, the hole is left behind in the valence band. The electron carries a negative charge. However, the hole it leaves behind in effect carries a *positive* charge, because there is now one more proton in the nucleus than there are electrons around the nucleus. Since the hole in a covalent bond is shared by two atoms, this positive charge is spread over considerably more territory than the negative charge of the electron.

The thermal generation of the free electron and its corresponding hole, as shown in Fig. 5-3, is often referred to as *electron-hole pair generation*. Since they are *intrinsic*, or belong to the pure form of the element, the electron-hole pairs are called *intrinsic carriers*. Of these two intrinsic carriers, as indicated in Fig. 5-4, the electron in the conduction band has the higher energy, and hence the higher velocity. In contrast, the other half of the electron-hole pair, the hole in the valence band, can move only very slowly by comparison.

Hole movement The movement of the hole is a concept often difficult to grasp. Note that the hole in the valence band is a vacant orbit of some low energy. Suppose another valence orbit with a similar low energy is occupied by an electron. If this electron gains or loses the right amount of energy, it can move into the vacant hole. Note that

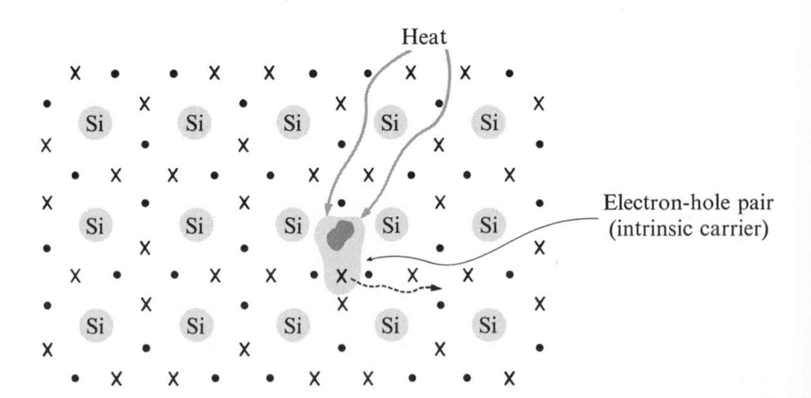

FIG. 5-3 Electron-hole-pair generation due to absorption of heat.

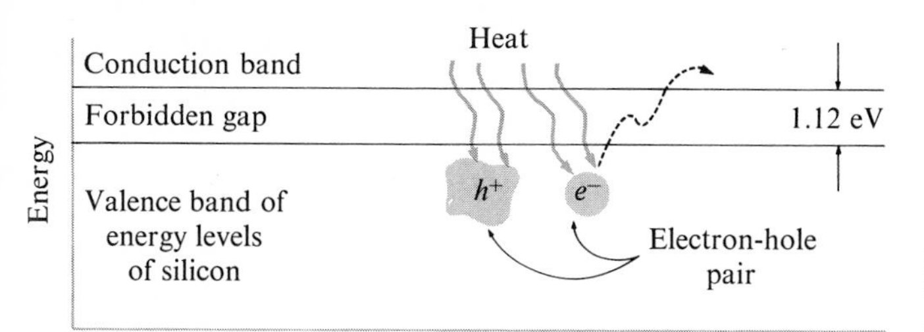

FIG. 5-4 Energy-level diagram showing the intrinsic hole in the valence band and the intrinsic electron moving to the conduction band.

this electron is moving within the valence band; it has *not* gone into the conduction band. The *hole* then moves to the place formerly occupied by the electron. This movement of the hole, which takes place entirely in the valence band, is very slow compared with movement of the electron in the higher-energy conduction band.

The hole continues to move in this fashion until an electron from the *conduction* band loses energy and returns to the valence band to recombine with it. Thus, although the mechanism of hole movement involves only the valence band of energy, the mechanism of *recombination* involves both the *conduction-band electrons* and the *valence-band holes*.

The current due to these intrinsic carriers is quite small. It is due almost entirely to thermally generated electron-hole pairs. Theoretically, at a temperature of zero it would cease altogether.

EXTRINSIC CARRIERS

As we have seen, the germanium or silicon crystal in its pure form has covalent bonds that produce a stable configuration of eight electrons around each atom in the crystal. Four of these electrons are owned and four are shared.

Few crystals, however, are completely pure. Most have impurities in their structure. Generally these impurities do not have exactly four electrons in their valence orbits; they may have three, or five, or some other number. If an impurity has five valence electrons, as shown in Fig. 5-5, four of these will bond as usual. The fifth is left free to wander about the crystal. The material is still electrically neutral in that the total number of electrons equals the total number of protons. However, since only four electrons are needed to fill the covalent bonds, there is one free electron that does not belong to any bond. This electron is known as an *extrinsic current carrier* because it is extra and does not belong to the original system.

If the impurity has only three valence electrons, when it bonds to the semiconductor atoms the bond will have one vacancy, an *extrinsic hole*. This hole, like the extrinsic electron, is an imperfection in the crystal. It is these imperfections that are the basis of the *extrinsic*

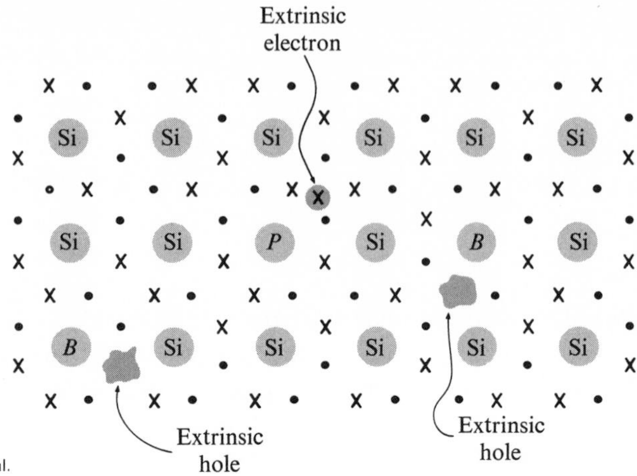

FIG. 5-5 Extrinsic carriers due to impurities in a crystal.

conductivity necessary for semiconductor action. All that is needed to produce a semiconductor is the controlled diffusion of the right impurity into the silicon or germanium crystal, a process known as *doping*.

DOPING

Pure silicon or germanium in its molten form can be doped with a controlled amount of the desired impurity. To make silicon with excess electrons, the impurity is generally phosphorus, arsenic, or antimony. Since the conductivity of the doped semiconductor material depends on the *degree* of impurity, this doping must be closely controlled. Impurities that come from group V of the periodic table introduce extra electrons to the semiconductor crystal. They donate electrons and therefore are known as *donor* impurities. Since the extrinsic carriers thus donated are negative electrons, the crystal becomes an *N-type* semiconductor. Although the carriers are negative, the material itself remains electrically neutral.

If the impurities are group III elements such as boron, aluminum, gallium, or indium, the material will be doped with excess extrinsic holes. Since the hole is an acceptor of electrons, the result is an *acceptor*, or *P-type*, material, because the hole has a positive charge. Again, the material is electrically neutral. It simply has vacancies in the bonds that are going to accept electrons.

FORBIDDEN ENERGY GAP

A semiconductor material that has been doped takes on the properties of both the doping agent and the original material. Recall that the forbidden gap in pure silicon is about 1.12 eV at room temperature. When silicon is doped with phosphorus, making it into *N*-type silicon, the forbidden gap is reduced to 0.039 eV. That is, only 0.039 eV of energy is needed to remove electrons from the valence orbit and put them into the conduction band. If the silicon is doped with arsenic, 0.049 eV is needed for the transition to the conduction band.

Similarly, silicon doped with boron, a group III impurity, has a forbidden gap of only 0.057 eV. However, in this case the mechanism of conduction is slightly different. When the impurity is from group III, instead of a filled electron energy state or orbit, an empty orbit is introduced. The electrons from the impurity atom lose energy and attach themselves to the silicon valence band. An intrinsic hole therefore exists slightly above the energy levels of the valence band. When a valence electron from some *other* covalent bond acquires a small amount of energy, this hole accepts the electron and another hole is generated—this time an extrinsic hole. The hole generated is in the valence band, and the electron that generated it by leaving is in an acceptor band just above the valence band.

Although we have gone into some detail here, an understanding of semiconductor action will stand us in good stead as we consider current flow in semiconductors. Figure 5-6 illustrates how the carriers are distributed in both *P*-type and *N*-type materials. Note that in no case are there holes in the conduction band. The *N*-type material has both intrinsic and extrinsic electrons in the conduction band, but only intrinsic holes in the valence band. In the *P*-type material the converse is true. There are only a few intrinsic electrons in the conduction band, but the valence band contains both intrinsic and extrinsic holes.

FIG. 5-6 Distribution of intrinsic and extrinsic current carriers on either side of the forbidden energy gap.

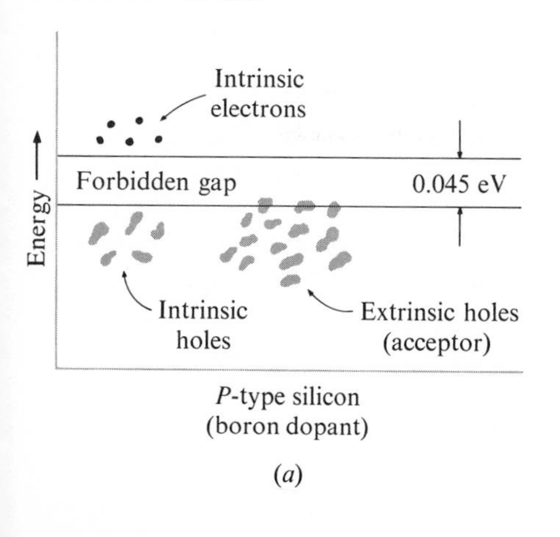

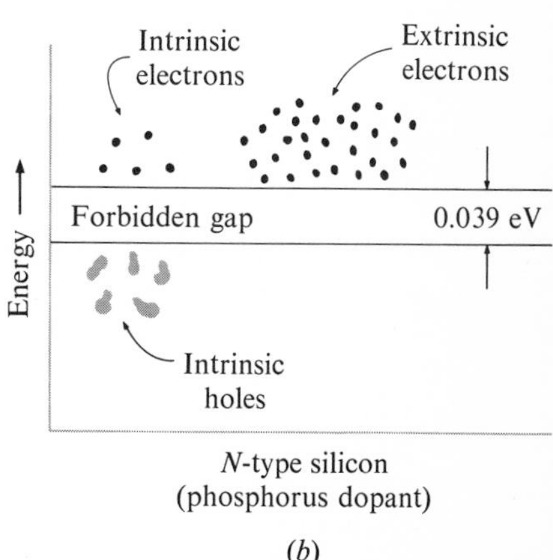

MINORITY AND MAJORITY CARRIERS

The extrinsic current carriers, electrons in the N-type material and holes in the P-type material, make up the largest number of carriers present in the crystals. For this reason they are known as the *majority current carriers*. In contrast, there are few intrinsic current carriers. The number changes with temperature, amount of light, and radiation conditions. The intrinsic carriers make up a minority of the carriers involved in the current and are referred to as *minority carriers*.

As indicated earlier, at a given temperature the number of minority carriers in germanium is about a thousand times more than in silicon. This is primarily because of the larger size of the germanium atom. The problems associated with minority carriers – those problems associated with the changes in temperature – are therefore a thousand times greater in germanium devices than in silicon devices. However, the energy gap for germanium is only about one-third that for silicon, and the mobility of the carriers in germanium is thus considerably greater. In most cases the problems associated with minority current carriers outweigh the need for a small forbidden energy gap and higher carrier mobility. For this reason silicon devices are in greater demand, especially where stability under temperature variation is a concern.

CURRENT-CARRIER DRIFT

Now that we have discussed majority electrons in the conduction band and majority holes in the valence band, let us consider the concept of drift current. If a piece of doped or extrinsic silicon of the N type is placed in a circuit as shown in Fig. 5-7, we have the following situation. The majority electrons in the conduction band are free to move under the influence of the electric field and will move or drift toward the positive terminal. Similarly, the minority or intrinsic electrons in the conduction band will also drift toward the positive terminal. However, the intrinsic *holes* in the valence band, which are positive, will move toward the negative terminal.

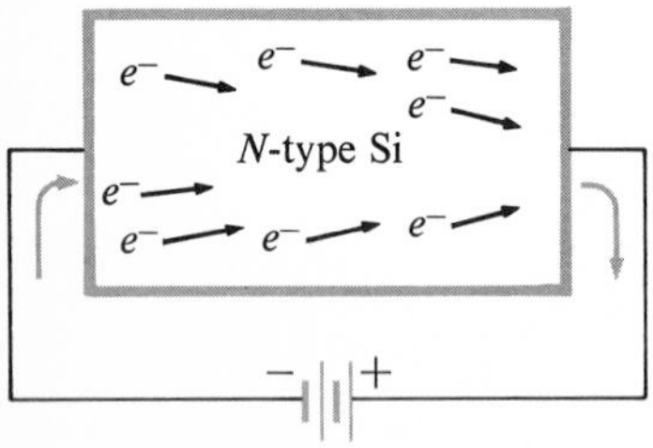

The electrons in the conduction band are able to move at much greater speed than the holes in the valence band. They are free to drift, whereas the holes may move only by the displacement of a nearby valence electron. Moreover, the bond electrons in the valence state are bound more tightly than the conduction-state electrons, so that the hole movement in N-type silicon is not only slower, but very small in amount. Under normal conditions the minority hole current in silicon is scarcely measurable in relation to the majority current.

FIG. 5-7 The majority carriers (electrons) in N-type silicon drift toward the positive terminal of the power supply.

INTRINSIC-CARRIER MOBILITY

The intrinsic electrons and holes released by thermal action both experience some difficulty in moving through the crystal lattice. However, since the electrons in the conduction band have higher energy, they are less impeded by the crystal than the electrons responsible for hole drift in the valence band. The net result is that the electrons in the conduction band are about three times as mobile as the holes in the valence band.

The mobility of intrinsic carriers is often expressed in terms of the *mobility factor* μ. Basically this means that the carrier will move at some velocity when accelerated by some potential acting through some distance of conductor. For example, in silicon at room temperature the minority electrons in the conduction band have a mobility factor of about 1,500 cm/s per V/cm. The holes in the valence band have a mobility factor of about 500 cm/s per V/cm.

The carrier mobility in germanium is considerably greater, since the valence electrons are already at higher energy than those of silicon. Typical mobility factors for germanium are 3,900 cm/sec per V/cm for electrons and 1,900 cm/s per V/cm for holes.

The approximate velocity of the carriers may be computed from the length of the crystal, the voltage, and the carrier mobility as

$$\text{Velocity} = \frac{\mu V}{l} \tag{5-2}$$

where μ = mobility, cm/s per V/cm
V = applied voltage
l = length of substance, cm

The conductivity of the semiconductor material depends, then, on the concentration of carriers, the charge (in Coulombs) of the carriers, and the sum of the mobilities. The mobility of the carriers generally decreases with a rise in temperature (more random motion), and the number of intrinsic carriers (electron-hole pairs) generated increases sharply with this same rise in temperature. The net result is an increase in minority drift current with an increase in temperature.

Sometimes the resistivity of a semiconductor material is given instead of the conductivity. The resistivity, expressed in Ohm centimeters, is the reciprocal of the conductivity.

THE *NP* JUNCTION

When semiconductor material of the *N* type is alloyed to the *P* type, the boundary, or *junction*, between them has special properties of its own. Current will pass in large quantities in one direction, but not in the other. This junction is not a simple mechanical junction, but is

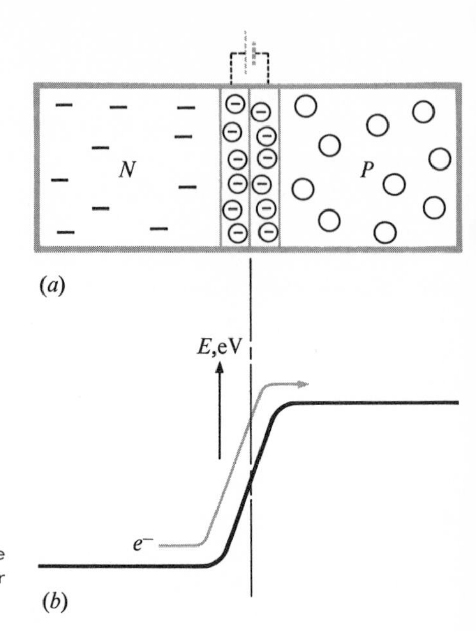

FIG. 5-8 The two depleted regions in (*a*) an *NP* junction and (*b*) a diagram of the energy (in electron Volts) required for an electron to cross the potential barrier of the junction.

formed by a special diffusion or alloying process. Figure 5-8*a* indicates the *NP* junction from its functional standpoint.

THE DEPLETED REGION

After the junction is formed, the majority electrons begin to diffuse across it from the *N*-type material into the *P*-type material. This is because charges tend to move from areas of high concentration to areas of low concentration. As the electrons start moving into the *P*-type material, they encounter large numbers of holes that immediately accept them. Diffusion continues until the number of electrons captured by the extrinsic holes of the *P*-type material is large enough to repel other electrons attempting to diffuse across the junction. Since the *P*-type material was neutral before the electron diffusion, the area near the junction takes on a negative charge as the holes become filled.

Similarly, some holes from the *P*-type material diffuse across the junction into the *N*-type material and find electrons with which to join. Thus on the normally neutral *N* side of the junction the collection of diffused positive holes contributes a small positive charge which repels further diffusion of holes.

Once the small area on either side of the junction has been depleted of uncovered carriers, diffusion appears to cease. In reality, a condition of equilibrium is established across the junction. For every majority electron that crosses into the *P*-type material an intrinsic or minority electron crosses from the *P*-type material into the *N*-type material. Similarly, for every majority hole that diffuses across the junction an intrinsic, or minority, hole crosses from the N-type material into the P-type material. Thus at the final equilibrium stage the recombination current of majority carriers exactly equals the minority or thermal current.

THE POTENTIAL BARRIER

The effect of the depleted region established in this fashion is to set up a potential barrier, a small voltage barrier, to oppose further motion of majority carriers across the junction. Figure 5-8*b* depicts this potential barrier. It is sufficient at this point to state that the junction equilibrium voltage for a given charge density is covered by *Boltzmann's law*, which reduces to the relationship

$$V_{eq} = \frac{kT}{q} \ln \frac{N_n P_p}{n^2} \tag{5-3}$$

where k = Boltzmann's constant, 1.38×10^{-23} J/°K
T = absolute temperature, °K
q = charge on the electron, 1.6×10^{-19} C
N_n = electron density in N material
P_p = hole density in P material
$n^2 = N_n P_n = N_p P_n$

For a detailed development of this relationship, see Corning in the references at the end of the chapter.

If both sides of the junction are doped to the same degree, the depleted regions extend equally into both sides of the junction. However, if one side is doped more lightly, as it is in a transistor, the two depleted regions will not be the same; the region with the lightest doping will be depleted farther into the bulk of the material than the one with the heavier doping. It is interesting to note that the greater part of the potential barrier also lies in the region of least doping.

The potential barrier under forward bias When the NP junction is placed in a circuit and connected as in Fig. 5-9, it is considered to be *forward biased*. Note that the negative terminal of the supply voltage is connected to the N-type material and the positive terminal is connected to the P-type material. The net effect of this forward bias is to reduce the height of the potential barrier by narrowing the depleted region. Although the barrier cannot be completely overcome, it can be reduced sufficiently for large amounts of current to pass over it.

Under the influence of this forward-bias voltage the electrons in the N-type material are repelled toward the junction. The holes in the

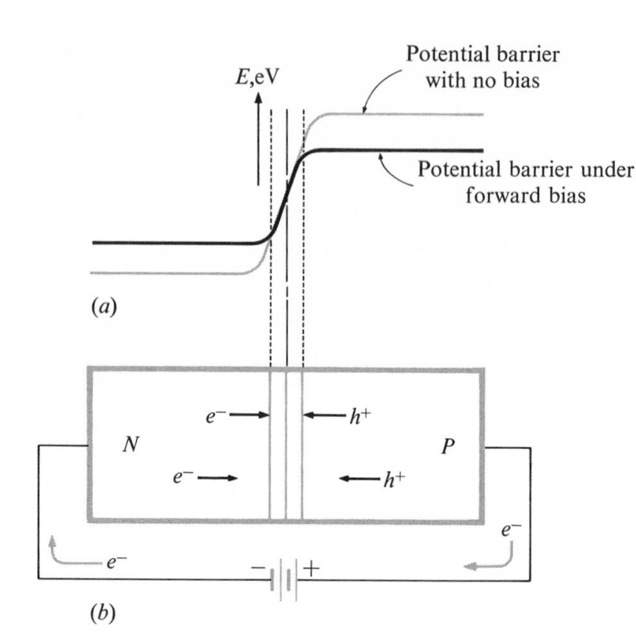

FIG. 5-9 (*a*) Potential-energy diagram indicating how the potential barrier is reduced when the junction is forward biased, and (*b*) direction of carrier movement in an *NP* junction diode under forward-bias conditions.

P-type material are also repelled toward the junction by the positive supply terminal. When the electrons and the holes come together at the junction, recombination occurs. That is, a conduction-band electron combines with a valence-band hole to reestablish the covalent bond. Both these carriers are thus removed from the scene. At the same time, however, an electron is injected into the *N*-type material by the negative terminal and an electron is removed at the positive terminal. Under forward-bias conditions, then, reduction of the potential barrier allows recombination, which in effect allows current to drift through the crystal.

Recombination may not occur right at the junction, but the probability is greatest very near the junction and falls off as the distance from the junction increases.

The potential barrier under reverse bias When the *NP* junction is connected in the circuit as shown in Fig. 5-10, the junction will be *reverse biased*. That is, the positive terminal of the voltage supply is connected to the *N*-type material and the negative terminal is connected to the *P*-type material.

The effect of a reverse bias on the potential barrier is predictable from basic principles. Electrons are drawn away from the junction by the influence of the positive terminal, and the holes are drawn away by the influence of the negative terminal. The depleted region widens, and the potential barrier becomes higher and hence more difficult to cross.

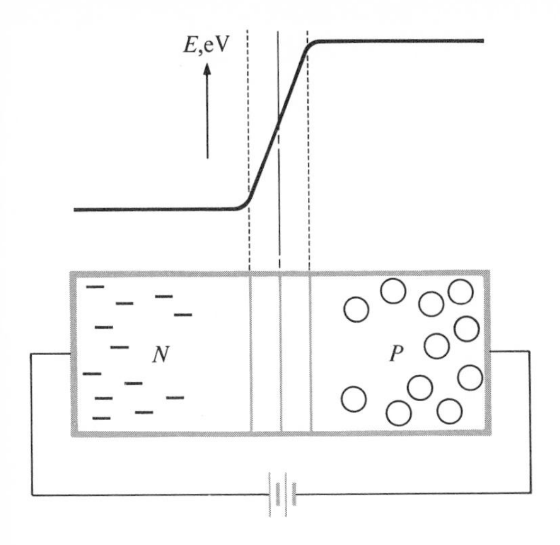

FIG. 5-10 The reverse-biased *NP* junction showing the movement of carriers away from the junction. Note the widened depleted regions and the increased electron-energy difference across the junction.

Recombination ceases, and for all practical purposes forward current also ceases.

LEAKAGE CURRENT

Under reverse-bias conditions the ideal result would be absolutely no current in the reverse direction. This situation, however, implies infinite resistance. Obviously a condition of infinite resistance cannot be achieved. There is always some infinitesimal amount of leakage. The difficulty is that it increases with temperature.

As might be expected, the leakage, or reverse current, is due to the thermal generation of intrinsic minority carriers (electron-hole pairs). The number about doubles with every 10°C rise in temperature. As discussed previously, the problem of leakage current is generally quite severe in germanium devices and tends to be negligible in most silicon devices.

When the junction is reverse biased, the minority electrons in the *P*-type material are in the conduction band and are attracted toward the positive terminal, which is on the other side of the junction. In effect there is no potential barrier for the minority carriers in a reverse-bias situation. For the minority carriers the reverse-bias condition is actually forward bias. Therefore the electrons in the *P*-type and the holes in the *N*-type materials move toward the junction, and recombination occurs. This leakage current is of the order of nanoamperes for silicon junctions and remains essentially constant with changes in voltage.

The reverse-current characteristics are indicated in Fig. 5-11. Note that the leakage current does increase with temperature but does

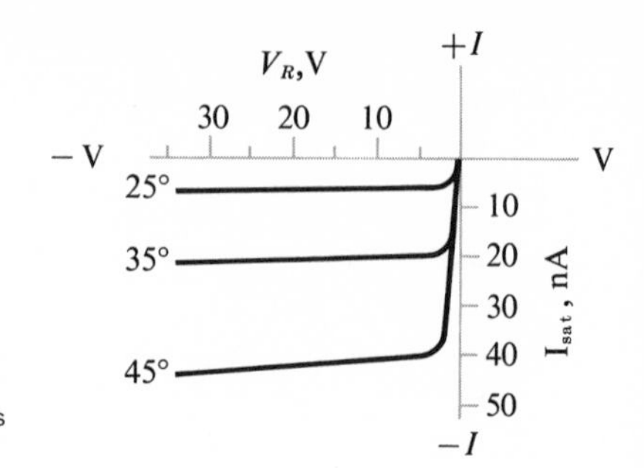

FIG. 5-11 Leakage current in a reverse-biased silicon diode at various temperatures.

not increase significantly as the reverse voltage is increased. When all the thermally generated minority carriers are participating in the leakage or reverse current, it is usually referred to as the *saturation current*.

THE *NP* JUNCTION AS A DIODE

Most single-junction devices, called *diodes*, display *rectifying* action; that is, they allow current to pass in only one direction. This is, of course, the ideal, since, as we have seen, there is always some leakage.

The forward current of a diode increases almost exponentially once the forward potential barrier is overcome. This means that for a silicon diode the current will begin to flow when the potential difference across the diode is about 0.5 V. This 0.5 V represents the approximate height of the potential barrier at room temperature. The potential may be lower or higher, depending on the amount of doping, the geometry of the device, and other factors. After the barrier has been partially overcome, the current begins to increase rapidly, as shown in Fig. 5-12. As the temperature increases, the random mobility of the carriers decreases, and the junction potential decreases. Remember that the forward current is due primarily to majority carriers, and so the decrease in their mobility overshadows the increase in minority carriers. Thus there should be at least some decrease in junction potential with a rise in temperature.

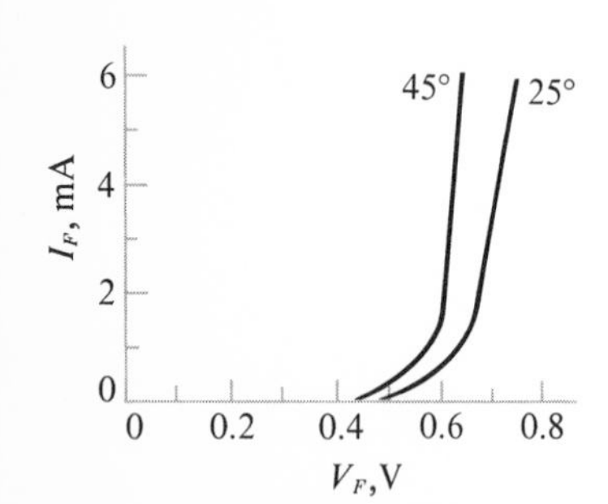

FIG. 5-12 The forward characteristics of a silicon diode at different temperatures.

If a circuit is connected as shown in Fig. 5-13*a*, the change in junction potential with a rise in temperature will be obvious. Be certain the diode used is a germanium diode and that the voltmeter is on the 0- to 0.5-V scale. The junction potential is only about 0.2 V for germanium, and a less sensitive meter would not indicate the significant change. Heat the diode with a hot light bulb and observe the increase in current as well as the decrease in junction voltage.

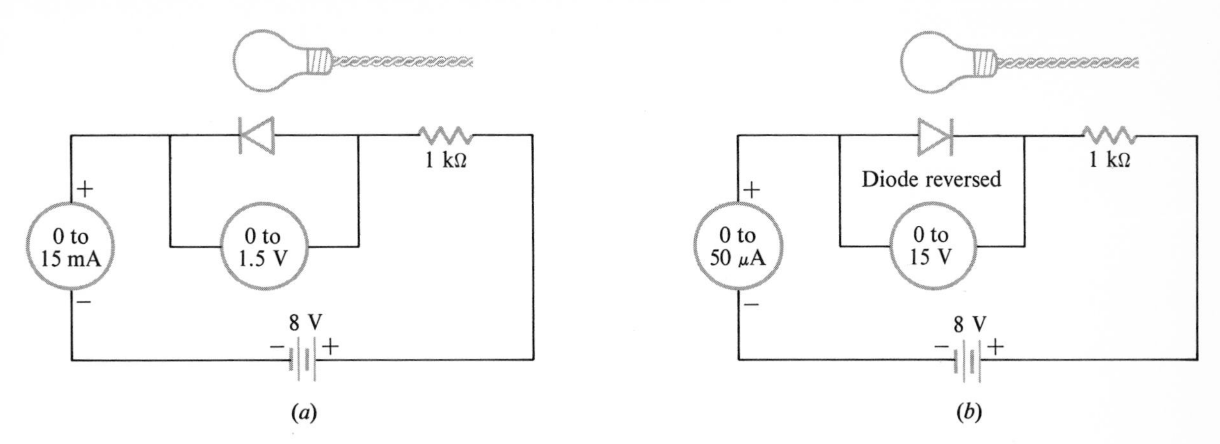

FIG. 5-13 (*a*) Forward-bias junction voltage with changes in temperature; (*b*) reverse-bias junction voltage and leakage current with changes in temperature.

Next, replace the germanium diode with some silicon diode of the glass-case variety. Set the point of operation exactly as with the germanium diode, except change the voltage scale to the 0- to 1.5-V range. Bring the hot light bulb near to the silicon diode, as before, and note any changes in junction voltage and current.

The increase in current for the silicon diode may not be measurable unless the scale is very large. The increase in current for the germanium diode, however, generally shows up rather well.

Now reverse the diodes, as in the circuit of Fig. 5-13*b*, and repeat the experiment; remember to change the voltmeter scale to the 0- to 15-V range. Note that the reverse current is almost negligible until the germanium diode becomes quite hot.

SUMMARY OF CONCEPTS

1. Electrons in the atom occupy discrete orbits, or energy levels, and may move to other orbits by gaining or losing energy.
2. Electrons in their highest energy levels are the most likely to leave the atom and become current carriers.
3. The outermost electrons in the lower-energy levels are known as valence electrons. Their various levels make up the valence band.
4. Electrons in the valence band may absorb sufficient energy from heat or light to escape to the conduction band.
5. Materials such as germanium and silicon form crystals by means of covalent bonding.
6. When covalent bonds are broken through the addition of energy, electron-hole pairs are formed. These electrons and holes are the intrinsic current carriers.
7. Intrinsic electron-hole pairs are those native to the pure crystal.
8. Extrinsic current carriers, which are not native to the pure crystal, may be introduced by diffusing carefully controlled impurities into the material.
9. Silicon or germanium doped with group III elements, such as boron or aluminum, forms a *P*-type crystal.
10. Silicon or germanium doped with group V elements, such as phosphorus or antimony, forms an *N*-type crystal.
11. Doping a pure semiconductor material reduces the energy difference between the valence band and the conduction band, the forbidden energy gap.
12. The majority current, or extrinsic conductivity, in a semiconductor is due to the extrinsic current carriers produced by doping.
13. The minority current, or intrinsic conductivity, is due primarily to the intrinsic electron-hole pairs generated by temperature increase.
14. Minority carriers are present in larger quantities in germanium than in silicon because of its larger atomic size and higher electron energies.
15. Random mobility of carriers is a function of temperature.
16. The resistivity of a semiconductor depends on the amount of doping.
17. A depleted region is formed across every *NP* junction when carriers cross the junction to fill bonds.

18. When a depleted region has developed across an *NP* junction, a potential barrier of a few tenths of a Volt develops.
19. The potential barrier is narrowed by application of forward bias, a positive voltage to the *P*-type material and a negative voltage to the *N* type.
20. The potential barrier is widened by application of reverse bias across the junction.
21. Recombination of holes and electrons occurs at the *NP* junction when the forward bias exceeds the potential barrier.
22. There is a leakage current across an *NP* junction under reverse-bias conditions because no potential barrier exists for the minority carriers.
23. The leakage current approximately doubles for every 10°C rise in temperature.
24. The forward voltage across a diode drops as the temperature increases because of decreased mobility of the carriers.
25. Holes have effective movement in the valence band through the displacement of electrons, while electrons move freely with higher energy in the conduction band.

GLOSSARY

Boltzmann's constant = 1.38×10^{-23} J/°K

Bond The means by which one atom is bound to another in a material. The two most common types of bonds are the ionic bond, and the covalent bond.

Carrier An electron or hole that, when in motion, constitutes current. Electrons are negative carriers and holes are positive carriers.

Conduction band The range of energy that a current carrier must have in order to move freely through a semiconductor crystal.

Covalent bond A bond in which valence electrons are shared by neighboring atoms to form a stable configuration for each, as in germanium and silicon crystals.

Depleted region The area on either side of the junction of *P*-type and *N*-type materials joined to form a diode. This region is almost devoid of free carriers.

Diode A functional device made by diffusing *P*-type silicon or germanium into *N* type to produce an *NP* junction with a high resistance in one direction and a low resistance in the other.

Doping The introduction of a small amount of some trivalent or pentavalent element such as boron or phosphorus to change pure silicon or germanium into *P*-type or *N*-type crystals.

Drift current The zigzag motion of current carriers through a crystal in the general direction of the positive (or negative) potential of the source voltage.

Electron-hole pair The spontaneously generated intrinsic current carriers in a pure or doped semiconductor due to the release from a bond of an electron as the temperature rise supplies sufficient energy.

Electron Volt That energy represented by an electron accelerating through a potential difference of one Volt, about 1.6×10^{-19} J.

Extrinsic carriers The electrons or holes that are placed in a semiconductor crystal by the process of doping. They play the major role in semiconductor conductivity.

Forbidden energy gap That range of energy between the valence band and the conduction band which an electron must acquire in order to move from the crystal bond to the free, or conduction, state.

Forward bias The electrical connection between the *NP* junction and a power source needed to cause majority current, e.g., the positive terminal to the *P*-type material connection.

Free electron An electron that has acquired sufficient energy to escape from an atom or bond.

Hole An absence or vacancy left in the valence bond when an electron escapes to the conduction band. It has all the characteristics of a positive charge or particle, including motion through the crystal lattice.

Insulator A material whose valence electrons are bound very tightly, so that few free electrons exist to provide a means of conduction.

Intrinsic conductivity The property of a semiconductor material to conduct small amounts of current without doping, due primarily to thermally generated electron-hole pairs.

Ion An atom that has taken on a charge by a loss or gain of electrons.

Junction Usually the *NP* junction, the border region formed when *P*-type material is alloyed to or diffused into *N*-type material.

Leakage The tiny current that passes through a reverse-bias *NP* junction owing to the presence of minority carriers; it increases with an increase in temperature.

Majority carriers Extrinsic electrons in *N*-type and holes in *P*-type semiconductor materials, due primarily to doping.

Minority carriers Intrinsic holes in *N*-type and electrons in *P*-type semiconductor materials, due primarily to thermal generation.

Mobility Generally the ability of electrons and holes to move in a semiconductor material, greater in germanium than in silicon.

Mobility factor μ A mathematical expression of a carrier's velocity when it is acted upon by some potential difference through some distance.

Photoelectric The release of electrons from the surface of a conductor upon exposure to light owing to their absorption of light energy.

Planck's constant h A constant used in several physical formulas, having a value of 6.56×10^{-27} erg-s.

Potential barrier The barrier to forward current across an *NP* junction, amounting to a few tenths of a Volt, as a result of the charge produced by carriers in the depleted region near the junction.

Quantum A basic discrete quantity of energy that may be absorbed or released by an electron.

Recombination The process by which a hole is reunited with or filled by an electron.

Resistivity The resistance offered by a unit cube of a certain material, a measure of how tightly bound the carriers are in a crystal of semiconductor material.

Reverse bias The electrical connection between the *NP* junction and the power source needed to prevent majority current, e.g., the negative terminal to the *P*-type material.

Semiconductor That group of elements and compounds whose resistivity lies between that of insulators and conductors, with a forbidden energy gap that may be bridged by means of doping.

Valence Generally the number of outermost electrons that an atom may give up or acquire in order for a chemical reaction to occur.

REFERENCES

Brazee, James G.: *Semiconductor and Tube Electronics*, Holt, Rinehart and Winston, Inc., New York, 1968, chap. 1.

Corning, John J.: *Transistor Circuit Analysis and Design*, Prentice-Hall, Inc., Englewood Cliffs, N.J., 1965, chap. 1.

Cutler, Phillip: *Semiconductor Circuit Analysis*, McGraw-Hill Book Company, New York, 1964, chaps. 1–2.

G. E. Transistor Manual, 7th ed., General Electric Company, Syracuse, N.Y., 1964, chap. 1.

Lenert, Louis H.: *Semiconductor Physics, Devices, and Circuits*, Charles E. Merrill Books, Inc., Columbus, Ohio, 1968, chaps. 1–4.

Malvino, Albert Paul: *Transistor Circuit Approximations*, McGraw-Hill Book Company, New York, 1968, chaps. 1–2.

Seidman, A. H., and S. L. Marshall: *Semiconductor Fundamentals Devices and Circuits*, John Wiley & Sons, Inc., New York, 1963, chaps. 1–2.

Shockley, William: *Electrons and Holes in Semiconductors*, D. Van Nostrand Company, Inc., Princeton, N.J., 1950.

Surina, Tugomir, and Clyde Herrick: *Semiconductor Electronics*, Holt, Rinehart and Winston, Inc., New York, 1964, chap. 1.

REVIEW QUESTIONS

1. Why are no two electron orbits exactly the same?

2. What conditions are necessary for an electron to change from one orbit to another or leave the atom?

3. What is the shape of most electron orbits?

4. What kind of bonding is found in a pure silicon crystal?

5. What distinguishes a semiconductor from a conductor? From an insulator?

6. Why is it necessary to dope pure silicon or germanium in order to make diodes or transistors?

7. How does current-carrier drift differ from diffusion?

8. What is meant by the forbidden energy gap? In reality, what is it?

9. In what way is electron-hole pair generation related to intrinsic conductivity?

10. Do holes move in the valence band or the conduction band? Why?

11. Describe the function of a donor atom in the formation of *N*-type silicon.

12. What happens to the forbidden energy gap when silicon is doped?

13. What is the difference between a majority carrier and a minority carrier?

14. What are the electron mobility factors for silicon and germanium at room temperature? What are the hole mobility factors?

15. Describe an *NP* junction in terms of the depleted regions.

16. In what way is the depleted region a factor in the junction potential barrier?

17. What is the effect on the potential barrier of a forward bias? A reverse bias?

18. What is the cause of leakage current?

19. In what way does the temperature affect the amount of leakage current?

20. What is the approximate forward bias needed to pass some small amount of current through a silicon diode at room temperature?

21. As the temperature increases, what will happen to the amount of forward bias in Quest. 20?

PROBLEMS

1. If the leakage current of a germanium diode is 20 μA at 30°C, what would it be at 40°C? At 50°C?

2. From the graph of Fig. 5-11, what is the forward voltage of the silicon diode for 2 mA of current at 25°C? At 45°C?

3. If the mobility factor for electrons in silicon at room temperature is 1,500 cm/s per V/cm, what is the velocity of the electrons if the length is 1 cm and the applied voltage is 10 V?

CHAPTER 6

OBJECTIVES

1. To develop a clear understanding of the silicon diode and its dynamic characteristics
2. To explore the Zener, or reverse-breakdown, phenomenon of the silicon diode
3. To explore the significance of maximum forward and reverse voltages and currents
4. To compare the silicon diode and the ideal diode
5. To describe the actions and applications of half-wave, full-wave, and bridge rectifiers
6. To compare the capacitor input filter, the π-section filter, and the choke input filter
7. To develop a simple experimental regulated power supply
8. To investigate the temperature-compensation properties of the diode

THE SILICON DIODE AND ITS APPLICATIONS

Without the semiconductor diode and its multitude of applications, modern electronics would no doubt still be in the era of the vacuum tube. Its tiny size has made it indispensable in computers, radar, sonar, telemetry, integrated circuitry, power supplies, and a host of other applications. With new manufacturing techniques many parallel diodes can be produced on the same chip to control vast amounts of current that formerly required huge vacuum tubes. In this chapter only a few applications pertinent to power supplies and detection will be reviewed, but a wealth of information is available for other applications.

THE SILICON DIODE

The silicon diode, unlike the germanium diode, may be operated under extremes of temperature without the problem of reverse leakage current. Leakage may be several microamperes with germanium diodes, but it is only a few

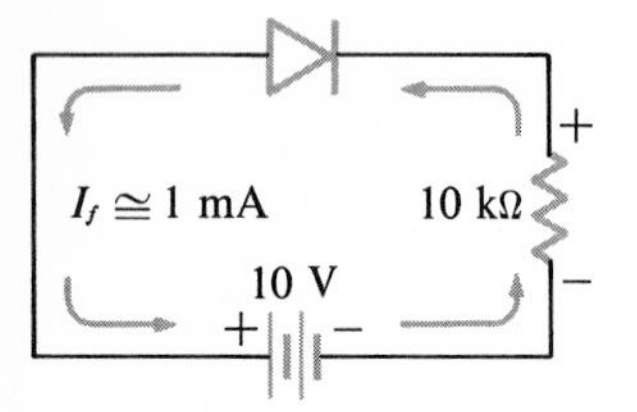

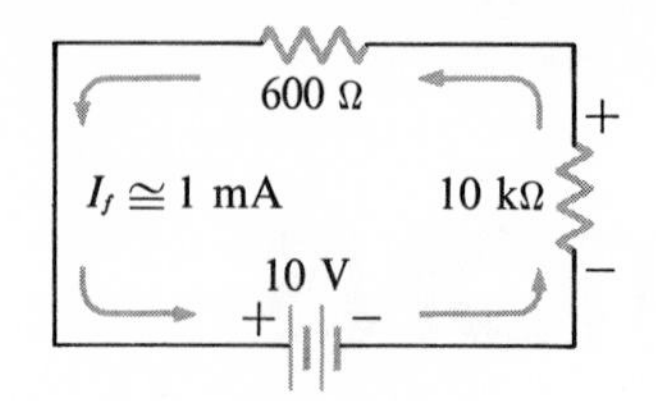

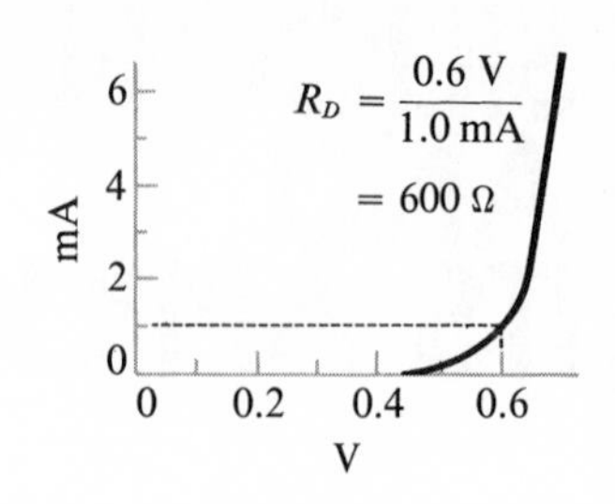

FIG. 6-1 Forward-bias silicon-diode resistance characteristics.

nanoamperes with the better silicon diodes. For this reason the bulk of our discussion will be limited to silicon diodes.

Figure 6-1 illustrates the forward-bias resistance characteristics of a typical silicon diode. Note that very little current flows until the potential barrier of the *NP* junction has been lowered to about 0.5 V. At about 1.0 mA the forward voltage across the diode is approximately 0.6 V. From Ohm's law, the resistance of the diode at 1 mA of current must be about 600 Ω. Although this value may vary from one type of diode to the next, it generally ranges from about 0.5 to 0.7 V. Thus the forward resistance of the real diode is between 500 and 700 Ω.

The reverse-bias characteristics show a very high resistance, generally many millions of Ohms. As Fig. 6-2 indicates, even though some current does leak through the reverse-biased junction, it is a very small amount, corresponding to a very high reverse resistance. Note that if a sufficiently high voltage is applied in reverse, the diode will break down, and large amounts of current will flow. At this voltage, referred to as the *peak reverse voltage* (PRV), the diode is likely to be destroyed. This much voltage should never be applied to a diode in the reverse direction.

THE ZENER EFFECT

There are some diodes that are manufactured specifically to be operated in reverse. These diodes are known as *Zener diodes* or, more technically, *breakdown diodes*. Under certain doping conditions, as the potential across the junction approaches this Zener voltage, instead of going over the potential barrier, which is very high, the current car-

FIG. 6-2 Reverse-bias silicon-diode resistance characteristics.

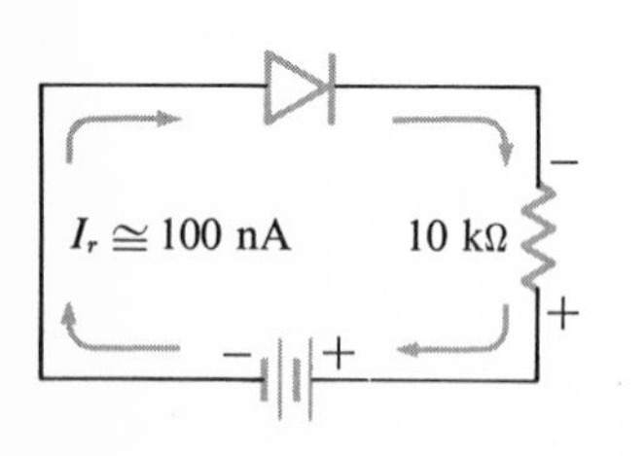

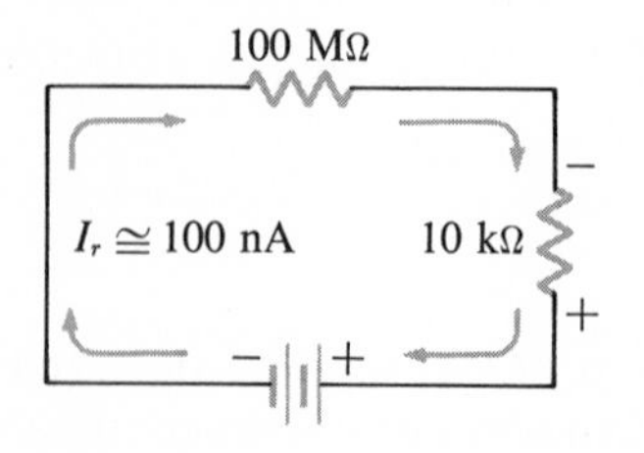

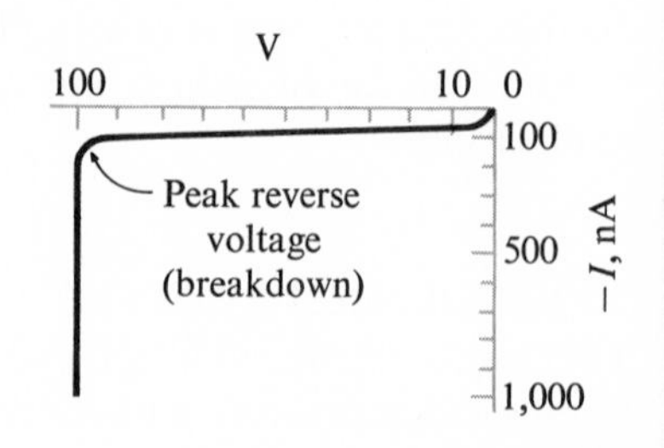

riers tunnel through it. At about the same voltage *avalanche* occurs. The avalanche condition is just what its name implies. Electrons that are knocked loose have so much energy that they knock others loose, and breakdown occurs, with large amounts of reverse current as the result.

The Zener effect occurs at low voltages, usually less than 8 V. The avalanche effect occurs at only a slightly higher voltage, and so the two effects are generally lumped together. However, it is the avalanche effect that is used in the breakdown diodes of greater than 8 V. For our purposes both effects will be lumped together under the term *breakdown diode*.

Note in Fig. 6-3 how sharp the knee of the curve is. It is this property that makes the breakdown diode useful for application as a voltage regulator. The two symbols shown in Fig. 6-3 are those commonly used in schematic drawings.

The normal operation of the breakdown diode is just the reverse of that for the regular silicon diode. There must be sufficient current flow to assure that the voltage will not drop below the knee. An increase in voltage beyond the breakdown voltage will cause more current to flow, but the diode voltage will remain almost the same. Some breakdown diodes have a softer knee than the one illustrated; that is, it is more rounded. Sometimes when the diode has been damaged slightly the knee becomes so softened that it is no longer usable, since there is no definite break point.

A typical circuit for holding a voltage constant by means of a breakdown diode is shown in Fig. 6-4.

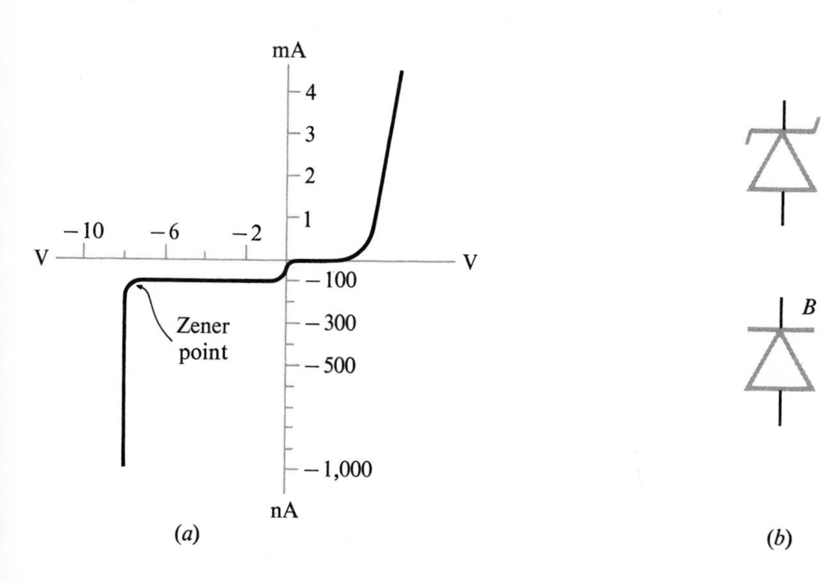

FIG. 6-3 (*a*) Forward and reverse characteristics of the Zener diode, and (*b*) common symbols for the Zener and breakdown diodes.

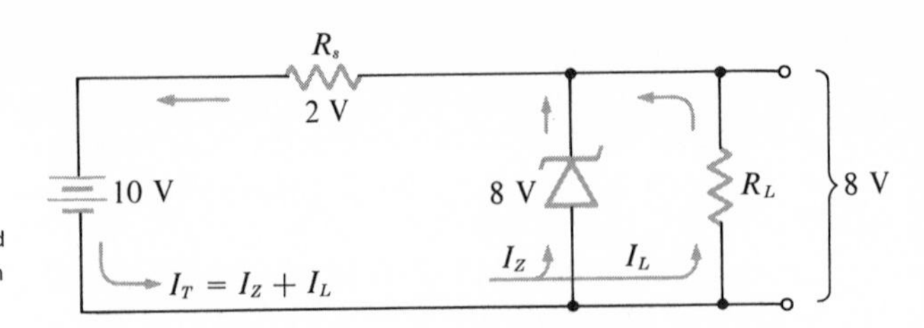

FIG. 6-4 A breakdown diode used to regulate the load voltage to 8 V when the supply voltage is greater than 8 V.

THE IDEAL DIODE VERSUS THE SILICON DIODE

If the circuit designer had his way, all components would be ideal in that a switch would be open or closed. The open switch would be equivalent to infinite resistance, allowing no current to pass, and the closed switch would be a short circuit of zero resistance, allowing infinite current to pass. In the ideal diode the reverse resistance would be infinite—an open switch—and the forward resistance would be like a short circuit. Unfortunately, this ideal condition does not exist. As we saw earlier, the diode conducts little current in the forward direction until the junction barrier potential is reduced. The barrier cannot be entirely overcome, but for all practical purposes the diode acts as a low-value resistor after a few milliamperes of current is flowing.

The resistance of a forward-biased silicon diode is made up of two resistances: the junction resistance and the bulk resistance of the diode itself. The junction resistance accounts for the curvature of the characteristic curve near the origin. The curves for these two resistances are shown in Fig. 6-5.

The resistance of the junction is affected by the current passing through the junction as well as its temperature. The ideal diode junction has a *temperature voltage* V_t of about 26 mV at 25°C, which varies according to the formula

FIG. 6-5 A comparison of the characteristics of the (*a*) ideal diode and (*b*) the real diode, and (*c*) the forward characteristic curve of a real diode indicating the bulk resistance.

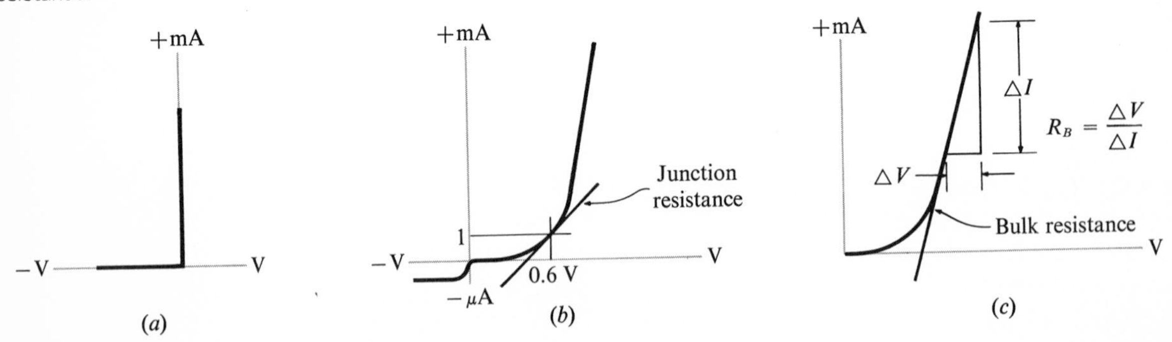

$$V_t = \frac{kT}{q} = 0.026 \text{ V} \qquad (6\text{-}1)$$

where k = Boltzmann's constant, 1.38×10^{-23} Watt-s/°K
T = absolute temperature, °K (room temperature = 300°K)
q = charge on the electron, 1.6×10^{-19} C

From this temperature voltage, or voltage that will vary with the temperature, the junction resistance R_j may be computed for the temperature at which the diode is used. For example, at room temperature and a forward current I_f of 1 mA,

$$R_j = \frac{kT/q}{I_f}$$

$$= \frac{0.026 \text{ V}}{1.0 \text{ mA}} = 26\ \Omega \qquad (6\text{-}2)$$

where T = 300°K
I_f = 1.0 mA

Although the junction temperature is usually higher than room temperature when the diode is in operation, this figure is a good one to remember. Let us, however, simplify the problem by using 30 Ω at 1 mA, which is closer to the actual operating value.

It can be seen, then, that as the current is reduced to 0.1 mA, the resistance increases to 30 mV/0.1 mA = 300 Ω, and so on. Note that the slope of the curve approaches the horizontal—that is, infinite resistance—as the amount of current decreases. Of course, this value is never actually reached. Conversely, as more current is allowed to flow, the curve becomes steeper, approaching 0 Ω. The limit to this is the bulk resistance and the resistance of the connections to the diode. The typical lowest resistance is generally between 2 and 10 Ω.

THE SILICON RECTIFIER

Because of its characteristic stability with temperature, the silicon diode makes a good *rectifier*. That is, it can be used to allow current to flow in one direction while blocking it in the reverse direction. To rectify really means to correct. Since most early electrical machines were operated from dc current sources such as batteries, alternating current had to be corrected, or rectified, to direct current before it could be used. The term rectifier now generally refers to a diode used to convert alternating current to direct current in a power supply. The same action is used in detecting a radio signal, but small-signal applications use signal diodes rather than power-type rectifiers. Figure 6-6 illustrates the rectifying action of the diode in an experimental circuit.

THE HALF-WAVE RECTIFIER

Connect the circuit as indicated in Fig. 6-6. If a signal generator is not available that will deliver 18 V *p-p*, substitute a 6.3-V transformer. The 6.3-V secondary is actually 6.3 V rms, which is about 18 V *p-p*.

Connect the leads from the load resistor R_L to the vertical input on the oscilloscope. Be sure to connect the negative terminal of the circuit to ground on the scope. All ground leads on all measuring instruments should be kept attached to the same common point in the circuit.

When the circuit is powered, the output signal observed on the oscilloscope should resemble the waveform of Fig. 6-6. Since only half of the total sine wave is left, it is known as *half-wave rectification*. Note that the trace descends below the reference line during the *off* half-cycle. This is because of the small amount of leakage current that passes through the load resistance.

If a dc coupled scope is available, you can observe the leakage part of the curve. Temporarily remove the vertical lead from the scope and align the trace to some convenient line on the screen. When the scope is reconnected to the circuit, the waveform should appear as indicated.

Since the sum of the voltages around a circuit must at all times equal the magnitude of the source or supply voltage, there will be about 0.6 V across the diode. The remainder will appear across the load resistance. Unless the load resistor is in the circuit, all the voltage will be across the diode, and it will rapidly conduct sufficient current to destroy itself.

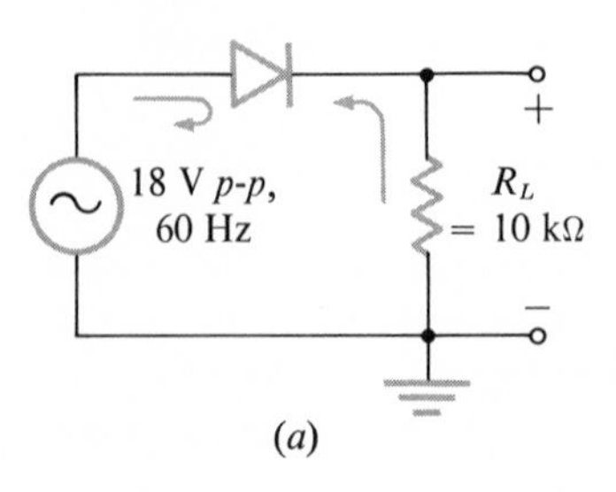

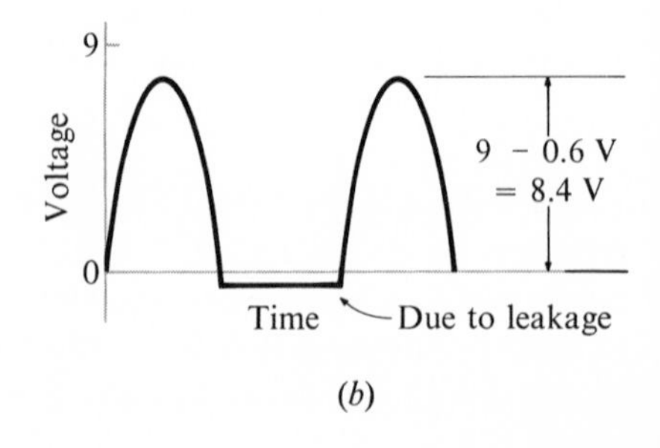

FIG. 6-6 (*a*) The diode acting as a rectifier in a circuit and (*b*) the half-wave rectified waveform.

After measuring the rectifier output or load voltage on the oscilloscope in peak-to-peak units, measure the same voltage with an electronic voltmeter. If the meter is calibrated in peak-to-peak units, compare the two readings. They will probably not be the same. The voltmeter actually responds to the average voltage, but it is calibrated to show rms voltage in most instances. Some sophisticated electronic voltmeters actually show rms voltage, but most meter manufacturers either rescale the meter or use some simple corrective measures. If the ac voltage being measured is not a sine wave, the reading will be in considerable error because rms voltage is a measure only of the *sine* wave.

Since most meters do show rms, or effective, voltage, the reading for the output voltage from a half-wave rectifier should be about one-half the peak value of the waveform as viewed on the scope. This is because we are measuring *the effective value of the effective value of the whole sine wave*. Stated another way, the rms value of one-half the sine wave is 0.707 times the rms value of the full sine wave, or

$$V_{hw,\,\mathrm{rms}} = 0.707 \times 0.707 \times V_p = 0.5 \times V_p$$

$$= 0.5 \times 8.4 = 4.2\ \mathrm{V} \qquad (6\text{-}3)$$

where V_p = peak voltage

This is not as involved as it seems. Remember that the power delivered to the load is being delivered for only one-half the time. Thus if the dc effectiveness of the full ac wave is only 0.707 times the peak value, then for half the full wave it should be 0.707 times the dc effectiveness of the full wave. This indeed proves to be the case.

THE FULL-WAVE RECTIFIER

The inefficiency of the half-wave rectifier may be easily remedied by using two half-wave rectifiers back to back. Instead of a single winding for the secondary of the power transformer, a center-tapped transformer must be used. Figure 6-7 illustrates how the full-wave rectifier is related to the half-wave rectifier. Note that for both halves of the cycle the current flows through the load in the same direction. Thus the full sine wave is corrected, or rectified.

The full-wave rectifier circuit of Fig. 6-7 uses a 12.6-V center-tapped transformer. Since only half of it is used for each half-cycle, the output voltage is 6.3 V rms. This corresponds to about 8.9 V peak, as will be measured on the oscilloscope. Again, remember that the various resistances and the transformer windings consume power, so that the voltage available to the load may not actually appear there if a great deal of current flows in the circuit.

The current in the circuit of Fig. 6-7 is direct current in that it always flows in the same direction. However, the amount of current is changing and is referred to generally as *pulsating direct current*. In order to smooth it out a filtering system must be used. This filter will be discussed later in the chapter.

Certain precautions should be taken in building a power supply with semiconductor diodes. The rectifier should never at any time during the cycle be subjected to a voltage greater than its peak reverse voltage. The problem will become clearer later, but generally it is safest to choose a diode with a PRV at least *three times* the rms voltage that is to be used. If, for example, the transformer secondary is rated at 120 V rms, the diode PRV should be no less than 360 V. Since the peak voltage of 120 V rms is about 170 V, the filter will charge up to this value. During the next half-cycle this voltage will add to the transformer voltage of 170 V peak, thus developing a voltage of about 340 V across the diode. If there is much fluctuation in the line voltage, this value could easily exceed 360 V.

The second safety factor to be considered is the maximum forward current the diode can pass safely. Although the load resistance may be such that only 0.5 A is being drawn from the circuit, the filter will have to recharge during each pulse of the input wave. These surge pulses may exceed by several times the rating of the diode. Most diodes will tolerate surges of current for short periods but will overheat and become damaged after a short time. The diode selected should be able to deliver the amount of current anticipated and handle surges with some degree of safety margin. For operation at high currents the diode rectifier must sometimes be mounted in a heat sink to prevent overheating.

In general any diode rectifier may be used in a power supply as long as the PRV and the maximum current will not be exceeded. For example, a diode specified for a PRV of 150 V at 50 mA should not be used where the 50 mA is to be delivered by a 120-V transformer. The safety margin is not great enough.

FIG. 6-7 The full-wave rectifier and waveform analysis.

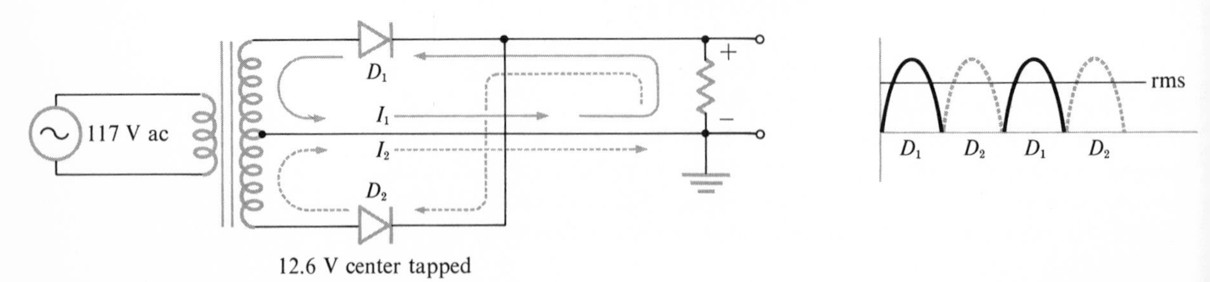

Compare the maximum ratings of the rectifier diodes listed in Table 6-1. Note that the 1N4001 diode has a PRV of only 50 V and the 1N4007 diode has a PRV of 1,000 V. Note also that although the average rectified forward current is only 1 A, the *repetitive* peak forward current is 10 A, and the peak surge current is 30 A. The lead temperature of 360°C may be tolerated at a short distance from the case for 10 s. However, the operating temperature range is −65 to +175°C.

THE BRIDGE RECTIFIER

In many applications involving low voltage and high current full-wave rectification is possible only with a full-wave bridge of four diodes rather than two. One reason is that center-tapped transformers of the ratings needed may not be available. Figure 6-8 shows how the four diodes are arranged in the bridge circuit so that during each half-cycle two of the diodes are conducting. As before, current always passes through the load resistance in the same direction.

One common problem in using the bridge rectifier is getting each of the diodes in the right direction in the circuit. Recall that electron current must pass into the diode through the cathode terminal. Hence all diodes must be placed in the circuit with their cathodes closest to the point at which the electron current reenters the bridge. It sometimes helps to remember that all the arrowheads should point in the direction of the positive terminal of the output. Manufacturers have tried to overcome this problem by printing a plus sign on the cathode end of the diode. This does not mean that the cathode is the positive end of the diode, but that this end goes into the circuit closest to the positive terminal of the supply. As a matter of fact, the positive terminal is the terminal that attracts the negative electron current.

The circuit and waveforms in Fig. 6-8 show that during the first half-cycle, when the top of the transformer secondary is positive, diodes 1 and 2 will conduct. Current coming out of diode 1 (D_1) will pass up through the transformer (minus to plus) from bottom to top, down through diode 2 (D_2), out to and up through the load resistance,

FIG. 6-8 The bridge rectifier and the input-output waveforms.

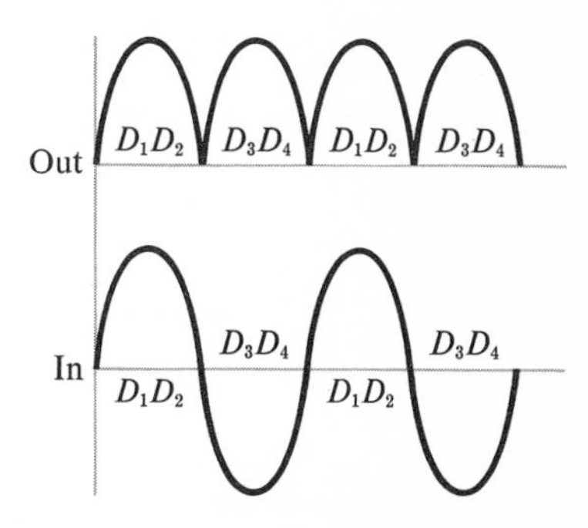

TYPES 1N4001 THROUGH 1N4007
DIFFUSED-JUNCTION SILICON RECTIFIERS

TYPES 1N4001 THROUGH 1N4007
BULLETIN NO. DL-S 657366, FEBRUARY 1965

50-1000 VOLTS • 1 AMP AVG

- **MINIATURE MOLDED PACKAGE**
- **INSULATED CASE**
- **IDEAL FOR HIGH-DENSITY CIRCUITRY**

***mechanical data**

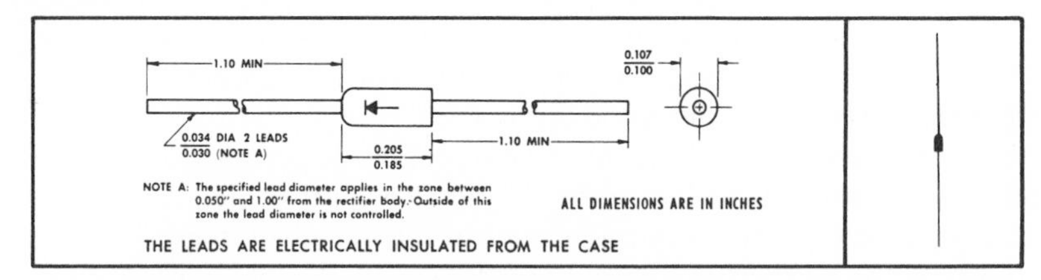

***absolute maximum ratings at specified ambient† temperature**

		1N4001	1N4002	1N4003	1N4004	1N4005	1N4006	1N4007	UNIT
V_{RM}	Peak Reverse Voltage from −65°C to +175°C (See Note 1)	50	100	200	400	600	800	1000	v
V_R	Steady State Reverse Voltage from 25°C to 75°C	50	100	200	400	600	800	1000	v
I_O	Average Rectified Forward Current from 25°C to 75°C (See Notes 1 and 2)	1							a
$I_{FM(rep)}$	Repetitive Peak Forward Current, 10 cycles, at (or below) 75°C (See Note 3)	10							a
$I_{FM(surge)}$	Peak Surge Current, One Cycle, at (or below) 75°C (See Note 3)	30							a
$T_{A(opr)}$	Operating Ambient Temperature Range	−65 to +175							°C
T_{stg}	Storage Temperature Range	−65 to +200							°C
	Lead Temperature ⅜ Inch from Case for 10 Seconds	350							°C

NOTES: 1. These values may be applied continuously under single-phase, 60-cps, half-sine-wave operation with resistive load. Above 75°C derate I_O according to Figure 1.

2. This rectifier is a lead-conduction-cooled device. At (or above) ambient temperatures of 75°C, the lead temperature ⅜ inch from case must be no higher than 5°C above the ambient temperature for these ratings to apply.

3. These values apply for 60-cps half sine waves when the device is operating at (or below) rated values of peak reverse voltage and average rectified forward current. Surge may be repeated after the device has returned to original thermal equilibrium.

* Indicates JEDEC registered data.

† The ambient temperature is measured at a point 2 inches below the device. Natural air cooling shall be used.

TEXAS INSTRUMENTS
INCORPORATED
POST OFFICE BOX 5012 • DALLAS, TEXAS 75222

25401

TABLE 6-1 *Courtesy Texas Instruments Incorporated.*

TYPES 1N4001 THROUGH 1N4007
DIFFUSED-JUNCTION SILICON RECTIFIERS

***electrical characteristics at specified ambient† temperature**

	PARAMETER	TEST CONDITIONS		MAX	UNIT
I_R	Static Reverse Current	V_R = Rated V_R,	T_A = 25°C	10	μa
		V_R = Rated V_R,	T_A = 100°C	50	μa
$I_{R(avg)}$	Average Reverse Current	V_{RM} = Rated V_{RM}, f = 60 cps,	I_O = 1 a, T_A = 75°C	30	μa
V_F	Static Forward Voltage	I_F = 1 a,	T_A = 25°C to 75°C	1.1	v
$V_{F(avg)}$	Average Forward Voltage	V_{RM} = Rated V_{RM}, f = 60 cps,	I_O = 1 a, T_A = 25°C to 75°C	0.8	v
V_{FM}	Peak Forward Voltage	V_{RM} = Rated V_{RM}, f = 60 cps,	I_O = 1 a, T_A = 25°C to 75°C	1.6	v

*Indicates JEDEC registered data.

THERMAL INFORMATION

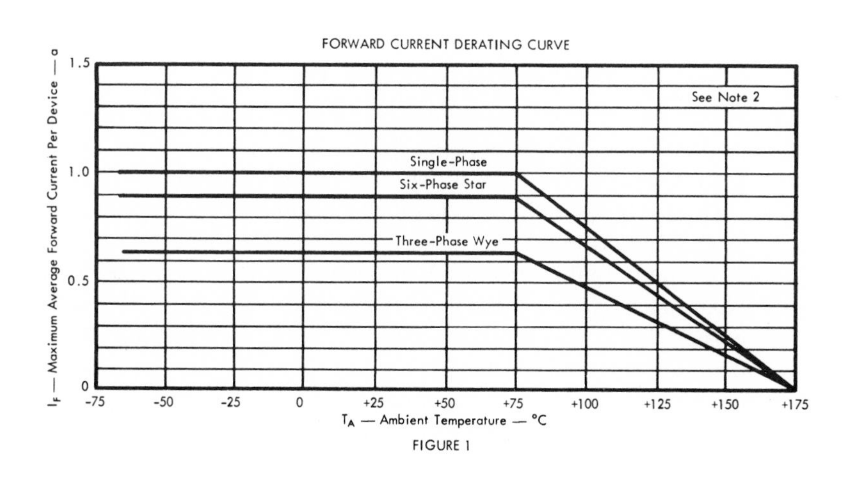

FIGURE 1

†The ambient temperature is measured at a point 2 inches below the device. Natural air cooling shall be used.

NOTE 2: This rectifier is a lead-conduction-cooled device. At (or above) ambient temperatures of 75°C, the lead temperature 3/8 inch from case must be no higher than 5°C above the ambient temperature for these ratings to apply.

and back to diode 1 to complete the circuit. When the transformer reverses the current direction, diodes 1 and 2 become reverse biased, but diodes 3 and 4 are forward biased. This allows current to pass up from the bottom of the transformer to the junction of diodes 1 and 2, down through diode 4, and out through the load resistance, as with the previous half-cycle. From the load resistance the current passes back through the forward-biased diode 3 and back through the transformer to complete the circuit. At no time does the current through the load change direction.

Although the bridge rectifier circuit requires four diodes instead of two, it has certain advantages. Note that there are two diodes in series in the circuit at all times. If both are the same type, only one-half the voltage is across one diode at any time. Thus lower-PRV diodes may be used than with the two-diode center-tapped full-wave rectifier circuit. Since the same current passes through each diode, the same current rating must be applied as before. Diodes of a similar type may also be paralleled, depending on the circuit, for currents higher than either diode alone can handle.

THE CAPACITOR INPUT FILTER

When the rectifier itself delivers current to a load, there is always a large *ripple voltage* present. If the circuit making up the load must have a smooth voltage supply, this is not satisfactory. Some circuits, such as a battery charger, need no smoothing, but amplifier circuits will amplify the ripple as well as any signal. The capacitor is a ready-made container capable of absorbing charge when it is not needed and discharging when the charge is needed by the circuit. Recall from our earlier discussion of basic electricity that the voltage of a capacitor depends on the number of charges stored and the capacity of the capacitor itself.

As the voltage of the output waveform begins to rise, the voltage on the capacitor also begins to rise, because the electrons encounter less opposition from an empty capacitor than from the load resistance. The greater the number of charges absorbed, the greater is the capacitor voltage. After a short time the capacitor will have charged up to very nearly the peak voltage of the waveform. Unless the load requires significant current from the filter, the capacitor will remain at essentially the peak voltage of the input waveform.

If much current is needed by the load, the capacitor discharges back into the circuit as the voltage from the transformer begins to drop. When the capacitor is discharged its voltage gradually drops until the voltage from the transformer again rises to the value of the voltage

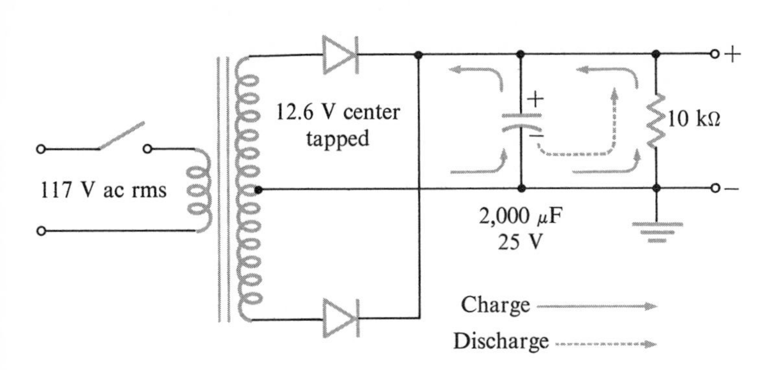

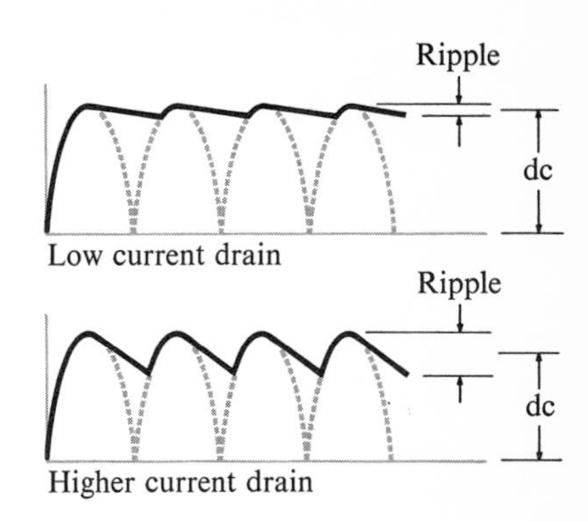

FIG. 6-9 Capacitor input-filter power supply and ripple waveforms.

stored in the capacitor. At this point the capacitor voltage again rises to its peak value. This change, or ripple voltage, in the capacitor is thus a function of the amount of current used. The greater the current, the greater the ripple. Figure 6-9 shows the capacitor input to the load as a supplier of current as well as a filter of the pulsating voltage from the output of the rectifier.

The bleeder resistance When a power supply is in operation there is generally some external circuit or load connected to its output. That is, current is being delivered to a real load. Sometimes, however, a power supply may be turned on without a connected load. Unless there is some internal path through which a small output current may pass, the power supply may not function properly. This auxiliary path is called a *bleeder resistance* and is shown in parallel with the load resistance in Fig. 6-10.

The term bleeder resistance stems from the early use of this circuit element. Note that without some attached load the capacitors in the filter will remain charged. The charge will leak off very slowly through the reverse-biased diodes. However, this may take several days, and it creates a serious safety hazard in the meantime, especially if high voltage is present. Some of the integrated-circuit voltage regulators, such as the Fairchild μA723C, are so constructed that unless a small bleeder current is used at all times, the circuit will malfunction.

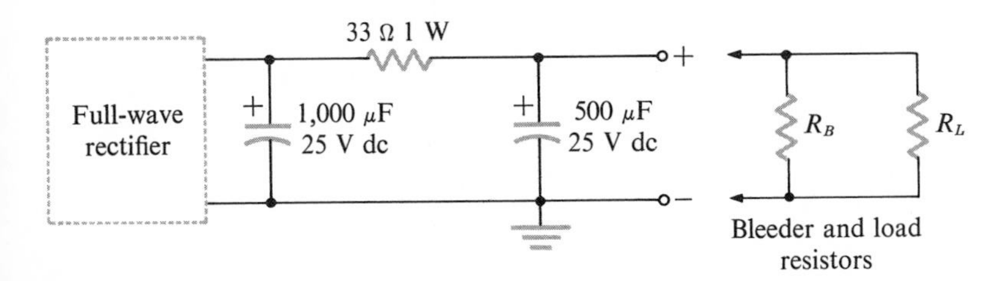

FIG. 6-10 The π-section (capacitor input) filter.

Surge currents Generally speaking, the larger the capacitor used to filter the pulsating voltage, the better the filtering and the smaller the ripple voltage. However, there is a practical limit to this ideal. Very large capacitors have very low impedance when they are empty. When the power supply is first turned on, the surge of current may burn out the diodes. Under certain load conditions the capacitor voltage differs from or is lower than the transformer voltage for only a short part of the cycle, and it must charge to peak value in this short period of time. This surge of current must also be tolerated by the diodes in the circuit. For this reason large-current power supplies often use small values of resistance in series with the diodes to limit the surge to some safe value. As a result, transistor-circuit power supplies often use capacitors for filtering in the 4,000- to 6,000-μF range.

THE π-SECTION FILTER

The capacitor acting alone as a filtering agent does not do as good a job of filtering as might be expected. However, there are some simple, as well as some very complex, ways to increase the filtering adequacy. Perhaps the simplest method is a form of the *π-section filter*, so named because it resembles the Greek letter pi (π). There are two basic forms of this filter, but we are concerned with the resistance-capacitance form shown in Fig. 6-10. Replacing the resistor by a choke coil may improve the filtering, but at great cost in money, space, and weight.

The action of the filter is as follows. Any current that passes through the load resistance is supplied in part from the discharge of electrons from the capacitor input of the filter. Because the charging current is a pulsating direct current, there is a small ripple voltage superimposed on top of the dc output voltage. As this pulsating ripple current passes through the resistance of the filter, it develops a ripple voltage as well. This ripple voltage, however, is absorbed, or shunted to ground, by the low reactance of the second capacitor in the filter. The reactance of the capacitor is, you will remember, a function of the ripple frequency. Since the input sine wave of 60 Hz was fully rectified, two pulses resulted for every cycle rectified. This means that the ripple frequency must be 120 Hz, just twice the fundamental frequency. Technically the ripple frequency is expressed as 120 *pulses per second (pps)*, and not in Hertz (cycles per second). The reactance of the capacitor, however, is a function of the rate of change of its charging voltage, and not whether the voltage is a changing pulse or a sine wave.

If the second capacitor is to do its job well, its reactance at 120 pps should be less than 10 percent of the value of the series resistance of the filter. For example, if the series resistance is 33 Ω, the capacitor reactance X_C at 120 pps must be less than 3.3 Ω, given by

$$X_C = \frac{1}{2\pi fC} = \frac{0.159}{fC} \qquad (6\text{-}4)$$

$$\text{since } \frac{1}{2\pi} = 0.159$$

Thus

$$C = \frac{0.159}{fX_C} = \frac{0.159}{120 \times 3.3} = \frac{0.159}{396}$$

$$= 0.0004 \text{ F} = 400\ \mu\text{F} \qquad (6\text{-}5)$$

A capacitor having more than 400 μF of capacitance should do the job for currents less than 20 mA, since current at this level will produce a ripple of only a few millivolts at most.

THE CHOKE INPUT FILTER

The use of a choke coil as an input to the capacitor in the filter arrangement has certain properties that make its mention worthwhile. It is more expensive, and is larger and heavier as well, but it does provide better voltage regulation than the capacitor input filter. This is because its high inductive reactance coupled with its low resistance result in an inherent opposition to current changes.

The dc resistance of the choke coil reduces the peak voltage that reaches the capacitor, thus limiting the voltage to which the capacitor may charge. This output voltage does, however, remain more constant than with the capacitor input filter. The peak surge currents tend to be limited by the reactance of the coil, but the maximum direct current is somewhat improved over that available from the capacitor input filter.

The choke input filter was frequently used with vacuum-tube circuitry. However, because of its size and weight it has given way to

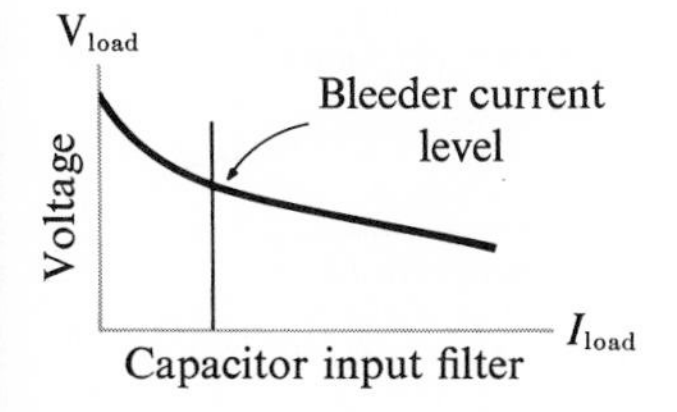

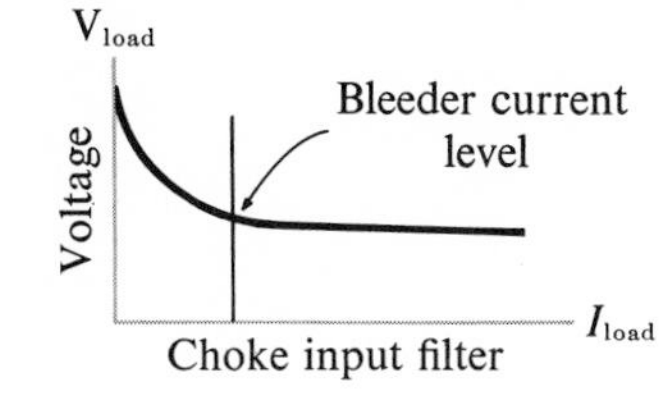

FIG. 6-11 Change in load voltage with increase in current.

the electronic filter or the semiconductor voltage regulator and current regulator or limiter.

THE DIODE AS A VOLTAGE REGULATOR

In addition to the current drawn through the load to be attached to the power supply, a small *bleeder current* is usually allowed. This requires a bleeder resistor, as indicated in Fig. 6-10. It serves two functions: to remove the charge from the capacitors after the power supply is turned off and to improve regulation of the output voltage with variations in current.

Regulation is a measure of how well the power supply is able to maintain an output voltage without change. For example, if the full- or rated-load voltage is 16 V and the no-load voltage is 18 V, the percentage of regulation is

$$\text{Percent reg.} = \frac{V_{\text{no load}} - V_{\text{rated load}}}{V_{\text{rated load}}} \times 100$$

$$= \frac{18\text{ V} - 16\text{ V}}{16\text{ V}} \times 100$$

$$= 0.125 \times 100 = 12.5 \text{ percent} \qquad (6\text{-}6)$$

Obviously, if the voltage varied 12.5 percent from no load to rated load, regulation of the power supply would be poor. However, if the voltage change were only 0.2 V instead of 2.0 V, the regulation would be 1.25 percent. This amount of regulation is not bad for certain applications, such as the experiments in this text. Industrial applications frequently require 0.001 percent regulation.

If the power supply is allowed to operate at no load, the output voltage will always be higher than with a load. As a small permanent load (a small current) is added to the output, the regulation will improve as indicated in Fig. 6-12. The small load is developed by means

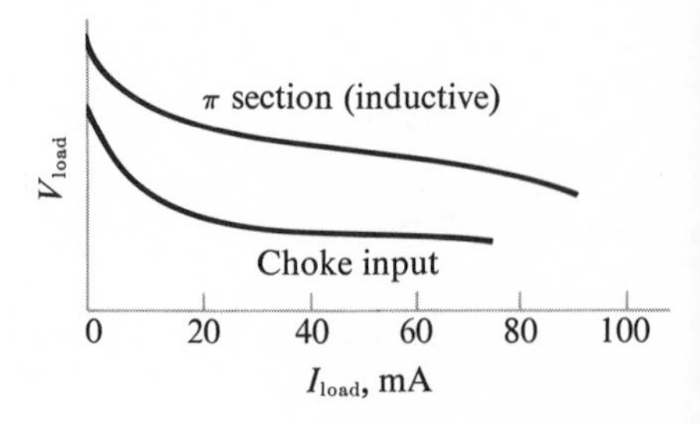

FIG. 6-12 The choke input filter and its voltage-regulation curves.

of the same type of bleeder resistor used to bleed off the capacitor charge when the supply is not in use.

THE BREAKDOWN DIODE AS A REGULATOR

Perhaps the most useful solid-state device for voltage regulation is the breakdown diode. As we saw in Fig. 6-4, the breakdown diode can be used to maintain a voltage within narrow limits as long as the amount of current is small. The Zener point, indicated on the characteristic curve in Fig. 6-3, is difficult to establish exactly, since the breaking point is actually a knee or curve. However, if a small amount of current is allowed to pass through the diode as it is connected, the voltage across it will be stable at some Zener voltage V_z for current I_z. As long as more than this amount of current is passed through the diode in this breakdown condition, the voltage will remain essentially constant, generally rising only a few millivolts.

The circuit of Fig. 6-13 should be entirely clear from Ohm's law and the principles of series-parallel circuits. Note that the current coming from the negative terminal of the rectifier may pass through both the breakdown diode and the load resistance. This same total current must pass through the series resistance of the π-section filter. Therefore the voltage at the output of the power supply must be $V_{out} = V_{in} - V_{\pi}$.

Since the output voltage is the same across the load resistance and the breakdown diode (they are in parallel), the current through these two branches must divide in some way inversely proportional to their resistances. In other words, the larger current will pass through the smaller of the two resistances. If the output voltage is to remain constant, the sum of these two currents must remain constant, since the total current must pass through the series resistance of the filter.

Observe that the output voltage from the rectifier is about 16.5 V peak. If the output voltage of the supply is to be 15 V dc, the voltage across the filter must be 1.5 V and must remain 1.5 V within the expected current range of the load resistance. This means that as the load current changes, the current through the breakdown diode must change accordingly. This is exactly what happens. An example will illustrate the point.

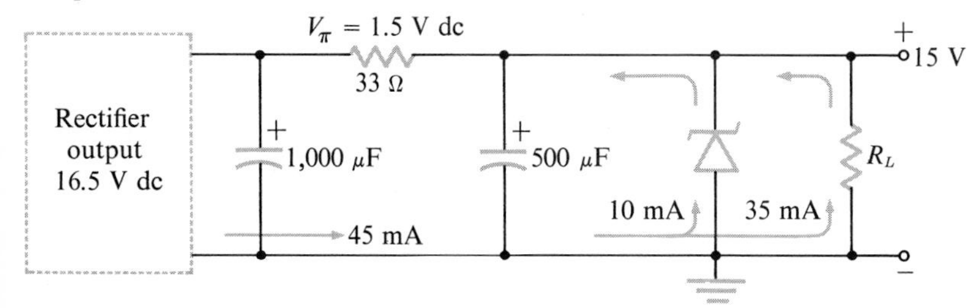

FIG. 6-13 A low-current power supply using a capacitor input π-section filter and a Zener voltage regulator.

If the total current I_T needed is 45 mA, the series resistance of the filter, R_π, must be

$$R_\pi = \frac{V_\pi}{I_T}$$

$$= \frac{1.5\ \text{V}}{45\ \text{mA}} = 33\ \Omega \tag{6-7}$$

Since this current must be constant, if the load current drops from 25 to 15 mA, the diode current must increase from 20 to 30 mA. This is because as the load current begins to drop, the voltage across the series resistance begins to drop too, but the load voltage in series must rise. However, since the breakdown diode is operating at breakdown, its voltage cannot rise, and so its current goes up instead (see Fig. 6-3*a*).

There must always be some minimum current passing through the breakdown diode, say, 10 mA for any operation beyond the knee. If the total current is 45 mA, the load may use up to 35 mA before the diode current drops below its reference current. When more than 45 mA of total current passes through the series resistor, the voltage across it will exceed 1.5 V, leaving less than 15 V for the output. At this point the output voltage is no longer regulated, since no current is passing through the breakdown diode.

In reality, of course, the diode breakdown curve is never completely vertical. As more current passes through the diode, the voltage tends to increase slightly. It will remain within a few millivolts, however, if a high-quality breakdown diode is used. For most low-current applications a 1-W breakdown diode is adequate. For operation at 15 V this will allow a maximum current through the diode (no current

FIG. 6-14 A power supply to power low-current experimental circuits. Output voltages are 8.2 and 16 V. Switch S_1 is set for 16 V in position 2.

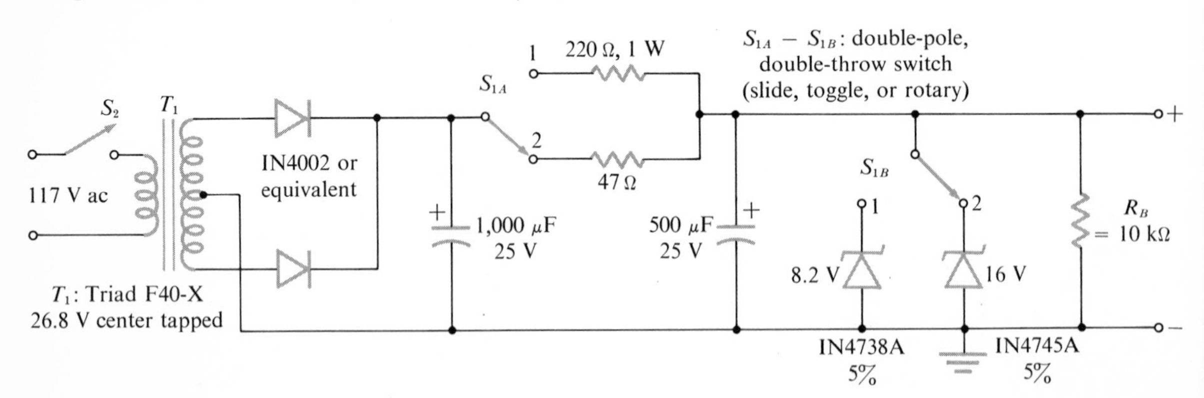

through the load) of about 60 mA. The diode should not be operated near this maximum capacity, however, since it may overheat if the heat is not dissipated by some means.

FIG. 6-15 A constant-voltage–constant-current regulated power supply (*courtesy Hewlett-Packard*).

AN EXPERIMENTAL POWER SUPPLY

As a simple power supply adequate for most of the experiments in the text, the following circuit will operate well and give two fixed voltages. A switching arrangement is needed to change from the low voltage to the high voltage. This may be accomplished by means of a slide, a toggle, or a rotary switch that will operate with two poles on a double throw.

The power supply is designed to deliver up to 30 mA to the load, with 15 mA minimum current passing through the breakdown diode to maintain regulation. The output voltages may change slightly with temperature, but the regulation and ripple control are more than sufficient for the very-low-current circuits.

This power supply is not designed to develop low ripple at the high currents needed to run the power amplifier circuits. The transformer will deliver up to 1 A and may be used in an unregulated arrangement similar to that in Fig. 6-9 for the power amplifier.

DETECTION OF RADIO WAVES

Basically the same circuit as used for the rectifier in the half-wave circuit is used to detect modulated radio waves. When an amplitude-modulated (AM) radio-frequency (RF) wave is transmitted, the waveform, or envelope, resemble Fig. 6-16. This waveform is the result of adding a low-frequency sound wave to a high-frequency radio wave. When the two waves are added, they add algebraically. Thus the amplitude varies both above and below the zero reference point. The frequencies also combine to yield the sum and the differences of the

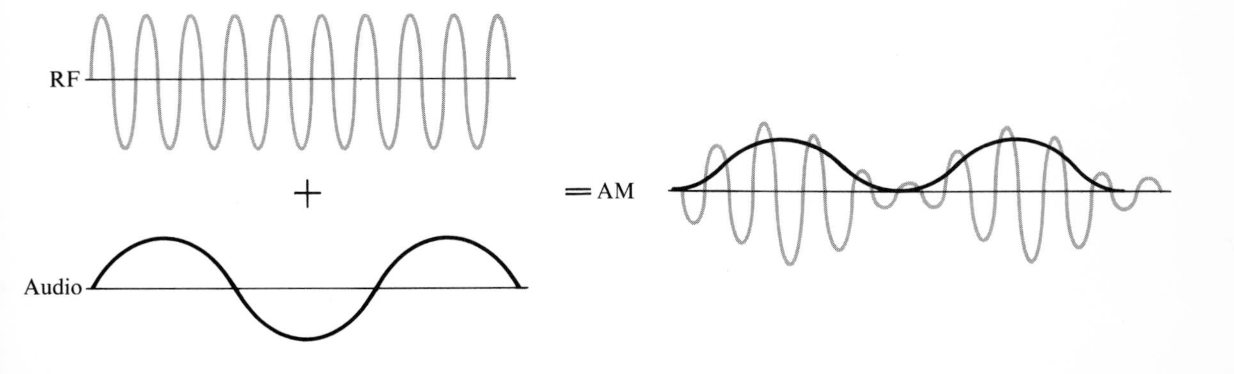

FIG. 6-16 Amplitude-modulated RF waveform generation.

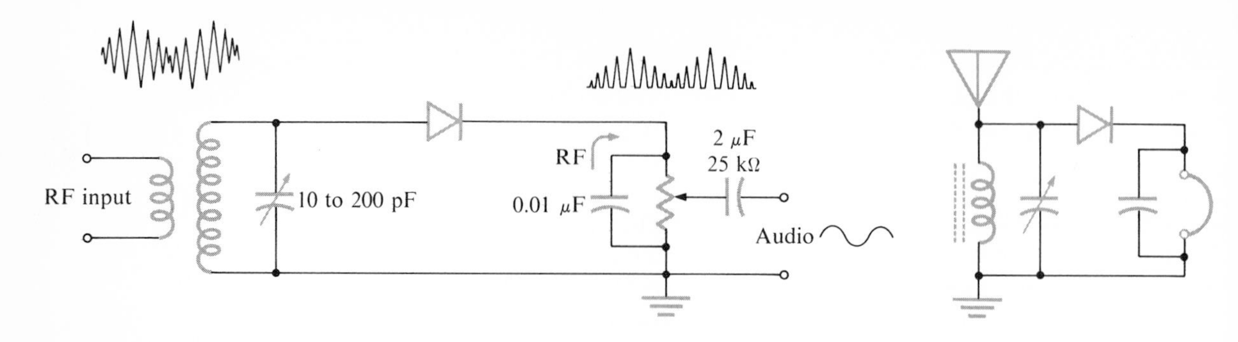

FIG. 6-17 Amplitude-modulated RF detector circuits.

two frequencies in addition to the two original frequencies. Of course, only the three frequencies that are in the RF range are actually transmitted by the radio.

The theory of detection and recovery of the information in the two sidebands (sum and difference) is rather complex, but the circuitry is quite simple. The circuit consists of a tuned RF transformer, a half-wave diode rectifier, and a load resistor which is bypassed at the radio frequency. The output voltage recovered across the load resistor is literally the average of the peaks of the waves that charge up the capacitor. It is the changing voltage on the capacitor, varying at the audio rate, that is found across the load resistor (see Fig. 6-17). This sound voltage is to be amplified by an audio amplifier, which develops sufficient power to drive the speaker.

Although full-wave rectification or detection is possible, it is rarely used except in some experimental "super" crystal receivers. This is because the voltage developed by the half-wave detection process is quite adequate.

TEMPERATURE COMPENSATION

The silicon diode is frequently used as a temperature-compensating element. When the temperature increases, the voltage across the diode actually decreases. Conversely, the rise in temperature across such devices as the Zener diode is followed by an increase in voltage. For this reason the silicon diode may be used in conjunction with other devices in extreme temperature conditions to counteract the effects of voltage change.

Figure 6-18 shows how the silicon diode is connected to compensate for the change in voltage across the Zener. As the Zener voltage begins to increase, the diode voltage begins to decrease, with the result that the voltage remains more nearly at its proper operating point.

Because of its temperature-compensating properties, the diode is frequently used instead of resistors when the voltage must remain

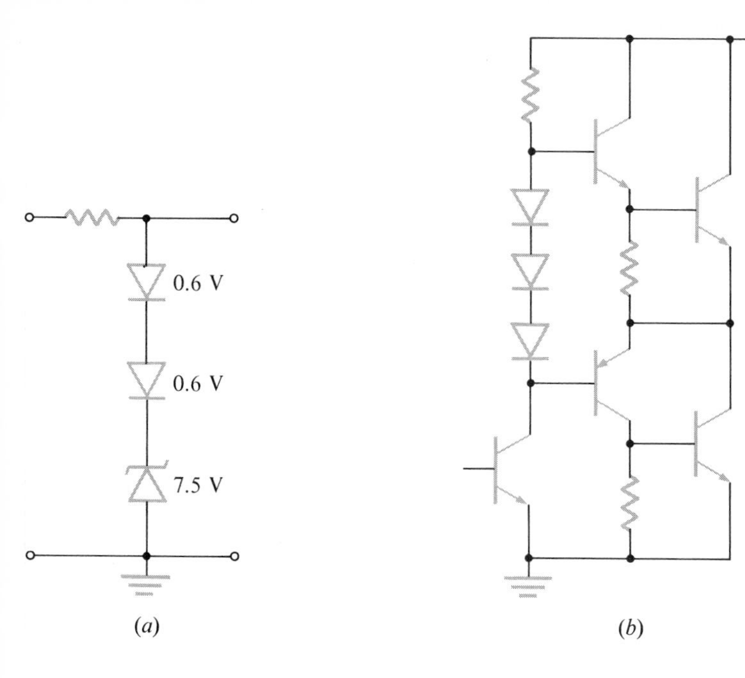

FIG. 6-18 The silicon diode as a temperature-compensation device. Diode voltage *decreases* as the temperature increases.

essentially constant with wide ranges in current but must decrease with a rise in temperature. This situation is illustrated in Fig. 6-18*b*, where the two transistor bases must be at temperature-compensated voltage differences. Each diode provides about 0.6 V of potential difference over a wide range of current while providing temperature compensation.

SUMMARY OF CONCEPTS

1. The forward characteristic of a diode is due to the combined action of the junction impedance and the bulk resistance.
2. The silicon diode only approximates the ideal diode, which would have no resistance in the forward direction and infinite resistance in the reverse direction.
3. The peak reverse voltage for the silicon rectifier represents the maximum reverse voltage, which is the point of operation of the breakdown, or Zener, diode.
4. The breakdown diode must operate beyond the knee of the curve to be used as a voltage regulator.
5. Junction resistance is a function of the junction temperature voltage kT/q and the forward current flow, about 26 to 30 Ω at 1 mA.
6. The output voltage of the half-wave rectifier in rms units is one-half the peak voltage, not one-half the full-wave rms voltage.
7. The output voltage of a half-wave rectifier can be made to approach the peak voltage by means of a capacitor input filter.
8. The full-wave and the full-wave bridge rectifier have output voltages that are easier to filter because the ripple or pulse frequency is twice the line frequency.
9. The choke input filter gives better regulation than the capacitor input filter because of its reactance.
10. The series-resistance π-section filter is superior to the capacitor filter because of the capacitor-bypass action of the ripple voltage built up on it.
11. The bleeder resistance in a power supply improves voltage regulation by preventing a no-load condition and by bleeding off the capacitor charge when the power supply is turned off.
12. An AM radio wave may be detected by half-wave rectification and filtering of the RF voltages, which leaves the audio voltage.
13. The junction temperature-voltage changes make it possible to use the diode as a temperature-compensating element in a circuit.

GLOSSARY

Breakdown diode A diode designed to operate with reverse bias and in the breakdown condition for use as a voltage reference. It is similar to a Zener diode, but makes use of the avalanche effect and operates at voltages greater than about 8 V.

Bridge rectifier A set of four diodes placed in a bridge formation, with all cathodes on the ends closest to the positive end of the load, used to give full-wave rectification.

Choke An inductor or coil, usually wound on some form of a core, used to limit alternating current or smooth out current ripple in a power supply.

Detection A process of rectifying an RF ac voltage for recovery of the modulated RF waveform.

Filter Generally an *RC* or an *RCL* combination of components placed in a circuit in such a way as to remove variations of voltage or current of a certain frequency, such as a 60-Hz power-supply ripple.

Full wave A term generally applied to a rectifier and referring to the output voltage having a positive pulse for each negative input pulse while allowing the positive pulses to go through unchanged.

Half-wave Refers to a simple rectifier or detector output where only the positive half-cycle is allowed to pass through the circuit to the output.

Heat sink A piece of heat-conducting material, such as aluminum or copper, on which transistors or diodes may be mounted to dissipate their heat of operation.

Ideal diode A theoretical diode that would allow current to pass in the forward direction as soon as the forward bias exceeded 0 V and would allow no current to pass under reverse-bias conditions.

Peak reverse voltage The amount of reverse bias that may be applied to a diode without breakdown occurring.

π-section filter An *RC* or *LC* filter laid out so that the resistance or inductance is in series with the circuit current, with the filter capacitors connected at each end of the resistance or inductance to ground. The circuit diagram resembles the Greek letter π.

Power supply A circuit that can supply the desired voltage and current to an active circuit. Generally the power supply changes alternating current to regulated smooth direct current.

Rectifier Some form of diode used to convert alternating current to direct current by allowing current to pass in only one direction.

Regulation The percentage of change in output voltage from a power supply as input voltage changes or as output load current changes.

Ripple The small sawtooth-shaped voltage present in most power supplies after the rectified ac voltage wave has been filtered.

Surge current The large current that surges through a circuit when the power is first connected and before the capacitors have had a chance to charge up and limit further current.

Temperature voltage That part of the diode junction voltage, about 26 to 34 mV, which is a function of the absolute temperature and from which the dynamic junction resistance may be computed ($V_t = kT/q$).

Zener effect The carrier tunneling effect that accounts for the reverse-breakdown phenomenon in Zener diodes, the primary property of Zener diodes with breakdowns of less than 8 V.

REFERENCES

Amos, S. W.: *Principles of Transistor Circuits*, 3d ed., Hayden Book Company, Inc., New York, 1965, chap. 1.

Brazee, James G.: *Semiconductur and Tube Electronics: An Introduction*, Holt, Rinehart and Winston, Inc., New York, 1968, chap. 2.

Corning, John J.: *Transistor Circuit Analysis and Design*, Prentice-Hall, Inc., Englewood Cliffs, N.J., 1965, chap. 2.

Cutler, Phillip: *Semiconductor Circuit Analysis*, McGraw-Hill Book Company, New York, 1964, chap. 2.

Leach, Donald P.: *Transistor Circuit Measurements*, McGraw-Hill Book Company, New York, 1968, experiments 4–6.

Lenert, Louis H.: *Semiconductor Physics, Devices, and Circuits*, Charles E. Merrill Books, Inc., Columbus, Ohio, 1968, chap. 4.

Malvino, Albert Paul: *Transistor Circuit Approximations*, McGraw-Hill Book Company, New York, 1968, chaps. 2–4.

Preferred Semiconductor and Components from Texas Instruments, Texas Instruments, Catalog CC202, Dallas, Tex., 1970, pp. 25401–25402.

Surina, Tugomir, and Clyde Herrick: *Semiconductor Electronics*, Holt, Rinehart and Winston, Inc., New York, 1964, pp. 12–20.

REVIEW QUESTIONS

1. What properties of the silicon diode make it more useful in some applications than the germanium diode?

2. What is the forward-bias voltage needed for current to pass through a silicon diode?

3. What will happen if a diode is placed in a circuit so that the reverse bias is greater than the peak reverse voltage of the diode?

4. What is meant by the Zener effect? What is its use?

5. Are all Zener diodes breakdown diodes? Why?

6. In what way does the silicon diode differ from the ideal diode?

7. What is meant by rectification? How does a silicon diode rectify?

8. What is the difference between half-wave and full-wave rectification?

9. What safety factors should be observed in using silicon diodes in a power supply?

10. Why is a bridge rectifier sometimes used in place of a two-diode full-wave rectifier?

11. In what way does a capacitor input filter smooth out the pulsations in rectified ac voltage?

12. What is meant by surge current?

13. In what way does a π-section filter do a better job of filtering than a simple capacitor filter?

14. What is the function of a bleeder resistor in a power supply?

15. What is the ripple frequency of a full-wave rectifier with 60-Hz input voltage? Why?

16. Why is regulation of a power supply so important?

17. Which is better—a regulation of 1.5 percent or 0.5 percent? Why?

18. What is the advantage of a choke input filter over the capacitor input filter? The capacitor input over the choke input filter?

19. Describe the action of a Zener, or breakdown, diode as a regulating element in a power supply.

20. Why must there be some current through the Zener, or breakdown, diode if it is to operate properly?

21. How may a breakdown diode in series with a silicon diode be used to increase the regulation voltage available from a breakdown diode alone?

PROBLEMS

1. In a circuit like that of Fig. 6-1, what would be the voltage across the load resistance if it were 5 kΩ instead of 10 kΩ?

2. What is the reverse resistance of a diode if 50 nA of leakage current is observed when 100 V of reverse bias is applied?

3. In the voltage-regulator circuit of Fig. 6-4, what is the value of the series control resistance if the Zener current must be at least 10 mA and the load current is 30 mA?

4. From the data in Prob. 3, what would happen to the Zener current if the load current were reduced to 20 mA?

5. If the diode's dynamic junction resistance is about 30 Ω when the forward current is 1.0 mA, what is its approximate resistance at 2.0 mA?

6. If a 12.6-V rms transformer is used to supply the power to a half-wave rectifier, what is the peak output voltage from the rectifier? The dc voltage measured on a meter?

7. In Prob. 6, if the output of the transformer-rectifier is filtered by a capacitor, what is the approximate output voltage?

8. From Prob. 7, since a diode has a voltage of 0.6 V under load, what is the output voltage at 20 mA?

9. In a full-wave rectifier circuit such as that of Fig. 6-7, what is the peak output voltage if the transformer is a center-tapped 12.6-V type? What is the dc output voltage?

10. If the forward diode voltage of the four diodes in a bridge rectifier circuit is 0.65 V each at 40 mA, what will the unfiltered peak voltage output be at that current if a 12.6-V-rms untapped transformer is used? (Remember that two diodes are used for each half-cycle.)

11. What is the regulation of a power supply whose no-load output voltage is 35 V but whose voltage drops to 33.5 V under load.

12. Draw the circuit arrangement that might be used for a breakdown-diode regulator to supply about 18 V output if only 8.2-V breakdown diodes and regular silicon diodes are available.

CHAPTER 7

OBJECTIVES

1. To introduce the transistor as a transfer resistor
2. To describe the transistor in terms of the resistance network
3. To relate transistor structure to the junction diode
4. To describe the diffusion process of constructing transistors and integrated circuits
5. To discuss the operation of the emitter-base junction
6. To describe the basics of transistor action
7. To relate transistor action to diode action
8. To introduce the concept of current gain, or amplification
9. To consider various input and output conditions or parameters
10. To compare basic transistor circuit arrangements

TRANSISTOR ACTION

Transistor action was a natural outgrowth of research on diode actions. It was not until December 23, 1947, that three Bell Labs scientists, William Schockley, John Bardeen, and Walter Brattain, demonstrated to the world that a tiny piece of germanium could be made to amplify voice signals about 40 times. The first transistors, patented by Bardeen and Brattain, were point-contact structures. The junction transistor was patented the following year by Schockley. In recognition of their work Schockley, Bardeen, and Brattain received the Nobel Prize in 1956.

THE TRANSISTOR

The *NP*-junction diode, which we have discussed at length, served as the basis for development of the *transistor*—originally an acronym for *trans*fer res*istor.* The name arose from the nature of the three-terminal device, since it seemed to function somewhat as indicated in Fig. 7-1. Of course, there is much more

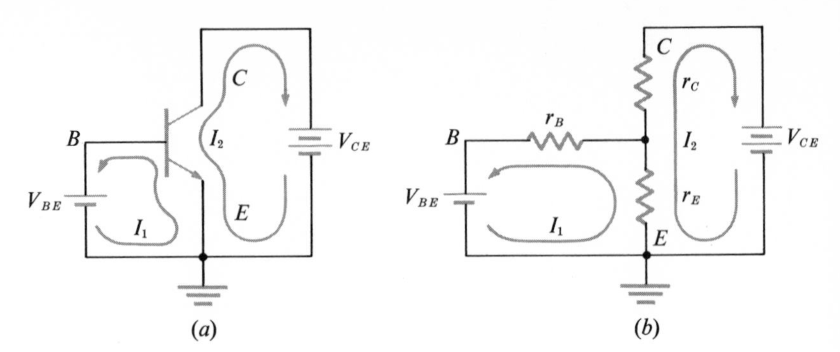

FIG. 7-1 (*a*) The transistor symbol with proper bias and (*b*) the transistor resistance equivalent circuit.

to the transistor than just three resistors. The circuit of Fig. 7-1 will not actually function as a transistor, but it does serve to illustrate the action.

The three terminals, denoted as *E*, *B*, and *C*, were named the *emitter*, the *base*, and the *collector*. The emitter seemed to emit the current carriers into the base. The base was the physical base on which the contacts for the emitter and collector were placed. The collector collected the current carriers from the base region.

The *NPN* transistor more nearly matches the usual orientation to electron current, and so our discussion will be in terms of this device, rather than its counterpart, the *PNP* transistor.

The structure of the transistor is such that a small voltage connected between the base and the emitter will cause current to flow in that circuit. With the emitter-base current flowing, if a supply voltage V_{CE} is connected between the collector and the emitter as shown in Fig. 7-1, a much larger current will flow in that circuit.

It may be further determined that if the supply voltage V_{BE} is removed, stopping the current I_1, then I_2 will also cease, except for a tiny leakage current, even though the second supply V_{CE} is still connected.

When the V_{BE} supply is reconnected, allowing current I_1 to flow once again, current I_2 immediately resumes in the collector-emitter circuit. An increase in I_1 is followed by an increase in I_2, a much larger current than I_1. A decrease in I_1 results in a large decrease in I_2. Since a small variation of current I_1 seems to control a large variation of I_2, the transistor does indeed act as a *transfer resistance*. It is this phenomenon of apparent control of a large current by a small current that makes the transistor useful as an amplifying device.

TRANSISTOR CONSTRUCTION

EARLY PROCESSES

The structure of the transistor actually consists in much more than these three resistances connected together. However, for our purposes

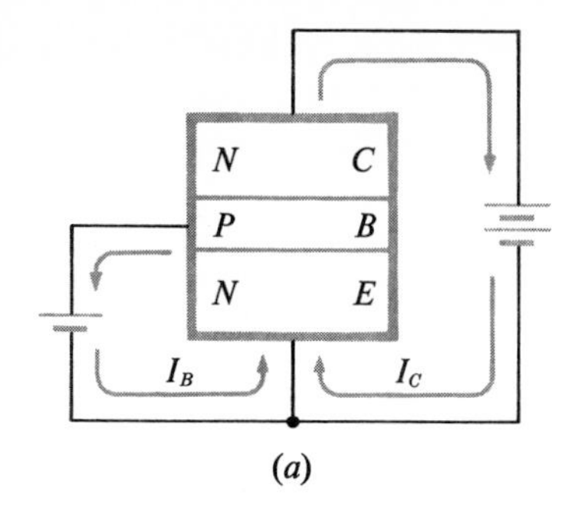

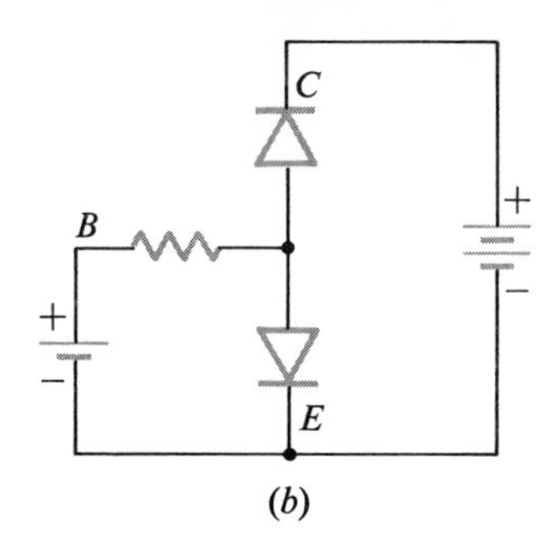

FIG. 7-2 (*a*) The layered structure of the *NPN* transistor and (*b*) transistor-diode structural similarity.

certain minor effects may be disregarded for the time being. We are more concerned here with the major effects.

The method of constructing transistors has changed many times over the years, but they are still basically three-layer devices. An *NPN* transistor looks much like an *NP* junction and a *PN* junction with a shared *P* layer, as shown in Fig. 7-2*a*. In operation this transistor resembles the two diodes connected as shown in Fig. 7-2*b*. Placing two diodes back to back will not, however, make one transistor.

As can be seen in Fig. 7-2, electrons enter into the transistor at the emitter terminal. Some of the electrons are allowed to leave the device at the base terminal, while most of the electron current proceeds on to the collector, where it is collected from the transistor by the positive terminal of the power supply. The actual process will be developed later in the chapter.

One of the major difficulties in the early methods of constructing transistors was that base regions had to be very thin to allow current carriers to pass through in very short periods of time. However, as the bases were made thinner and thinner to meet circuit requirements, their physical strength was impaired.

Partly because of the types of materials available and partly because of the processes in use at the time, the *PNP* transistor was the easiest and first to be manufactured. Figure 7-3 shows one type of *PNP* alloy-junction transistor. The alloy-junction transistor actually consisted of a thin layer of *N*-type germanium as the base, with the emitter and the collector alloyed into it. These emitter and collector layers consisted of small amounts of *P*-type materials or impurities such as indium, alloyed directly to the base. Such *PNP* transistors were somewhat simpler to fabricate than the *NPN* types. This accounts in part for their extensive use in most early circuits.

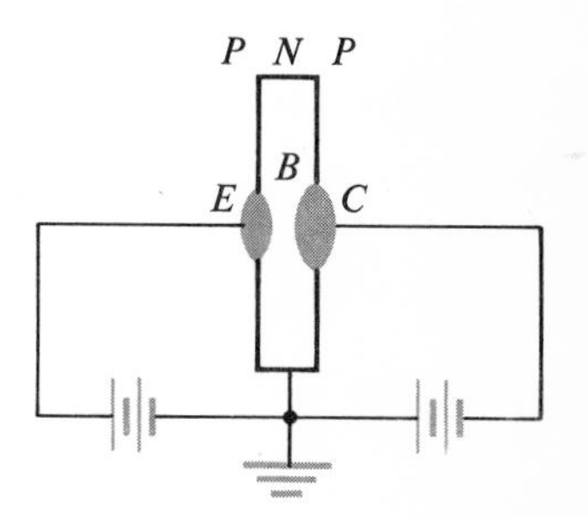

FIG. 7-3 A diagram of the *PNP* alloy-junction transistor.

THE DIFFUSION PROCESS

In 1960 Fairchild Semiconductor Company developed the *Planar diffusion process*. With this new development the *NPN* transistor manufactured from silicon became the more common type. The superiority of the silicon transistor in controlling leakage currents and its lower sensitivity to temperature changes have made it the choice for most circuits where extremely high frequencies or ultrafast switching times are not required.

The Planar process is essentially a photographic process. A thin wafer of *N*-type silicon is covered with a thin layer of silicon dioxide (glass); it is then masked, photographically exposed to imprint hundreds of transistors on the same wafer, and then etched. Figure 7-4 illustrates the basic steps. With the glass layer thus etched through, exposing the silicon, the wafer is placed in a furnace and heated in an atmosphere of gasses containing *P*-type material from some group III element such as boron. This heating process causes the impurity to diffuse into the *N*-type wafer substrate, thus producing one of the *NP* junctions.

Silicon dioxide is then redeposited over the diffused areas, and the wafer is exposed for the next mask, developed, and etched. The second diffusion uses gasses from the elements of group V, such as antimony, arsenic, or phosphorus. This time an *N*-type region is diffused into the *P*-type material previously diffused into the *N*-type substrate. This completes the second junction, which is to be the emitter-base junction.

The process is completed by again applying a glass layer, remasking, and etching, now at the places where external conducting wires are to be connected. The etched areas allow a metallized area to be alloyed to the exposed emitter and base layers. One feature of this method is that the thinness of the base layer can be controlled by controlling the time allowed for the diffusion of the emitter into the base layer.

FIG. 7-4 Some of the important steps in the Planar diffusion process for manufacturing transistors.

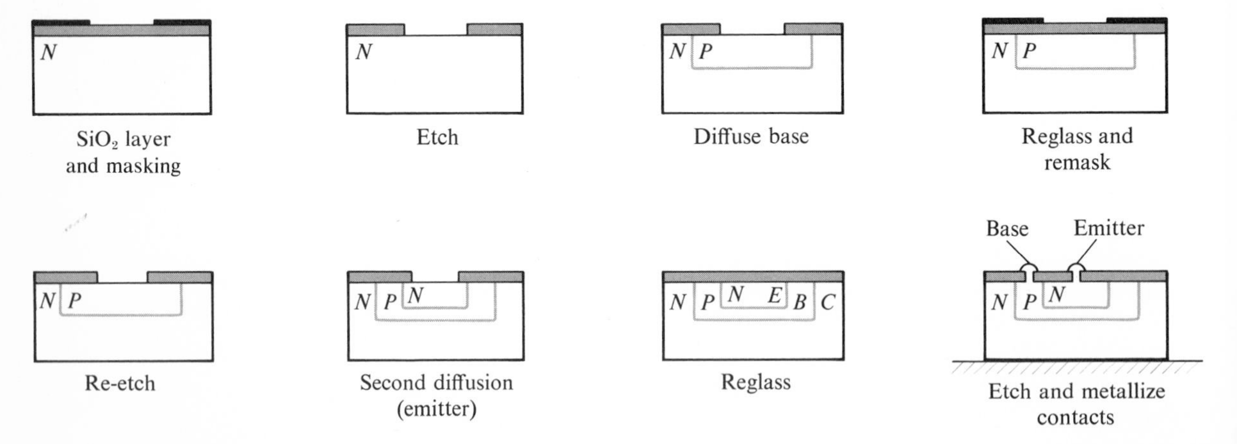

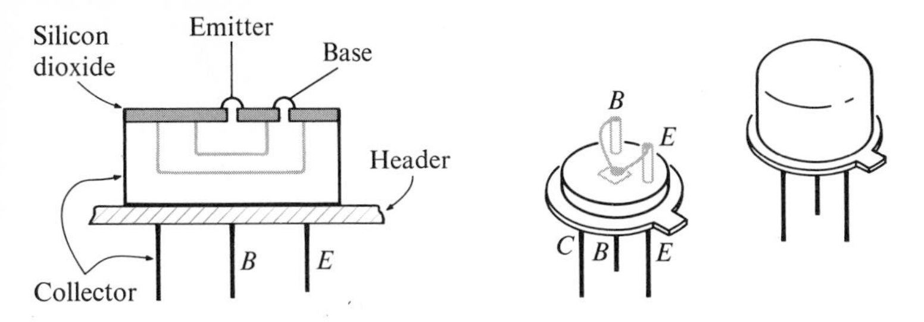

FIG. 7-5 The Planar process transistor and its case mounting.

TRANSISTOR MOUNTING

After it is completed, the tiny transistor chip is separated from the rest of the chips on the wafer and welded to the header of the transistor case, as shown in Fig. 7-5. The header acts both as the collector contact and as a heat sink to carry away the heat generated in the operating transistor. Tiny wires are then ultrasonically bonded to the metallized contacts of the base and emitter, and connected to the proper post terminals. This completes the internal transistor wiring. The final part of the fabrication consists of placing a cover on the case and filling it with dry nitrogen or some other inert gas.

Most companies market, in addition to their metal-cased transistors, a line of devices cased in plastic epoxy or ceramic. For most applications not requiring extremes of temperatures and moisture the plastic- or ceramic-cased transistors may be preferred because of their price and their electrical isolation from the case. They are much cheaper to produce, and a nonmetal case removes any concern that the case may short out to some other component. (A metal case is connected to the header, which is connected directly to the collector of the transistor.)

THE TRANSISTOR JUNCTION

An understanding of the operation of the emitter-base junction is basic to an understanding of transistor operation. The junction is a diode and acts like a diode, but it differs in some ways from the rectifier diode. The base region, for example, is more lightly doped than the emitter (the reasons for this will be evident shortly).

In normal operation the emitter-base junction is forward biased, as indicated in Fig. 7-6. Recall from our earlier discussion of the silicon diode that a forward bias of approximately 0.6 V is needed to allow any significant current to pass. The forward bias lowers the potential barrier and narrows the depleted region near the junction. At room temperature a forward bias of 0.6 V will allow about 1 mA of current to pass through the junction. Beyond this point a slight increase

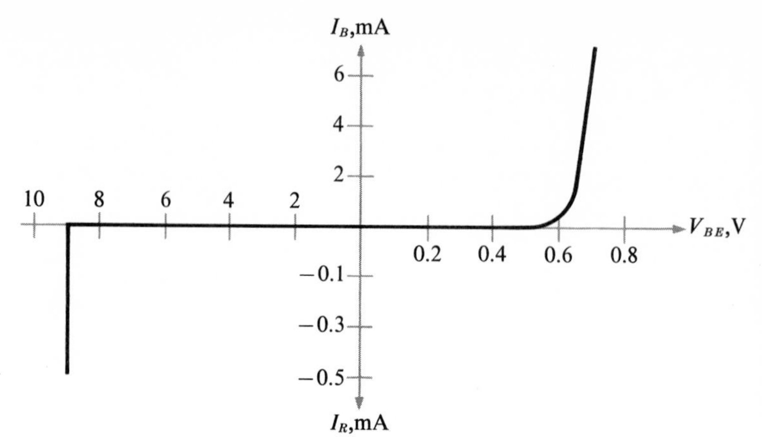

FIG. 7-6 Forward and reverse characteristics of the 2N4400-transistor emitter-base junction.

in forward bias is accompanied by a large increase in current. However, as indicated in Fig. 7-6, for currents less than 10 mA the forward voltage across the emitter-base junction will not exceed about 0.75 V. As the temperature increases, the forward voltage across the diode decreases.

When the emitter-base junction is reverse biased, it will break down at a relatively low voltage. For example, the 2N4400 device shown in Fig. 7-6 is an *NPN* silicon transistor mounted in a plastic case and has a breakdown emitter-base voltage of 9 V. Other transistors have emitter-base junctions that display breakdown characteristics at voltages from 4 to about 9 V. This sharp breakdown point allows the emitter-base junction of a transistor to be used as an inexpensive low-power Zener diode in some applications.

The collector-base junction of the 2N4400 transistor displays a forward-biased curve similar to that of its emitter-base junction. However, when the collector-base junction is reverse biased, its breakdown voltage exceeds 60 V, and for some devices may be as high as 120 V. Since the transistor is normally operated with the collector-base junction reverse biased, this means that supply voltages of up to 60 V may be used under certain conditions without damage to the transistor.

TRANSISTOR ACTION

The action of a transistor results from the combined actions of the emitter-base junction, which is forward biased, and the collector-base junction, which is reverse biased. The total picture of transistor action is based on the following principles:

1. Current flow is due to forward-bias conditions.
2. The height of the potential barrier is a function of the bias voltage.

3. Reverse-bias conditions affect the width of the depleted regions and the leakage currents.
4. Minority current carriers make up the leakage current, which is a function of junction temperature.
5. Leakage current is a problem under reverse-bias conditions, since the minority carriers encounter no potential barrier at the junction (reverse bias for majority carriers is forward bias for minority carriers).
6. Current flow in the *NPN* transistor is due to the diffusion and drift of electrons across the base region from the emitter to the collector.
7. Gain, or amplification, in a transistor is the ratio of the output to the input current, voltage, resistance, or power.

THE DEPLETED REGIONS

A schematic representation of the junction transistor is shown in Fig. 7-7. Note the depleted regions around each junction. These are the regions where the bond vacancies or holes in the *P*-type material of the base have been filled by electrons which have *diffused* across the junctions from the emitter and the base. The depleted regions in the *N*-type materials of the emitter and base similarly result from the diffusion of holes through the junctions to accept free electrons in these areas.

These two depleted regions are almost without free carriers and thus present a rather high resistance to current. In addition to the apparent resistance of the depleted regions, there is another important effect. Recall that the *N*-type and *P*-type silicon is *electrically neutral*. As a result, when the depleted region is formed the holes that cross the emitter and collector junctions form a small positive potential. Similarly, the electrons that combine with the holes in the base to form the depleted regions form a negative potential. Therefore, without any external biasing at all, there is a potential difference, or forward potential barrier at each junction, as indicated in Fig. 7-7.

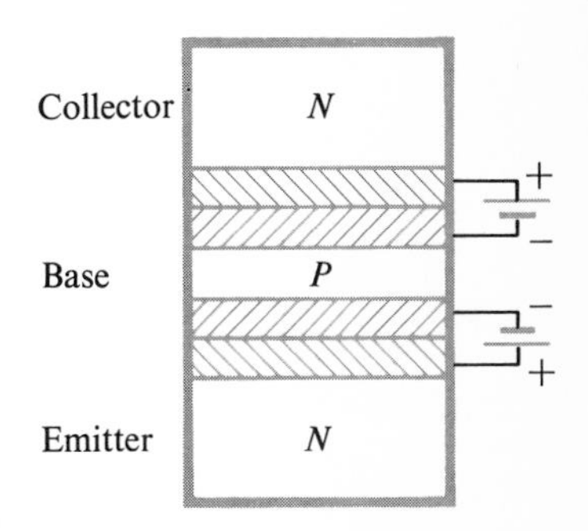

FIG. 7-7 A schematic representation of the transistor showing the depleted regions under no-bias conditions.

Since the base region is usually very lightly doped, when forward bias is applied to the emitter-base junction the depleted region becomes narrower and the potential barrier is lowered. However, the depleted region extends farther into the base than into the emitter. When the potential barrier has been sufficiently lowered as a result of the forward bias, to about 0.6 V, current begins to move over the potential barrier as shown in Fig. 7-8.

Because the collector-base junction is reverse biased, its depleted region becomes wider and extends farther into the base than into the collector. Note from the energy-level diagram in Fig. 7-8 that once electrons get into the base region, they no longer need added energy to complete the journey; they fall to a lower potential energy as they

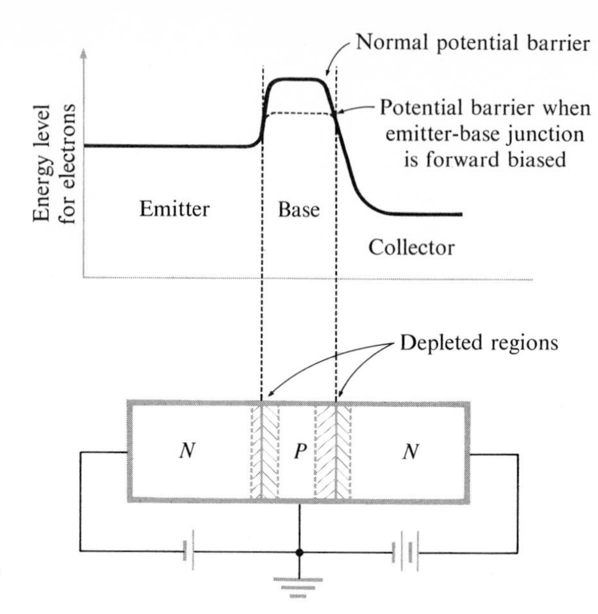

FIG. 7–8 The potential-energy diagram as it relates to the biased conditions of the transistor junctions.

drift into the collector region and are attracted by the positive electrostatic potential produced by the supply voltage.

THE COMMON-EMITTER CIRCUIT

Let us now examine the majority-current movement through the transistor in terms of the complete transistor circuit. Construction of a circuit such as that in Fig. 7-9 should clear up several details.

A forward bias of about 0.6 V between the base and emitter should allow about 1 mA of emitter current to flow into the emitter and over the potential barrier into the base. As the current diffuses into the base, most of the electrons diffuse through the base and into the depleted region near the collector. The positive potential of the collector attracts them at this point, and they drift across the junction and into the collector, from which they are returned to the power supply.

BASE CURRENT

A few electrons that leave the emitter do not get to the collector. About 1 to 5 percent of the electrons recombine with the available holes in the *P*-type material of the base. If equilibrium is to be maintained, for each electron-hole recombination in the base a new hole must be admitted into the base from the external bias supply. This is accomplished when an electron leaves the base to return to the power supply through the resistor R_B, which limits the base current (Fig. 7-9*a*). The base current is due primarily to this recombination effect. Some other

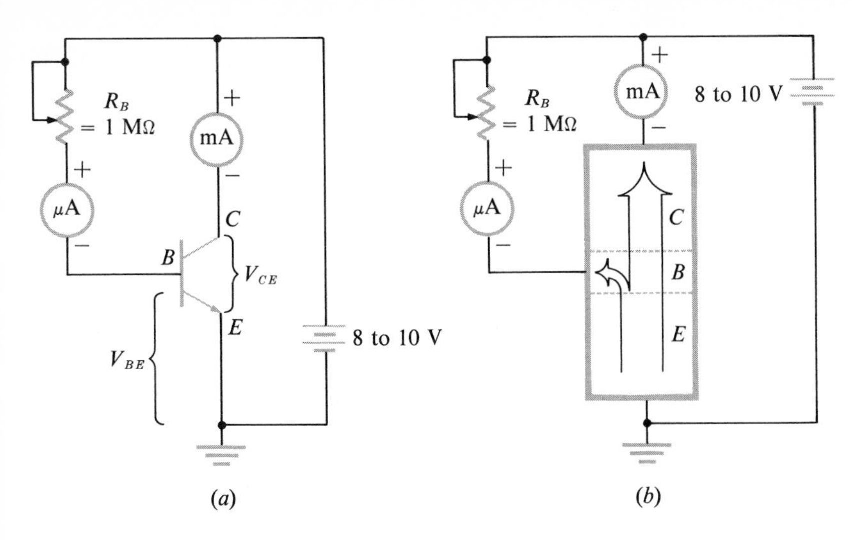

FIG. 7-9 *(a)* The biased transistor and *(b)* the base-emitter equivalent biased circuit with current flow schematic.

effects are present, but since they contribute very little to the function of the transistor, we shall disregard them at this point.

EMITTER CURRENT

If we explore this sequence of events again in terms of the circuit, it becomes apparent that the emitter current I_E is the sum of the base current I_B and the collector current I_C,

$$I_E = I_B + I_C \tag{7-1}$$

It will also be observed that changes in base current can control proportionally larger changes in collector current.

Notice in Fig. 7-9*a* that the forward bias for the emitter-base junction (positive voltage at the base) and the reverse bias for the base-collector junction (positive voltage at the collector) are both supplied from the same power source. The fact that only about 0.6 V is needed to forward bias the base presents no problem, since the remaining 9.4 V of the power supply is developed across the base-current-limiting resistor R_B as the base current passes through it.

CURRENT GAIN IN A COMMON-EMITTER CIRCUIT

Operation of the circuit of Fig. 7-9 requires that the base-current-limiting resistor R_B be adjusted until a collector current of 1.0 mA registers on the collector-current milliammeter. For the 2N4400 transistor suggested for this experiment, the base current indicated on the microammeter should be about 20 μA. The actual amount of base current observed will depend on the individual device, and not the transistor type. Since a collector current of 1.0 mA is produced when the base current is 20 μA, there is a current gain. This is because in a common-emitter circuit the input serves as the base current and the output serves as the collector. The output current is thus controlled by the input current. This current gain, denoted as β, is expressed mathematically as

$$\beta = \frac{I_C}{I_B} = \frac{1.0\ \text{mA}}{20\ \mu\text{A}} = \frac{1.0\ \text{mA}}{0.02\ \text{mA}} = 50 \tag{7-2}$$

In other words, the collector current is 50 times larger than the base current; that is, the current gain β is 50.

If R_B is adjusted to give a reading of $I_B = 40\ \mu$A, the collector current should increase to $50 \times 40\ \mu\text{A} = 2.0$ mA. Generally, as long as the collector current is small (between 0.5 and 5.0 mA), the forward current gain β should remain about the same, in this case about 50. In most transistors β will actually increase somewhat as the collector current increases. However, as long as currents near 1 or 2 mA are used, this change in β can generally be disregarded.

HYBRID PARAMETERS: INPUT VERSUS OUTPUT

Most manufacturers no longer refer to common-emitter current gain as β, although this term frequently appears in literature. Instead they are concerned with the relation of the output to the input. In some cases the output current is compared to the input voltage; in other cases output current is compared to input current. In one case the *change* in input voltage is compared to the change in output voltage. For this reason a *hybrid* measurement, denoted as h, has been developed, with a subscript to designate the particular parameter or condition being evaluated. For example, β is denoted as h_{FE}. Literally this means *the hybrid parameter expressing the forward-current transfer ratio, or current gain, in a grounded, or common-emitter, circuit.* Although this is quite a mouthful, for one symbol, it will become familiar with continued work with solid-state devices.

Let us review the hybrid-parameter description in terms of a circuit. To begin with, *ground* does not refer to the polarity of a terminal, but rather to that terminal of the transistor which is common to both input and output circuits of the device. Thus the F part of the subscript refers to forward current gain and the E refers to the common circuit terminal, which in this case is the emitter. If we were describing the forward current gain in a *common-base* circuit, the hybrid parameter would be h_{FB} instead of h_{FE}. By the same token, if it were the collector terminal that was common in the circuit, the forward current gain would be denoted as h_{FC}.

Note one other important point about this notation. *When capital letters are used as subscripts the current or voltage referred to is assumed to be dc*. When the subscripts are lowercase the parameter must be assumed to be a dynamic or ac voltage or current unless otherwise specified. Thus the hybrid parameter h_{fe} is still forward current gain in a common-emitter circuit, but it represents the amplification or control of *signal*, or *changing*, currents. Similarly, V_{ce} refers to an ac voltage, while V_{CE} represents a dc voltage.

THE COMMON-BASE CIRCUIT

Regardless of which terminal of the transistor is common to both the input and the output circuits, the forward bias must always be between the emitter and the base. Similarly, for most modes of operation the collector-base junction must be reverse biased; that is, for an *NPN* transistor the positive terminal of a power supply must be attached to the collector.

The common-base circuit is usually shown with two batteries or current sources, as in Fig. 7-10. The current distribution is the same as in the common-emitter arrangement; that is, for the same bias conditions there are the same base current and collector currents. The difference is that for the common-base circuit the input circuit (always shown on the left) is the emitter circuit, not the base circuit. The emitter current must be considered as the input current, with the collector current still the output current.

The emitter current from Eq. (7-1) was given as the sum of the base current and the collector current. If, as before, the base current is set at 20 μA and the collector current is 1.0 mA, the emitter current must be 1.02 mA.

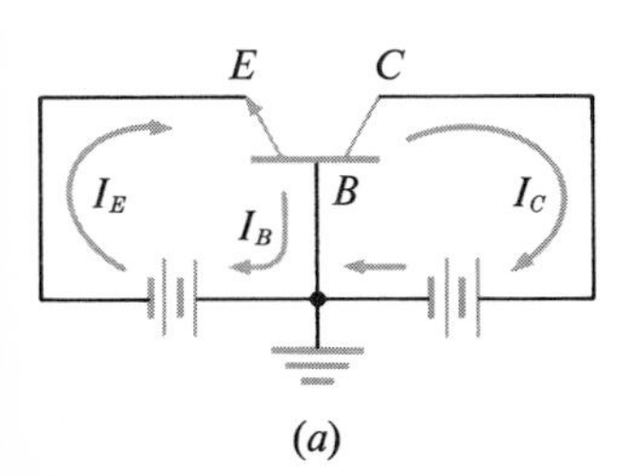

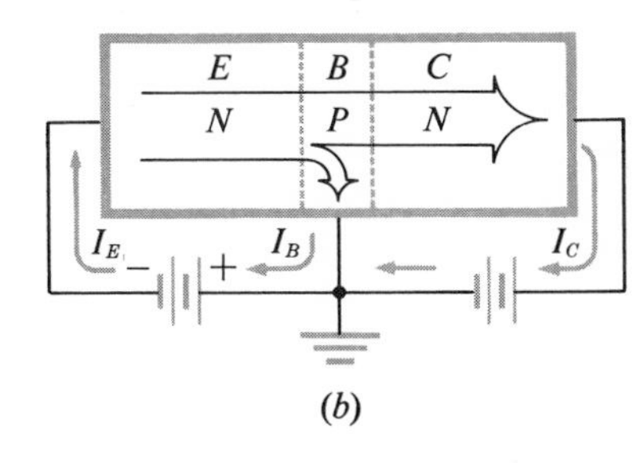

FIG. 7-10 (*a*) The common-base circuit arrangement and its (*b*) current-flow schematic.

CURRENT GAIN IN A COMMON-BASE CIRCUIT

The current gain of the transistor in a common-base circuit is sometimes called α, but it is now generally denoted by the hybrid parameter h_{FB}. Regardless of how it is designated, it is still determined by dividing the output current by the input current. In a common-base circuit the input current is the emitter current and the output current is the collector current. Hence the forward current gain for the circuit of Fig. 7-10 is

$$\alpha = h_{FB} = \frac{I_C}{I_E} = \frac{1.0 \text{ mA}}{1.02 \text{ mA}} = 0.98 \qquad (7\text{-}3)$$

Although this value of h_{FB} is less than 1 and represents an actual loss in current, the common-base circuit still has many uses.

Remember that the same transistor may be used in either a common-emitter or a common-base circuit. It is merely the comparison of different currents as output and input that yields the vast difference in magnitude.

VOLTAGE GAIN, OR AMPLIFICATION

To illustrate the usefulness of a common-base circuit and its current gain h_{FB}, consider how this circuit can produce a large voltage gain, or *amplification*, even though h_{FB} and h_{fb} are less than 1. Recall from Ohm's law that voltage is the product of the current and the resistance, $V = IR$. It follows, then, that the circuit voltage gain A_v should be the product of the circuit current gain A_i and the circuit resistance gain A_r.

The circuit current gain is very nearly the same as the transistor current gain, that is, $A_i = h_{fb}$. Hence the large voltage gain must be due to the large resistance gain of the circuit. The output resistance in a circuit such as this is in fact very high because the collector-base junction is reverse biased. However, the input resistance at 1.0 mA of current should be between 25 and 35 Ω. Thus if the output resistance to alternating current is 6,000 Ω and the input resistance, or impedance, is 30 Ω, the resistance gain will be about 6,000/30 = 200. Then, since the current gain is 0.98, the voltage gain will be

$$A_v = A_i A_r$$
$$= 0.98 \times 200 = 196 \quad (7\text{-}4)$$

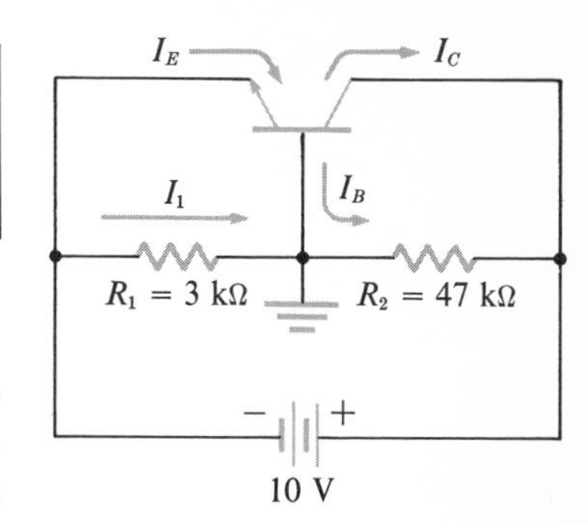

FIG. 7-11 The common-base circuit with voltage-divider bias.

Note that the lowercase subscripts indicate ac values.

The common-base circuit is not as popular as the common-emitter arrangement because of its very low input impedance. However, it is frequently used when a low-impedance circuit must be matched to a high-impedance circuit without loss of power. It is also used in very-high-frequency (VHF) and ultrahigh-frequency (UHF) amplifier circuits, since the common-base arrangement has a much higher cutoff frequency than the common-emitter arrangement.

The bother of using two power supplies can be eliminated by means of the voltage-divider arrangement of Fig. 7-11. Current I_1 passes through resistances R_1 and R_2 even if there is no transistor in the circuit because it is directly across the power supply. This current develops a voltage at the base of about 0.6 V, leaving about 9.4 V of the power supply's 10 V to be developed across the resistance R_2. Application of Ohm's law and a little bit of reasoning will remove any mystery from this simple arrangement.

It takes a forward bias of about 0.6 V at the base, with respect to the emitter, to lower the potential barrier enough to permit 20 μA of base current to flow out through R_2 to the power supply. If I_1 is at least 10 times as large as the base current, the voltage across R_1 and R_2 should remain essentially fixed even if no base current is used. Since $10 \times 20\ \mu\text{A} = 10 \times 0.02\ \text{mA} = 0.2\ \text{mA}$, let us assume this to be the value of I_1 and choose values for R_1 and R_2 accordingly to limit the current to about 0.2 mA. Thus

$$R_1 = \frac{V_B}{I_1}$$
$$= \frac{0.6\ \text{V}}{0.2\ \text{mA}} = 3\ \text{k}\Omega \quad (7\text{-}5)$$

$$R_2 = \frac{V_{CC} - V_B}{I_1}$$
$$= \frac{10\ \text{V} - 0.6\ \text{V}}{0.2\ \text{mA}} = \frac{9.4\ \text{V}}{0.2\ \text{mA}} = 47\ \text{k}\Omega \quad (7\text{-}6)$$

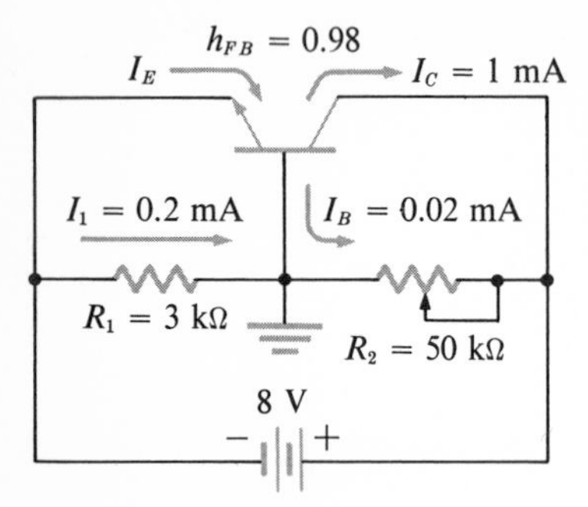

FIG. 7-12 The common-base experimental circuit with voltage-divider bias.

The experimental circuit of Fig. 7-12 may be used to verify the calculations for the single-power-supply common-base circuit. Of course, other supply voltages may be used, and a potentiometer may be substituted for R_2 to correct for minor differences in h_{FB} from one transistor to the next. For example, if an 8-V supply such as that of Fig. 6-14 is used, the potentiometer substituted for R_2 may be adjusted to a lower resistance to set the correct voltages and currents, since the sum of the voltages in the series $R_1 + R_2$ must equal the new supply voltage of 8 V.

Because the same transistor may be used in either the common-emitter or the common-base circuit, it is often convenient to be able to compute h_{FB} (or α) when only h_{FE} (or β) is known. One value is converted to the other as follows:

$$\beta = h_{FE} = \frac{I_C}{I_B} \qquad I_C = I_B h_{FE} \tag{7-7}$$

and

$$\alpha = h_{FB} = \frac{I_C}{I_E} \qquad I_C = I_E h_{FB} \tag{7-8}$$

Substitution of equivalent values does not change the accuracy of the equation, but it does change the form. Since $I_C = I_B h_{FE}$, substitution in Eq. (7-8) yields

$$h_{FB} = \frac{I_C}{I_E} = \frac{I_B h_{FE}}{I_E}$$

Then, since $I_E = I_B + I_C$,

$$h_{FB} = \frac{I_B h_{FE}}{I_B + I_C} = \frac{I_B h_{FE}}{I_B + I_B h_{FE}} \tag{7-9}$$

Factoring reduces the equation to

$$h_{FB} = \frac{I_B h_{FE}}{I_B\,(1 + h_{FE})} = \frac{h_{FE}}{1 + h_{FE}} \tag{7-10}$$

Thus the value of h_{FB} for the 2N4400 transistor is found to be

$$h_{FB} = \frac{50}{1 + 50} = 0.98 \tag{7-11}$$

This agrees with the value 0.98 determined experimentally. Develop the conversion formula of h_{FB} to h_{FE} for yourself from the facts that $h_{FE} = I_C/I_B$ and $I_C = h_{FB}I_E$.

THE COMMON-COLLECTOR CIRCUIT

The common-collector circuit, or *emitter follower*, as it is sometimes called, is not so frequently employed as the common-emitter circuit. However, its high input impedance and its low output impedance make it very useful for some applications. The basic biasing considerations are the same as those for the common-emitter and common-base circuits. The only difference is that the collector is the terminal that is common to both the input and the output circuits.

If the same transistor is used in this circuit that was used in the previous circuits, similar currents will result, since I_E differs from I_C by only the 2 percent which becomes the base current I_B. If I_C and I_E had been measured very accurately in the previous circuits, their measured values would have differed about 2 percent. However, because most meter movements and scales are within no more than 3 percent of the desired reading, actual readings taken with two different meters could be in error. If one meter read high and the other read low, the two readings might be the same or quite different, depending on which one was used in the collector circuit. If the meter is moved from the collector circuit to the emitter circuit to make the current measurements, *the readings will still be in error, because the meter itself has some resistance that will upset the voltage distribution when it is moved.*

These possible errors must be taken into consideration in any attempt to make accurate measurements. A 2 percent difference in currents cannot be measured accurately with ±3 percent meters. However, remember that all the current that goes into the emitter must come out, and that the sum of the base and collector currents is equal to the emitter current. The results may be no more accurate, but they are easier to deal with when one reading is taken on a microammeter and the other on a milliammeter. It is a simple matter to arrive at the emitter current, or to subtract the base-current reading from the emitter-current reading to arrive at the value of the collector current.

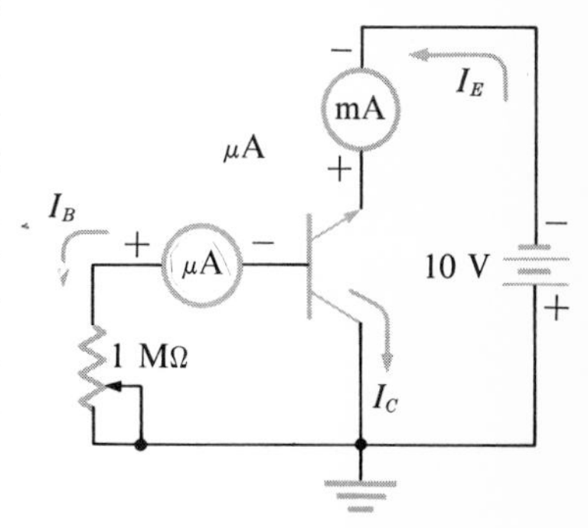

FIG. 7-13 Biasing arrangement for the common-collector circuit.

If, as before, the resistance of R_B in Fig. 7-13 is varied or adjusted to allow 20 μA of base current, the emitter current should be about 1.02 mA. The input current is the base current and the output current is the emitter current. The current gain of this common-collector circuit may thus be calculated as

$$h_{FC} = \frac{I_E}{I_B} = \frac{1.02 \text{ mA}}{0.02 \text{ mA}} = 51 \qquad (7\text{-}12)$$

Note that the value of h_{FC} is just one larger than that of h_{FE}. This is no coincidence. Since $I_E = I_B + I_C$,

$$h_{FC} = \frac{I_B + I_B h_{FE}}{I_B} \qquad (7\text{-}13)$$

By factoring we arrive at

$$h_{FC} = \frac{I_B (1 + h_{FE})}{I_B} = 1 + h_{FE} \qquad (7\text{-}14)$$

This value $1 + h_{FE}$ appears in many transistor equations and has caused much confusion among those who do not understand its origin. It is a good idea to commit it to memory.

SUMMARY OF CONCEPTS

1. The transistor is a three-terminal device with one terminal common to both the output and the input.
2. The transistor may be compared to two diodes connected back to back.
3. A forward potential difference of about 0.6 V must exist between the emitter and base of a transistor if the emitter-base diode is to conduct current.
4. The amount of collector current depends on the amount of base current, which in turn depends on the amount of forward bias between the base and the emitter of the transistor.
5. The ratio of output current (collector current) and input current (base current) is known as the forward current gain, denoted by h_{FE}.
6. The transistor is subject to the same leakage problems as the diode, except that the leakage may be amplified.
7. The collector-base junction is normally reverse biased, which places the correct-polarity charge at the collector in order to attract the carriers through the collector region of the transistor.
8. Emitter current divides such that a small amount (1 to 2 percent) becomes base current and the remainder goes on through the transistor to become collector current ($I_E = I_B + I_C$).
9. Three basic circuit arrangements of the transistor amplifier are possible, since one terminal is always common to both input and output: common emitter, common base, and common collector (emitter follower).
10. Regardless of the circuit arrangement, the current amounts and paths through the transistor remain unchanged.
11. The voltage gain, or amplification, is the product of the current gain and the resistance gain, according to Ohm's law.
12. Comparative measurements of input and output voltages and currents are known as hybrid parameters.
13. Parameters are described by subscripts such that the second letter in the subscript identifies the circuit arrangement being described; e.g., h_{FE} designates a common-emitter circuit.

GLOSSARY

Alloy junction An *NP* junction made by alloying *N*-type material to *P*-type material instead of using the diffusion process.

α Forward current gain in a common-base circuit, now referred to by the hybrid parameter h_{FB} for dc gain or h_{fb} for ac gain. It is the comparison of collector current to emitter current.

Base That region of the transistor that lies between the emitter and the collector, in an *NPN* transistor a very thin layer of *P*-type material.

β Forward current gain in a common-emitter circuit, now referred to by the hybrid parameter h_{FE} for dc gain or h_{fe} for signal or ac gain. It is the comparison of collector current to the base current.

Collector That region from which current passes out of the *NPN* transistor, that is, the region that collects the current coming through the base from the emitter.

Common-base circuit A transistor configuration in which the base is common to both the input and the output circuit.

Common-collector circuit A transistor configuration in which the collector is common to both the input and output circuits, sometimes called an emitter follower.

Common-emitter circuit The most common transistor configuration, with the emitter common to both the input and output circuits. It has both high voltage gain and high current gain.

Emitter That terminal into which current enters the *NPN* transistor.

Emitter follower The common-collector circuit, so named because the signal at the emitter follows in phase with the signal at the base or input.

Gain The amplification of a circuit, a comparison of output to input voltage (A_v), current (A_i), resistance (A_r), or power (A_p).

Hybrid parameter A descriptive term used to designate a particular property or action of the transistor. It is a ratio involving two or more different currents, voltages, or current and voltage, and involving both input and output functions of a circuit, for example, $h_{fe} = I_c/I_b$ (with V_{CE} held constant).

Parameter A descriptive property or condition, such as size, shape, temperature, or color.

Planar A diffusion process for making transistors and integrated circuits, developed by the Fairchild Semiconductor Company.

Ultrasonic bonding A process by which the tiny connecting wires are bonded to the silicon transistor or integrated circuit chip by means of the high-frequency sound vibrations.

REFERENCES

Amos, S. W.: *Principles of Transistor Circuits,* 3d ed., Hayden Book Company, Inc., New York, 1965, chap. 2.

Brazee, James G.: *Semiconductor and Tube Electronics: An Introduction*, Holt, Rinehart and Winston, Inc., New York, 1968, pp. 165–179.

Corning, John J.: *Transistor Circuit Analysis and Design*, Prentice-Hall, Inc., Englewood Cliffs, N.J., 1965, chap. 3.

Cutler, Phillip: *Semiconductor Circuit Analysis*, McGraw-Hill Book Company, New York, 1964, chap. 3.

Lenert, Louis H.: *Semiconductor Physics, Devices, and Circuits*, Charles E. Merrill Books, Inc., Columbus, Ohio, 1968, chap. 5.

Seidman, A. H., and S. L. Marshall: *Semiconductor Fundamentals, Devices, and Circuits*, John Wiley & Sons, Inc., New York, 1963, chaps. 4 to 6.

REVIEW QUESTIONS

1. What are the names of the three transistor terminals, and why are they so named?

2. What conditions are needed to cause base current to flow in a transistor?

3. What conditions are needed in order to have collector current?

4. Why is the collector current so large in comparison to the base current?

5. Why will two diodes back to back not produce a transistor?

6. Why is it necessary to have very thin base regions in transistors?

7. Describe the Planar process.

8. Why is the *NPN* silicon transistor considered better than the *PNP* germanium transistor for most applications?

9. Why are minority current carriers a problem under reverse-bias conditions?

10. In an *NPN* transistor, for every electron that recombines with a hole in the base region, how many electrons normally move on into the collector region?

11. What happens to the width of the depleted regions in a transistor under forward and reverse bias?

12. When a single power supply is used, what circuit component is used to control the amount of base current used?

13. What does the hybrid parameter h_{FE} denote?

14. What is meant by the terms "common" and "ground"?

15. What do capital letters used as subscripts refer to, as in V_{CE} or h_{FE}?

16. What is the difference between h_{FE} and h_{FB}?

17. In the circuit of Fig. 7-12, what is the function of resistance R_1?

18. Why is it difficult to measure emitter and collector currents accurately with one or two meters?

PROBLEMS

1. From the graph of Fig. 7-6, what is the approximate base-emitter voltage at $I_B = 3$ mA?

2. If the collector current of a transistor is 15 mA and the base current is 75 μA, what is the value of h_{FE}?

3. What is the base current of a transistor with $h_{FE} = 50$ when the collector current is 2.0 mA?

4. If the emitter current of a transistor is 2.06 mA and the collector current is 2.0 mA, how much is the base current?

5. From the transistor conditions of Prob. 4, what is the value of h_{FB}? Of h_{FE}?

6. If a forward bias of 0.65 V is required to allow 20 μA of base current, what size of current-limiting resistance must be used for a power source of 4.5 V?

7. In Prob. 6, what is the voltage across the current-limiting resistance?

8. In the circuit of Fig. 7-12, if the base current is increased to 0.03 mA, what is the new voltage across R_1 and R_2?

9. If the value of h_{FE} in a transistor is 120, how much is h_{FB}?

10. If h_{FE} in a transistor is 60, how much is h_{FC}?

CHAPTER 8

OBJECTIVES

1. To relate series resistances and their associated currents and voltages to the common-emitter circuit
2. To explore amplifier action and linear operation
3. To relate forward and reverse bias to control of transistor currents
4. To develop simple design concepts
5. To relate the amplifier to its characteristic curve

THE DC COMMON-EMITTER AMPLIFIER

The word "amplifier" usually brings to mind some electronic box which makes the sound or signal going into the microphone louder as it comes out of the attached speakers. This increase in loudness is indeed the result of amplification, but what conditions must exist inside the box for amplification to take place?

Just as a doorway must be designed to accommodate the largest person who will be using it, so an amplifier must be large enough for the largest signal to pass through. However, unless a large person goes through the doorway fairly close to its center, he must usually contort his body in some way to avoid bumping into the sides. The same is true with the amplifier. To allow large signal amplification without distortion, the dc voltages and currents must be set at the center point of operation to provide for equal changes of voltage or current in either direction.

DC CIRCUIT BIAS

As you have worked with resistors, batteries, and meters you have become familiar with Kirchhoff's voltage law, whether you knew it or not. This law means simply that the voltages measured across resistors in series must total that of the voltage supply or battery. For example, if the resistances and the supply voltage are as shown in Fig. 8-1, each resistor will have a voltage across it proportional to its resistance value, since the current is the same through each resistance. In other words, the 1.0 mA of current through the 1,000-Ω resistor will yield a voltage measurement across it of 1.0 V.

As a start in examining the amplifier, construct a circuit similar to that of Fig. 8-2, being careful to note the meter polarities. Adjust the potentiometer R_1 until 1.0 mA of current is flowing in the circuit. This means that the total resistance of the circuit will be 9.0 kΩ,

$$R = \frac{V}{I} = \frac{9\text{ V}}{1.0\text{ mA}} = 9\text{ k}\Omega \tag{8-1}$$

The voltmeter should read 4.3 V, about one-half the supply voltage.

If the size of the fixed resistor is 4.7 kΩ, which is a standard value, 1.0 mA of current through the potentiometer set at 4.3 kΩ will register 4.3 V on the voltmeter V_1. From Ohm's law, the total circuit resistance is 9.0 kΩ, and since the fixed resistance R_2 is 4.7 kΩ,

$$R_1 = 9.0 - 4.7 = 4.3\text{ k}\Omega \tag{8-2}$$

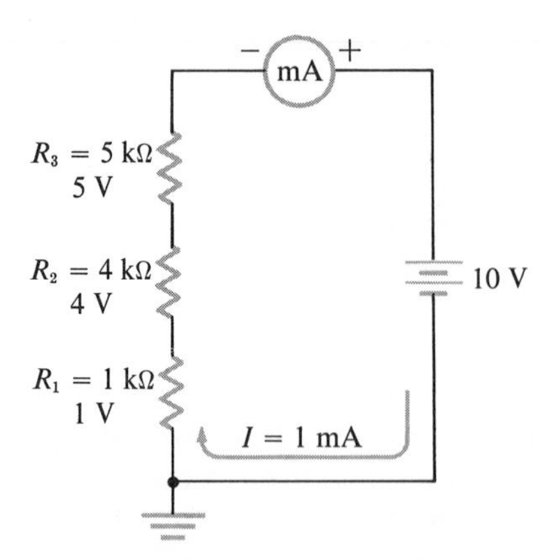

FIG. 8–1 Comparison of voltage, current, and resistance values in a series circuit.

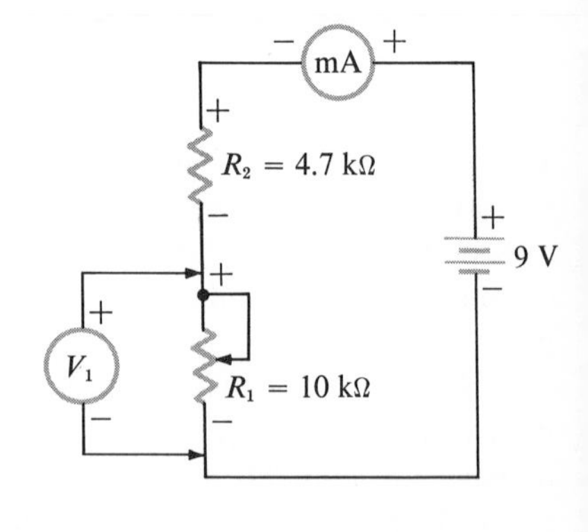

FIG. 8–2 Resistance equivalent circuit of a transistor and load in an amplifier circuit.

With 1.0 mA of current,

$$V_1 = IR_1 = 1.0 \text{ mA} \times 4.3 \text{ k}\Omega = 4.3 \text{ V} \quad (8\text{-}3)$$

As the potentiometer is reset to lower and lower values of resistance, the voltage across it will also drop. As the potentiometer is reset to larger and larger values of resistance, the voltage will increase. If this operation is alternated, the measured voltage will vary as shown in Fig. 8-3.

Note that as you move the potentiometer to 0 Ω, the voltage also drops to zero, and you can go no further, but as you again increase the resistance, the voltage once more increases.

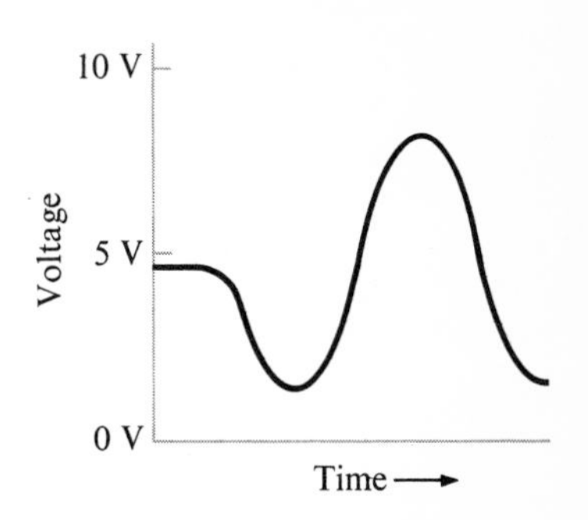

FIG. 8-3 Voltage variation with time as the potentiometer in Fig. 8-2 is varied.

THE COMMON-EMITTER CIRCUIT

The common-emitter amplifier circuit of Fig. 8-4 operates in much the same fashion as the circuit in Fig. 8-2. The transistor has the resistance between its collector and emitter of 4.3 kΩ when a voltage of 4.3 V is measured between the collector and emitter and there is 1.0 mA of collector current.

The transistor in Fig. 8-4 replaces the resistance in Fig. 8-2. However, the resistance of the transistor, R_{CE}, depends on how much base current is allowed. If there is no base current I_B, there can be no collector current I_C, and hence R_{CE} will approach infinite resistance. The voltage V_{CE} will therefore approach 9.0 V, since the load resistance of 4.7 kΩ in series with the transistor is so small in comparison. However, if I_B is allowed to increase to the point where the transistor becomes saturated with current, R_{CE} will decrease almost to zero, and therefore the voltage V_{CE} will also approach zero.

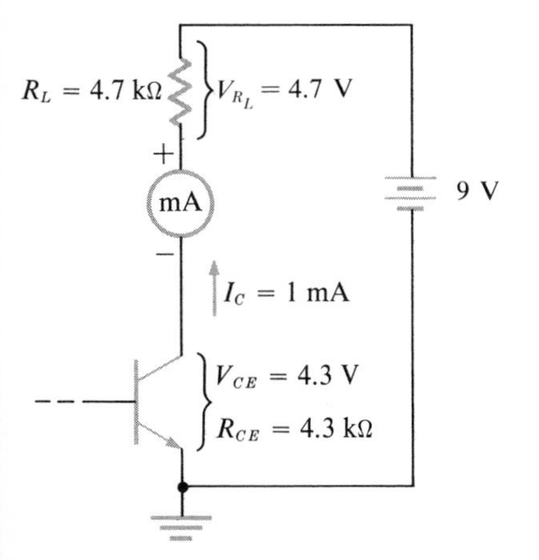

FIG. 8-4 Common-emitter amplifier equivalent of the circuit in Fig. 8-2.

It is apparent from this explanation that if the voltage across the transistor is to be able to vary an equal amount in either direction, the transistor resistance R_{CE} must equal the load resistance R_L, and the transistor must be biased so that its voltage V_{CE} is equal to one-half the supply voltage. In this case we used standard values of resistance and a standard-value 9-V battery. Ideally the value of V_{CE} should equal V_{R_L} in a good design and be 4.5 V when the supply voltage is 9.0 V.

If we complete the common-emitter circuit of Fig. 8-4 as in Fig. 8-5, to allow base current to flow, we can verify the preceding concept. The fixed 270-kΩ resistor is used simply to limit the amount of base current to a safe value in case the potentiometer is accidentally moved near to 0 Ω. Any value between 100 and 300 kΩ may be used. The transistor used in the circuit should be a silicon *NPN* transistor having a breakdown voltage greater than 20 V.

Adjust R_2 until V_{CE} is 5 V and note the amount of base current I_B and collector current I_C. This point of operation is known as the *quiescent point Q*. Measure V_{BE} for reference; it should be about 0.6 V.

Decrease the resistance of R_2 until V_{CE} drops to about 0.5 V. Note the new values of I_B and I_C. This is the upper limit of current needed, about 40 μA of base current and about 2 mA of collector current, if the transistor forward gain h_{FE} is about 50. If h_{FE} is 100, the base current needed to produce 2.0 mA of collector current is only 20 μA. Note the new value of V_{BE}; it should be slightly higher than the previously noted value.

FIG. 8-5 (*a*) An experimental dc amplifier and (*b*) sample data available from the circuit in operation.

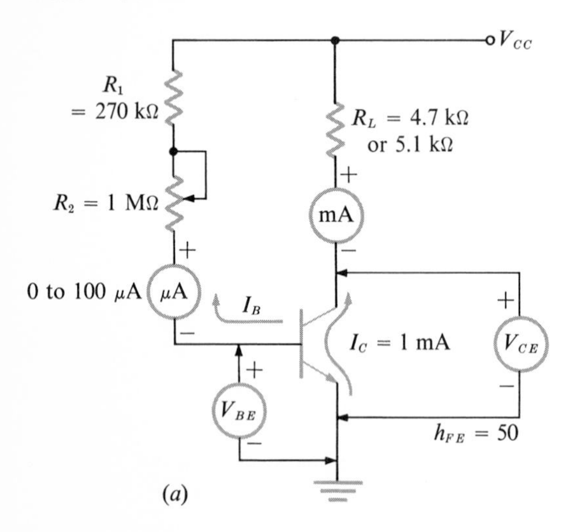

V_{CE}	V_{BE}	I_B	I_C
5 V	0.60 V	20 μA	1 mA
0.5 V	0.64 V	40 μA	2 mA
9.5 V	0.58 V	2 μA	0.1 mA

(*b*)

Now increase the resistance of R_2 until V_{CE} reaches about 9.5 V. Note the new values of I_B and I_C. They should be almost zero, since the transistor is approaching current cutoff as a result of the increasing collector-emitter resistance R_{CE}. V_{BE} should be slightly lower than before, less than 0.6 V. This is the lower limit of operation for this circuit design unless the signal is to be distorted.

This change in V_{CE} from 0.5 to 9.5 V corresponds to an ac voltage of 9.0 V peak to peak. Figure 8-6 represents these two extremes of variation in V_{CE}.

This change in V_{CE} was due to a change in collector current I_C from 1.0 mA, the reference or Q point, up to 2.0 mA, and down to 0.1 mA or near to 0 mA. This corresponds to a change in I_C of about 2 mA peak to peak. The change in I_C was produced by a change in base

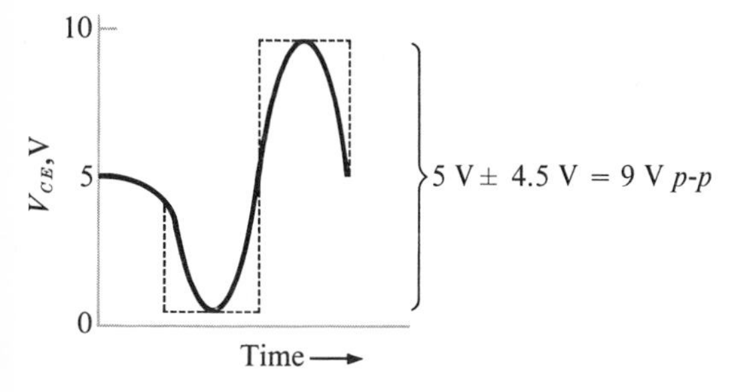

FIG. 8-6 Variation in the collector-emitter voltage resulting from variations in base current.

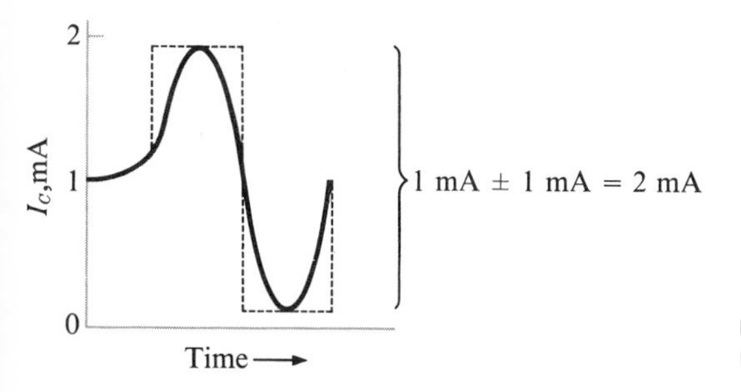

FIG. 8-7 Variation in collector current resulting from variations in base current.

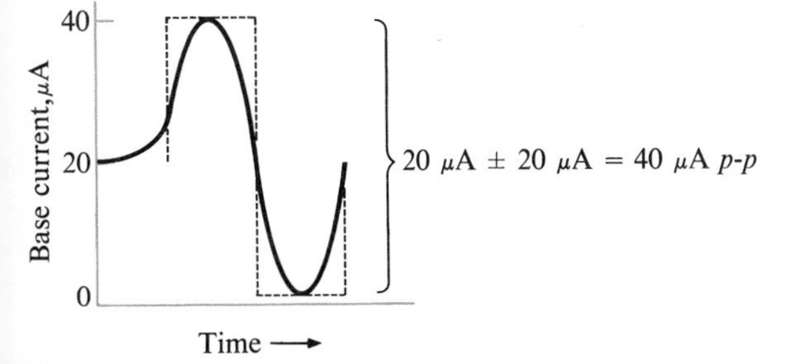

FIG. 8-8 Changes in base current used by the dc amplifier to produce changes in collector current and collector-emitter voltage.

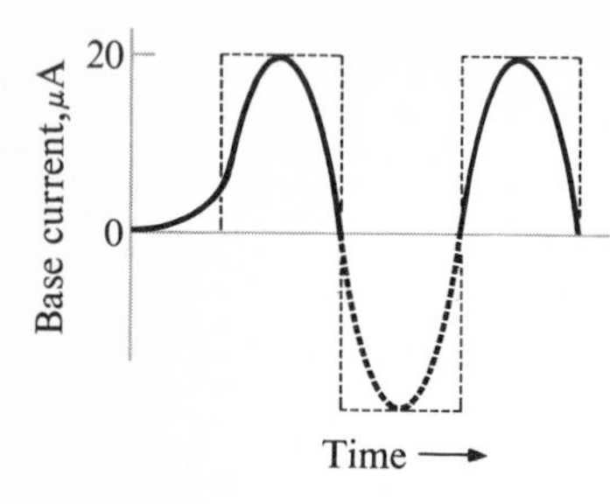

FIG. 8-9 Base-current variation of the dc amplifier biased at cutoff. Note that current passes in only one direction.

current I_B from the 20-μA Q point up to 40 μA and down to about 0 μA, or about 40 μA peak to peak.

Perhaps now it is apparent why there must be some base current, establishing the Q point, for the transistor to serve as an amplifying device without introducing distortion. If there is no base current, a signal voltage (or change in current) applied to the base will produce an output effect at the collector only as long as the base current is increased. A decrease in input voltage or a negative voltage will cut off the transistor, as shown in Fig. 8-9. Since the base-emitter junction is a silicon diode, when the forward bias drops below 0.6 V, the base current ceases. It will not reverse direction. It must always flow in the same direction, but it may increase or decrease if it starts somewhere in the middle of its range.

Note that when V_{BE} increases slightly V_{CE} decreases greatly, and when V_{BE} decreases V_{CE} increases. If the experiment is repeated and the voltage across the load resistance, V_{R_L}, is measured instead of V_{CE}, the opposite effect will be observed. This is because the increased current is passing through a fixed resistance, whereas the variable resistance of the transistor is decreasing.

Of course, nobody must sit and twist the shaft on a potentiometer to cause amplification. However, this concept illustrates the fact that slight increases or decreases in base-emitter voltage cause increases or decreases in base current, which cause greater increases or decreases in collector current. This *variation,* in collector current ΔI_C, yields a change in the resistance of the transistor ΔR_{CE}, which, from Ohm's law, must yield a change in V_{CE}, or ΔV_{CE}. The delta (Δ) before a symbol denotes change of a dc value.

The same procedure for biasing the transistor for dc amplification holds for any transistor. If a *PNP* type is used, however, all meters and the power supply must be reversed.

GRAPHICAL ANALYSIS

A more complete understanding of the operation of a transistor common-emitter amplifier requires further analysis of the current variations with changes in voltage. One can design a circuit that will function rather well without ever seeing a set of characteristic curves, but there is merit in knowing about the limits of the transistor.

If the emitter-base junction of a transistor is forward biased enough to allow 10 μA of base current while the collector-emitter voltage remains at 0 V, the collector current will be zero. As the collector-emitter voltage is increased in 0.1-V steps to 1 V, the collector current increases sharply up to a point and then levels off to some fairly constant value. As V_{CE} is gradually increased further to

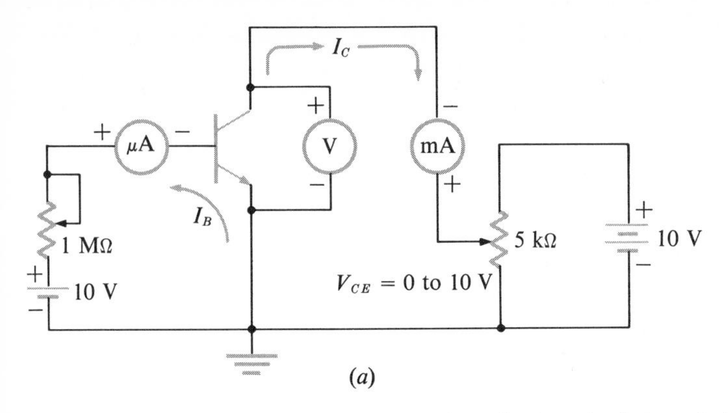

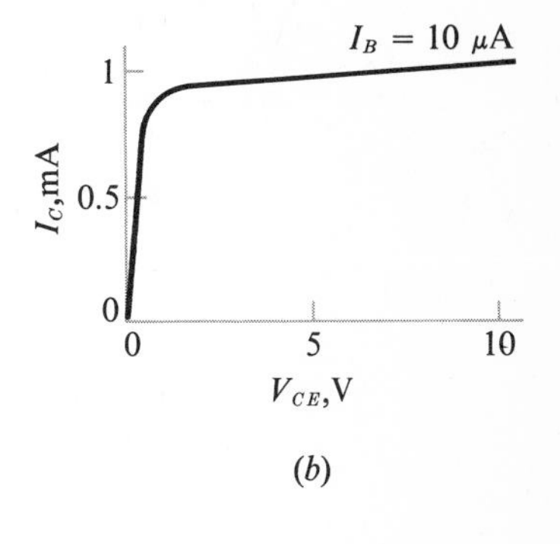

FIG. 8-10 (*a*) Experimental circuit for measuring collector-output characteristics and (*b*) a collector-output characteristic curve.

the limit of the power-supply voltage, there is only a slight increase in collector current until breakdown occurs. In other words, the collector current is controlled by the base current, and not by the collector-emitter voltage once V_{CE} exceeds the saturation voltage of about 0.5 V. This procedure may be observed and verified with the circuit of Fig. 8-10*a*, which results in the collector characteristic curve shown in Fig. 8-10*b*.

If the V_{CE} is again returned to 0 V and the base current is increased to 20 μA, a second collector characteristic curve may be developed as V_{CE} is again increased to its maximum value. In this manner a family of characteristic curves may be traced as shown in Fig. 8-11.

THE CURVE TRACER

Plotting a family of transistor collector characteristic curves is a time-consuming job when it is done by hand in the point-to-point fashion. Each point corresponding to some value of collector current, collector-emitter voltage, and base current must be individually determined, and then the points must be plotted.

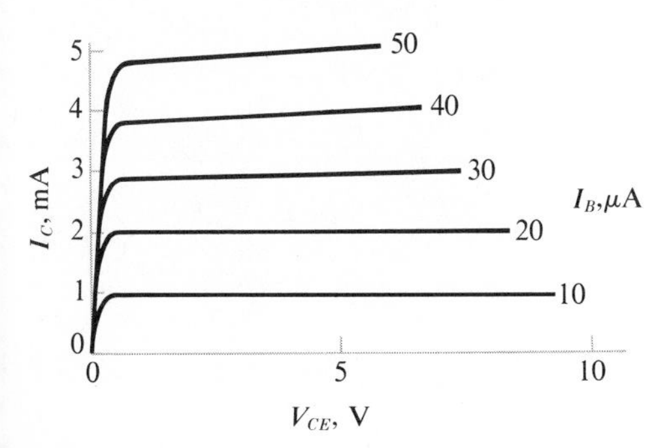

FIG. 8-11 Family of collector-output characteristic curves.

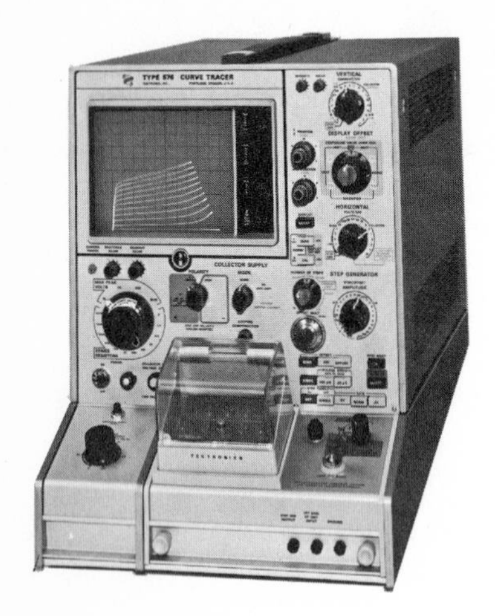

FIG. 8-12 The Tektronix Model 576 Curve Tracer (*courtesy Tektronix, Inc.*).

This cumbersome procedure may be simplified considerably by means of a curve tracer such as the Tektronix Model 576 shown in Fig. 8-12. Each variable can be set on the instrument so that the whole family of curves is displayed at once and can then be photographed, as shown in Fig. 8-13.

Although the curve tracer of Fig. 8-12 is indeed a tremendous aid in circuit design, it is quite an expensive laboratory instrument. If one is not available, the same results may be obtained by using an ordinary oscilloscope with the circuit shown in Fig. 8-14, which allows one curve to be traced at a time.

The size of the battery in the base circuit in Fig. 8-14 is not critical, since all voltage except the 0.6 to 0.7 V needed to forward bias the emitter-base junction of the transistor will be dropped across the series resistance. The function of the 47-kΩ base resistance is to limit the base current to some safe value, and it may be left out without affecting the circuit operation. A center-reading microammeter is employed, to allow for use with *PNP* transistors as well as *NPN* transistors.

The voltage V_{CE} must increase from zero to maximum for each curve. The 6.3-V rms transformer supplies a voltage which varies to about ±8 V, or 17.8 V *p-p*. When this sine wave is rectified, it provides the needed sweep voltage for the horizontal trace on the oscilloscope, about 0 to 8 V. As the collector-emitter voltage sweeps from zero to maximum, the collector current also increases from zero to the value limited by the series resistor and the base current. This produces a proportional voltage across the 100-Ω resistance. Since the oscilloscope is a voltage-measuring instrument, the change in voltage thus registered vertically on the scope screen represents the collector current.

With the circuit set up and connected to the oscilloscope as indicated in Fig. 8-14, set the sweep selector switch to *external horizontal*

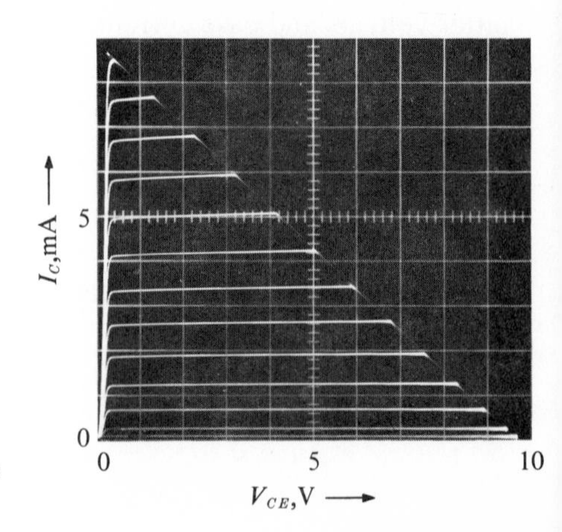

FIG. 8-13 A photograph of collector-output characteristics taken from a Tektronix Model 576. Base current is 10 μA per step.

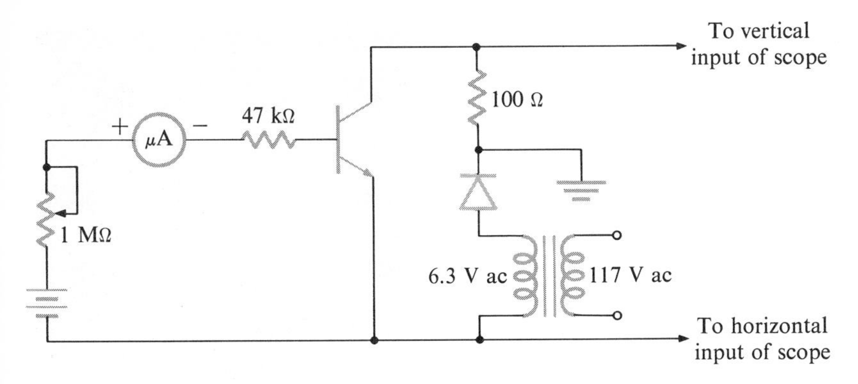

FIG. 8-14 An experimental curve tracer to display one characteristic curve at a time on a laboratory oscilloscope.

input and set the vertical attenuator to some range sensitive enough to display the vertical voltage variations. Increase the horizontal gain until the trace extends across the scope face. Set the base current for the first curve to be displayed, 10 μA. A display similar to that in Fig. 8-15 should be observed on the screen. The controls may have to be readjusted to center each curve at the same starting point. If a Polaroid oscilloscope camera is available, a multiple exposure should yield the family of curves, one for each base-current setting.

TRANSISTOR OPERATING REGIONS

If very precise measurements are taken, with V_{CE} between 0 and 1.0 V, the changes in collector current I_C appear more expanded, as in Fig. 8-16. This region of the characteristic curves is indicated on the manufacturer's data sheets as the *saturation region*. Saturation is that condition in an operating circuit in which the voltage is very low across the transistor (V_{CE}) when the current through the circuit is high. Recall from our analysis of dc amplifiers that the transistor is in series with a load resistance R_L. As the current increases to the maximum amount limited by the load resistor, the voltage developed across this resistance, V_{R_L}, approaches the supply voltage. That is,

$$V_{R_L} = I_C R_L$$
$$= 2 \text{ mA} \times 4.7 \text{ k}\Omega = 9.4 \text{ V} \tag{8-4}$$

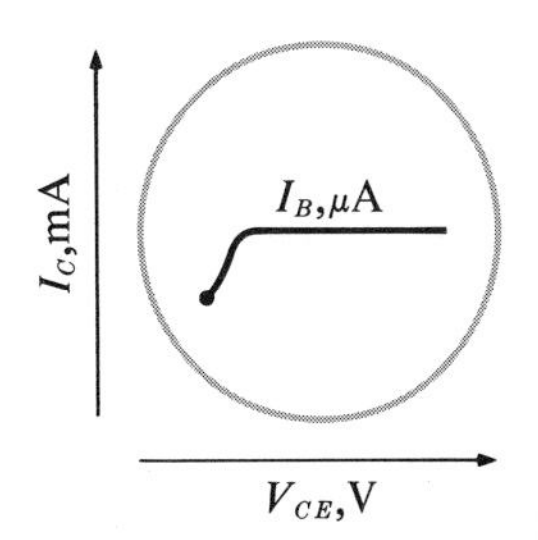

FIG. 8-15 The oscilloscope display observable with the curve-tracer setup of Fig. 8-14.

The limit of the voltage across the load resistance depends on the saturation voltage of the particular transistor used in the circuit. This saturation voltage in a particular amplifier design depends, in turn, on the maximum base current allowed. If a transistor is driven very hard by a high base current, the transistor will be driven rapidly

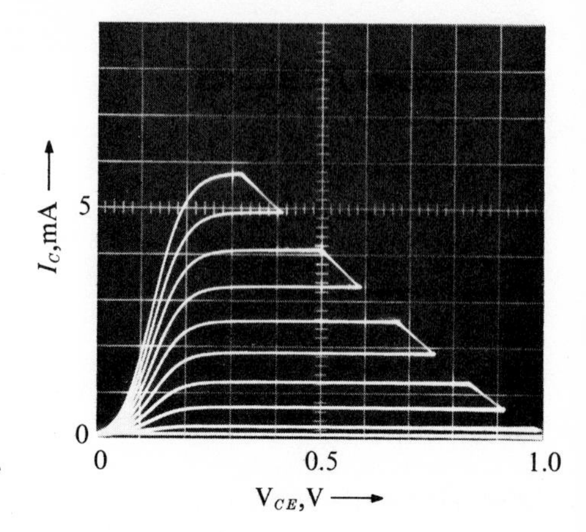

FIG. 8-16 The collector-output characteristics expanded to show the saturation region, $I_B = 10\ \mu A$ per step.

into saturation, and V_{CE} will drop to its lowest point; it cannot go to zero, however, since this would mean that R_{CE} had become 0 Ω. This is not possible, of course, because there is no such thing as a perfect conductor with zero resistance. Typical values of saturation voltage range from about 0.1 to over 1.1 V, depending on the amount of collector current used.

It is important to observe that if the collector-emitter voltage at saturation is less than the voltage between the ground and the base at that current, then both diodes of the transistor must be operating under forward-bias conditions. If, as in Fig. 8-17, the voltage between ground and the base, V_{BE}, is 0.65 V and the voltage between ground and the collector, V_{CE}, is 0.30 V, then

$$V_{BC} = V_{BE} - V_{CE}$$

$$= 0.65 - 0.30 = 0.35 \text{ V} \tag{8-5}$$

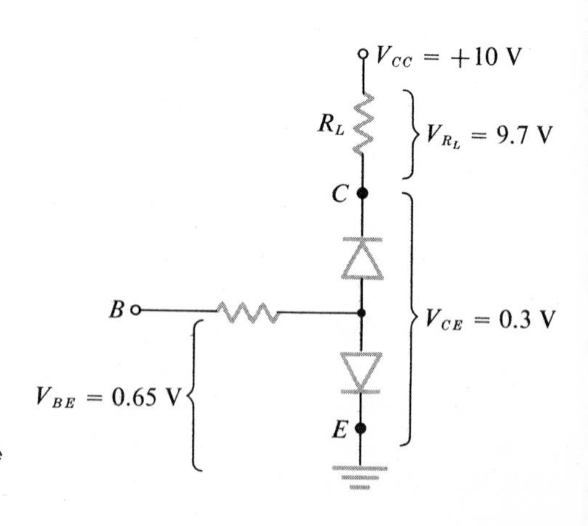

FIG. 8-17 The diode equivalent circuit showing bias conditions of the emitter-base and collector-base junctions during saturation.

That is to say, the base is 0.35 V more positive than the collector. This corresponds to a forward-bias condition between the collector and the base—a forward bias of 0.35 V.

Under saturation conditions the transistor acts as a switch in the *on* condition, a very valuable property. It does not allow operation as an amplifier, however, and so saturation conditions should be avoided in designing amplifiers for most purposes.

When the transistor is operated at currents somewhat less than needed for saturation, it is operating in what is called the *active region*. This means that any variation in base current will be accompanied by a correspondingly larger variation in collector current.

The various definitions of active region are often a source of confusion. It can be defined most easily in graphical terms, as the region to the *right* of the saturation region on the characteristic curves. It is the region in which V_{CE} is larger than the saturation voltage, generally greater than 1 V. This situation arises when the current is low enough so that it does not produce the large voltage drop across the load resistance that approaches the supply voltage.

One other extreme of operation is possible, that of the *cutoff* region. If base current is reduced to zero, the collector current also drops to zero. Of course, there is *leakage current* in all silicon transistors, even with no base current. This leakage current is generally only a few nanoamperes at room temperature. However, it doubles approximately every 10°C rise in temperature. This is not a significant factor in silicon transistors, but it is a serious problem in germanium transistors, which have up to 1,000 times more leakage than silicon transistors.

If a transistor amplifier is overdriven by a negative-going voltage or current, the signal may cause the transistor to cut off, introducing a distortion of the input and output signals for the time of cutoff.

Of the three possible operating conditions, we shall explore only the active region, or linear operation, at this point. Both cutoff and saturation conditions introduce distortion into the amplifier and are therefore to be avoided in its design.

INPUT-OUTPUT CHARACTERISTICS

Since the linear operation of an amplifier depends on how closely the changes in collector current follow, or track, the changes in base current, Fig. 8-18 should be of interest. Note that as the base current I_B begins to increase, up to about 5 μA the collector current does not track linearly. When I_B is between 5 and 40 μA the tracking is almost ideal. As the transistor approaches saturation in this hypothetical circuit, the rate of change of I_C slows down, becoming nonlinear once again. The linear operating range, or *gate*, for this transistor circuit, then, is about 5 to 40 μA of base current. The circuit should be biased

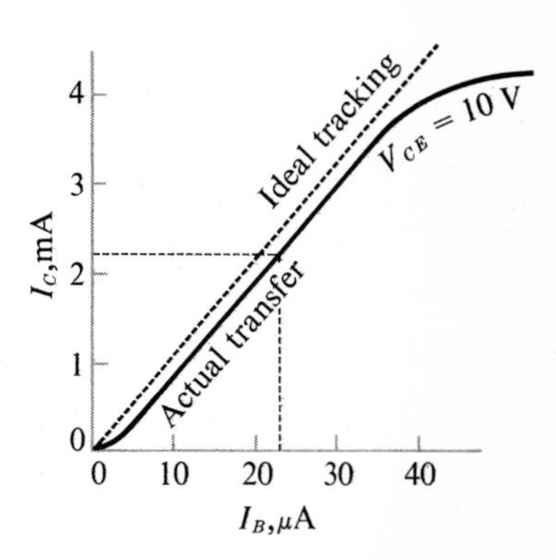

FIG. 8-18 Comparison of the input-output transfer characteristic curve with the ideal straight-line transfer.

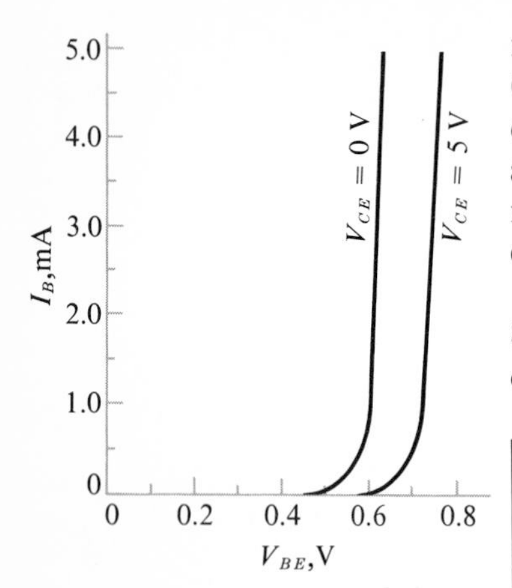

FIG. 8-19 Silicon-transistor input-impedance characteristic curves.

near the middle of this region for good design, with I_B about 22.5 μA. Of course, standard resistor values may produce a quiescent point of operation with I_B somewhere between 20 to 25 μA. Although this amount of off-center operation is not normally of any concern, it does not allow as large an input signal without some distortion as does centered operation.

The value of h_{FE} can be plotted directly from this curve of Fig. 8-18 by noting the amount of collector current for a particular value of base current, thus

$$h_{FE} = \frac{I_C}{I_B} = \frac{2 \text{ mA}}{0.02 \text{ mA}} = 100 \tag{8-6}$$

One more input characteristic curve will complete the picture for the common-emitter amplifier. This is the graph of changes in base-emitter bias voltage compared to changes in base current, with V_{CE} kept constant. The input characteristic curve of Fig. 8-19 strongly resembles that of the emitter-base diode with the collector-base junction open. Since the collector-base junction is reverse biased in normal operation, the input characteristic curve is shifted somewhat to the right. This indicates that the changes in output current, voltage, or resistance have *some* effect on the input characteristics. In this case the total input resistance, or input impedance, is increased owing to the reverse-bias condition of the collector-base junction.

An amplifier must be able to accept a signal from a signal source, such as a phonograph pickup, a microphone, or a radio tuner. Some signal sources are able to deliver lots of current and are referred to as low-impedance sources since they limit the current very little. Other sources are able to deliver only a voltage signal with very little current—that is, they are high-impedance sources. The input impedance of the amplifier, then, is of considerable importance in determining what kind of signal source to use. If the input impedance is too low, it will load down the signal source by forcing it to deliver too much current.

If a particular reference point of operation is chosen, the dc resistance may be determined from Ohm's law; since $R = V/I$,

$$R_{\text{in}} = \frac{V_{\text{in}}}{I_{\text{in}}} = \frac{V_{BE}}{I_B} \tag{8-7}$$

Suppose, for example, that $V_{CE} = 0$ V, $I_B = 0.5$ mA, and $V_{BE} = 0.62$ V. Then

$$R_{in} = \frac{0.62 \text{ V}}{0.50 \text{ mA}} = 1.22 \text{ k}\Omega \tag{8-8}$$

If $I_B = 1.0$ mA, then

$$R_{in} = \frac{0.64 \text{ V}}{1.0 \text{ mA}} = 640\ \Omega \tag{8-9}$$

For $I_B = 3$ mA,

$$R_{in} = \frac{0.70 \text{ V}}{3.0 \text{ mA}} = 232\ \Omega \tag{8-10}$$

This particular dc input resistance obviously changes rapidly with changes in dc base bias current. This is an important point and should not be confused with the dynamic or ac input impedance to an amplifier (which will be referred to later as the h parameter h_{ie}).

From these sets of curves, the dc operation of a transistor amplifier may be predicted. When the curves are compared at the Q point the predictions may be extended to the small-signal amplifier.

SUMMARY OF CONCEPTS

1. Amplification, whether it is voltage, current, resistance, or power, is merely a comparison of that output measurement to its input counterpart.
2. Linear amplification requires that the transistor be biased in the middle of its operating range.
3. Nonlinear amplification, or distortion, will result when the transistor is biased too near to saturation or cutoff.
4. The transistor is a variable resistor whose resistance depends on the amount of base current.
5. A fixed load resistance is needed in series with a transistor in order to divide the supply voltage and establish the point of operation known as the Q point.
6. Without a load resistor, the voltage across the transistor will remain the same as the power-supply voltage, regardless of its change in resistance and collector current.
7. The maximum change in V_{CE} that is possible in an amplifier is less than the power-supply voltage. The larger the supply voltage, the greater the possible output voltage change.
8. Characteristic curves allow graphical analysis and prediction of circuit behavior.

GLOSSARY

Active region That region of the transistor characteristic curves to the right of the saturation region, where any change in base current produces a correspondingly larger change in collector current.

Amplifier An active circuit capable of producing a gain in signal voltage, current, or power.

Characteristic curves A graphical presentation of changes in collector voltage, collector current, and base current.

Curve tracer An instrument or circuit capable of displaying the characteristic curve traces on an oscilloscope. The active device is generally plugged into the instrument that supplies the required sweep voltages and base-current steps.

Cutoff That condition of the operating transistor in which there is no base current, and therefore no collector current; the opposite condition from saturation.

Input impedance The input-signal current-limiting factor in an amplifier, a combination of circuit and device resistance and impedance.

Potentiometer A three-terminal variable resistance which allows a potential difference applied to the two ends to be divided by a movable center tap; commonly used in volume and tone controls.

Saturation The condition of a transistor circuit in which additional base current cannot produce further increases in collector current; that region near the left of the transistor characteristic curves.

REFERENCES

Amos, S. W.: *Principles of Transistor Circuits*, 3d ed., Hayden Book Company, Inc., New York, 1965, chap. 4.

Cutler, Phillip: *Semiconductor Circuit Analysis*, McGraw-Hill Book Company, New York, 1964, pp. 73–89.

Kiver, Milton S., and Bernard Van Emden: *Transistor Laboratory Manual*, McGraw-Hill Book Company, New York, 1962, experiments 4–7.

Leach, Donald P.: *Transistor Circuit Measurements*, McGraw-Hill Book Company, New York, 1968, experiments 8–9.

Malvino, Albert Paul: *Transistor Circuit Approximations*, McGraw-Hill Book Company, New York, 1968, chap. 6.

Seidman, A. H., and S. L. Marshall: *Semiconductor Fundamentals, Devices and Circuits*, John Wiley & Sons, Inc., New York, 1963, chaps. 7–8.

REVIEW QUESTIONS

1. Why does the collector-emitter resistance of a transistor depend on how much current is flowing?

2. What happens to the collector-emitter resistance of a common-emitter circuit such as that in Fig. 8-5*a* if the base current increases? Why?

3. What happens to the collector-emitter voltage as the base current is increased?

4. Approximately what voltage should be expected across a transistor when it is operating very near to cutoff?

5. What is meant by saturation?

6. Describe the process of amplification in a common-emitter dc amplifier.

7. What happens to the voltage across a transistor as the base current changes if there is no series or load resistance in the collector or emitter circuit?

8. What is meant by a family of characteristic curves?

9. What range of voltages might be expected across a saturated transistor?

10. How does the active region of the transistor characteristic curves differ from the saturation region?

11. Why is the input impedance of an amplifier important?

12. In what way does a transistor resemble a variable resistor?

PROBLEMS

1. In a circuit such as that of Fig. 8-4, if a collector current of 1.5 mA is observed through the 5-kΩ load resistance, what is the voltage across the resistor and the transistor?

2. What is the collector-emitter resistance of a transistor having 7.0 V across it when the collector current is 2.0 mA?

3. If a change in base-emitter voltage of 0.02 V causes the base current to change by 20 μA, what would be the change in collector current if h_{FE} were 50?

4. In Prob. 3, if the load resistance is 4.7 kΩ, what is the change in voltage across the load resistor due to the change in base current as indicated?

5. In Fig. 8-5, if $V_{CE} = 6$ V and $V_{R_L} = 4.5$ V, how much is the supply voltage?

6. If the base-emitter voltage is 0.63 V and the collector-emitter voltage is 0.5 V, how much is the collector-base voltage?

7. If the leakage current is 500 nA at 25°C, how much is the approximate leakage at 35°C?

8. From the curve for $V_{CE} = 5$ V in Fig. 8-19, determine the base-emitter resistance at $I_B = 2.0$ mA.

CHAPTER 9

OBJECTIVES

1. To examine the concepts that underlie the single stage ac amplifier
2. To explore the meaning of linear and nonlinear amplification
3. To examine graphically the difference between input signal voltage and signal current
4. To develop approximate formulas for voltage and current gain and for input impedance
5. To relate the input, transfer, and output characteristic curves

THE AC COMMON-EMITTER AMPLIFIER

The dc amplifier characteristics introduced in Chap. 8 set the stage for development of an amplifier to be used with the ac signals from such sources as microphones or phonographs. Comprehension of the more complex preamplifiers and power amplifiers used in high-fidelity systems depends on an understanding of the basic circuit underlying their design.

The quiescent point of operation of the dc amplifier with the common-emitter circuit will change if any voltage external to the circuit is applied directly. For this reason the dc amplifier must be modified for use in most ac applications.

THE SMALL-SIGNAL AMPLIFIER CIRCUIT

The simplest modification of the dc amplifier for alternating current is shown in Fig. 9-1. Note the addition of the input capacitor C_1 and the output capacitor C_2. The dc voltages V_{BE} and V_{CE} are critical for linear, or undistorted, operation of the circuit. For this reason capacitors C_1 and C_2 are used to prevent any external dc voltages from affecting the amplifier by blocking current flow. High-quality electrolytic capacitors are the most practical for this application. They are small and relatively inexpensive in comparison to other kinds. The correct polarity *must* be observed, however, because if they are placed into a circuit backward, the dielectric may break and begin to conduct current. This may cause severe damage to the circuit and may even generate sufficient heat to make the capacitor explode. Always be very careful to place the positive terminal of the capacitor in the direction of the more positive potential of the two points being connected.

The series resistance R_g, the other addition to the basic circuit, is not part of the amplifier at all. Its purpose is to properly load the signal generator so that the signal appearing at the base of the transistor will appear to come from a constant-current source rather than from a constant-voltage source.

Recall from the input characteristic curves developed in Chap. 8 that *the base-current change does not result from a linear change in the base-emitter voltage*. If a signal is delivered to the amplifier from the average signal generator used for vacuum-tube circuitry, the signal will be rather large, at least several tenths of a Volt. This type of signal will immediately drive the base of the transistor into its cutoff condition. However, if a large value of resistance is placed between the signal generator and the amplifier, this resistance R_g will appear in series with the input impedance of the transistor. Thus a large part of the signal voltage will appear across R_g, with only a small portion remaining to affect the base-emitter voltage of the transistor.

This is the condition required for linear operation. The same signal current passes through both the series resistance R_g and the input impedance Z_{in}, which in turn produces the change in base current of several microamps peak to peak. As we shall see later, the use of this resistance R_g is an application of Thévenin's theorem.

FIG. 9-1 The basic common-emitter small-signal amplifier.

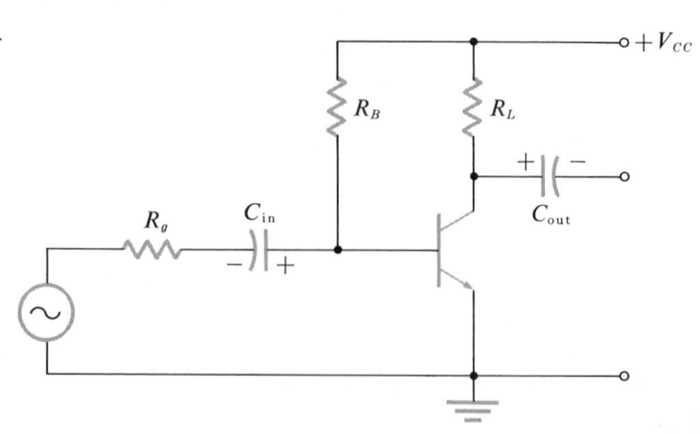

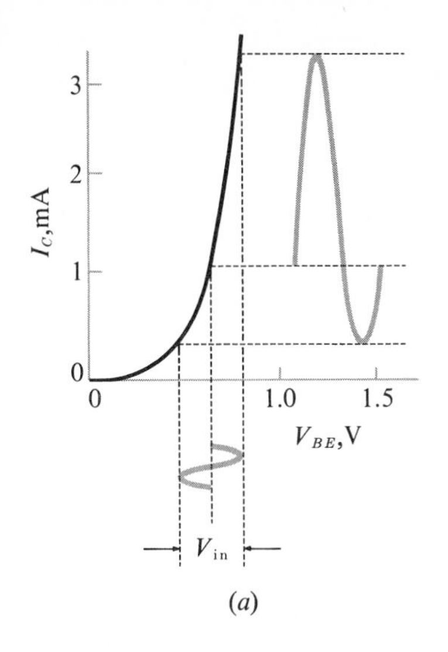

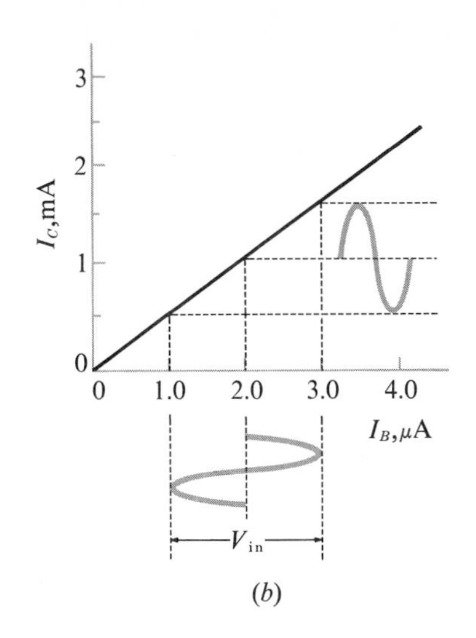

FIG. 9-2 (*a*) Voltage input signal with no resistance R_g in the input circuit, and (*b*) current signal input with R_g placed in series with the input signal current.

Figure 9-3 shows the actual input circuit to the amplifier as well as its equivalent circuit as seen by the signal generator. Note the resemblance to the Thévenin equivalent circuit. The resistance Z_{in} amounts to about 1.5 kΩ. Since $R_g = 100$ kΩ, the value of Z_{in} is only about 1.5 percent of the total series resistance R_T as seen by the signal generator. This may be expressed mathematically as

$$R_T \cong R_g + Z_{in}$$

$$Z_{in} \cong R_T \frac{Z_{in}}{Z_{in} + R_g}$$

$$= R_T \frac{1.5\,\text{k}\Omega}{1.5\,\text{k}\Omega + 100\,\text{k}\Omega} = R_T \times 0.015 \qquad \text{(9-1)}$$

FIG. 9-3 (*a*) The actual generator loaded-input circuit and (*b*) the equivalent input circuit.

100 kΩ
20 μF
$I_C = 1$ mA
$V_g = 2.0$ V *p-p*
1 kHz
R_{in}
(*a*)

$R_g = 100$ kΩ
$V_g = 2.0$ V *p-p*
1 kHz
R_{in}
= 1.5 kΩ
$V_{in} = 0.029$ V *p-p*
(*b*)

If the generator output voltage is known, the signal voltage input to the transistor can be determined easily by multiplying the generator voltage by the fraction or percentage determined in Eq. (9-1). Thus

$$v_{in} \cong v_g \frac{Z_{in}}{Z_{in} + R_g}$$
$$\cong 2.0 \text{ V } p\text{-}p \times 0.015 \cong 0.03 \text{ V } p\text{-}p \qquad (9\text{-}2)$$

As can be seen in Fig. 9-2*b*, this small variation in base-emitter voltage does generate a base-current signal of about 20 μA peak to peak, which is what is needed to produce a linear signal variation of collector current. It is difficult to measure this voltage, but it is not difficult to see that most of the signal voltage (about 98.5 percent) is lost across R_g. From Ohm's law, the base-current signal may be determined as

$$I_b = \frac{v_{R_g}}{Z_{in} + R_g}$$
$$= \frac{2.0 \text{ V } p\text{-}p}{101.5 \text{ k}\Omega} = 0.0195 \cong 20\ \mu\text{A } p\text{-}p \qquad (9\text{-}3)$$

MEASURING VOLTAGE GAIN

If the circuit in Fig. 9-4 is constructed, with all meters carefully placed with regard to range and polarity, the details of small-signal amplification may be verified. Set the supply voltage to some value between 8 and 10 V. The power supply indicated in Chap. 6 or a common 9-V battery will work well. Adjust the 1-MΩ potentiometer in the base circuit until about 1 mA of collector current is indicated. Connect the signal generator to the input circuit, and if one is available, connect an oscilloscope to the output circuit. Otherwise, use a high-input-impedance electronic voltmeter as an indicator of peak-to-peak voltage output.

Set the generator for a peak-to-peak output voltage of 2.0 V. An amplified version of the input signal should appear at the output terminals of the amplifier. If the output of the signal generator is now increased and decreased over a wide range of amplitude, you should see the distortion increase and decrease as the input signal drives the transistor into its nonlinear operating range.

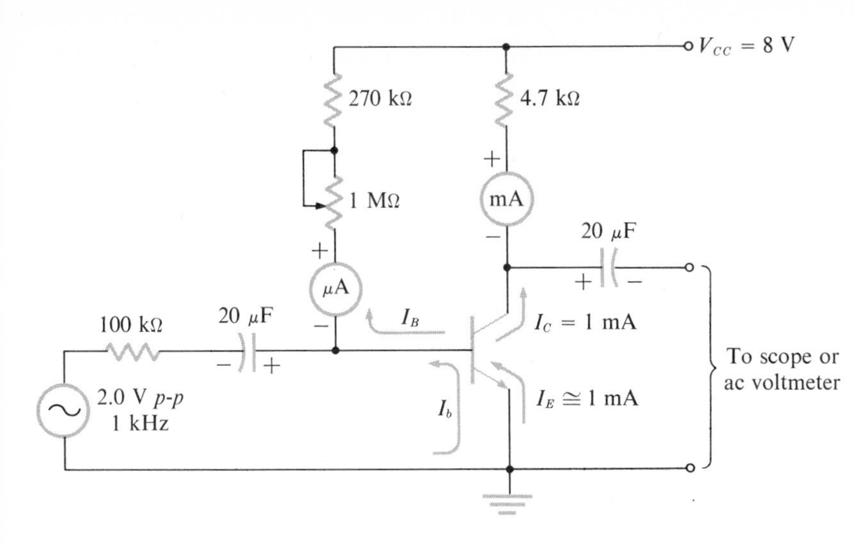

FIG. 9-4 An experimental small-signal common-emitter amplifier.

If an excessively large signal is used, the output signal will appear on the scope to be flattened on both the top and the bottom of the waveform. This indicates that the transistor is being driven into both saturation and cutoff. Saturation is indicated when the waveform is flattened on the bottom and cutoff is indicated when the waveform is flattened on the top.

Recall that an increase in base-emitter voltage is accompanied by a decrease in collector-emitter voltage. The limit of this decrease, however, is the approach to 0 V, the condition in which the transistor is saturated. When the base-emitter voltage decreases, the transistor begins to turn off (its resistance increases); this is accompanied by an increase in collector-emitter voltage. The limit of collector-emitter voltage, of course, is the supply voltage.

If the signal generator is reset to some very low value so as to give an undistorted output, the voltage gain of the circuit may be measured. For this experimental amplifier the voltage gain will be very high, about −150. This means that the output signal will be an inverted form of the input signal, but 150 times larger than the signal measured at the base. Stated another way, the signal voltage V_{ce} is 180° out of phase with the signal current and the input signal voltage.

If a sensitive scope is available, observe the input signal at the base of the transistor by connecting the scope between the base and ground. To avoid problems, make all measurements with reference to ground. Adjust the scope and signal generator until a usable waveform one space high on the most sensitive scale is observed. Check the output signal to determine if it is undistorted. This will require that the scope be set to a less sensitive range in order to display the entire signal.

To make the voltage-gain measurement, simply count the number of spaces used by the output waveform and multiply the number by the setting of the vertical attenuator. This system requires that the input be set at one-space height and that the attenuator for the range switching be fairly accurate. Once the input signal has been set, do not change the vertical gain setting, since this would alter the calibration of the scope.

CALCULATING VOLTAGE GAIN

It is possible to calculate the voltage gain of a circuit by means of a little reasoning based on Ohm's law. Being able to predict what should occur in a circuit can save a great deal of experiment time. Then when a difficulty does occur, it is easily spotted. The beginner is frequently more baffled by the equipment and the errors involved in experimentation than with the actual electronics theory. This may simply be because he has no idea what to expect.

Ohm's law established that

$$V = IR$$

and as we have seen, this relationship also holds for voltage gain A_v, current gain A_i, and resistance gain A_r, so that

$$A_v = A_i A_r \tag{9-4}$$

where $A_v = \dfrac{V_{ce}}{V_{be}} \qquad A_i \cong \dfrac{I_c}{I_b} \qquad A_r \cong \dfrac{R_L}{Z_{\text{in}}}$

If the current gain is considered to be about 50, and the resistance gain is about 4.7 kΩ/1.5 kΩ $\cong$ 3.1, then the voltage gain of the circuit should be near

$$A_v = -50 \times 3.1 = -155 \tag{9-5}$$

Although this is a straightforward development, it is not always possible to know Z_{in} accurately. Computation of the actual value will be discussed later in the chapter.

If the circuit of Fig. 9-4 is not changed, but the supply voltage is doubled, an interesting new phenomenon is observed. (When the supply voltage is changed, the base potentiometer should be readjusted to give one-half the supply voltage across the transistor and one-half across the load resistance.) Note that this doubling of the supply voltage also doubles the collector current and the base current, as should be expected from Ohm's law.

If the preceeding experiment concerning voltage gain is now repeated, it will be found that the voltage gain has also nearly doubled. Since the current gain has changed by only a small amount, the increase in gain can be accounted for only by a change in resistance gain. If the input current doubles, it is fairly safe to assume that the input resistance of the transistor must have decreased to about one-half its former value. As a matter of fact, this is a good approximation.

From an equation or relationship in solid-state physics concerning the mobility of current carriers, it can be shown that the impedance seen at the emitter terminal of the transistor is between 26 and 30 Ω when the emitter current is 1 mA dc [see Eq. (6-1)]. That is,

$$Z_{eb} = \frac{kT/q}{I_E} = \frac{0.026 \text{ V}}{1 \text{ mA}} = 26\ \Omega = h_{ib} \tag{9-6}$$

Actually, this value may be as low as 20 in some transistors and as high as 35 in others. When we study the common-base amplifier later, this impedance will be referred to as h_{ib}, *the hybrid parameter describing the input impedance to the common-base circuit.*

Let us assume that this emitter-base impedance h_{ib} is about 30 Ω, an easier figure to work with than 26. If this represents the input resistance to the common-base amplifier whose input current is the emitter current, then the input resistance (impedance) to the common-emitter amplifier should be about $1 + h_{fe}$ times larger. This is because the base signal current is h_{fe} times smaller than the collector current but $1 + h_{fe}$ times smaller than the emitter current. From application of Ohm's law, the input impedance must be $1 + h_{fe}$ times larger than h_{ib} for the common-emitter circuit.

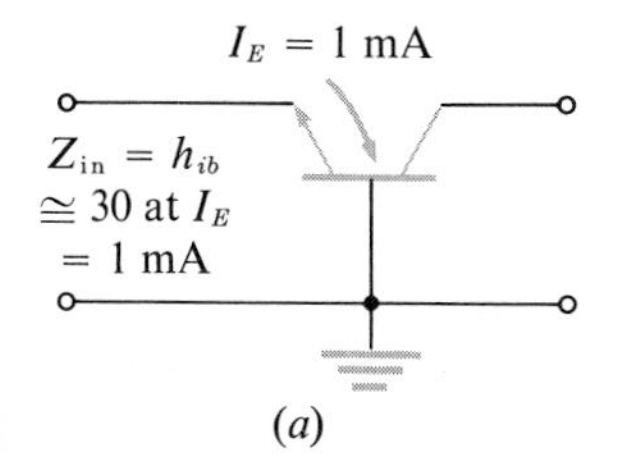

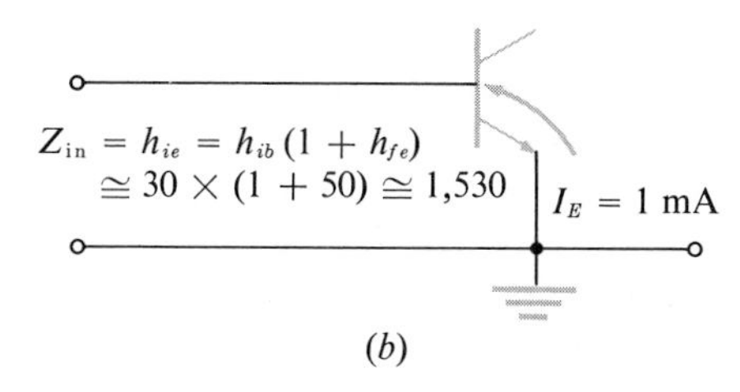

FIG. 9-5 The input impedance to (a) the common-base circuit and (b) the common-emitter circuit.

If h_{FE}, the dc gain of the transistor, is about 50 at 1 mA of emitter current, then h_{fe}, the ac gain, may also be about 50 at low frequencies. With this approximation the input impedance to the common-emitter amplifier using 1 mA of emitter current should be about 1,500 Ω and is often denoted as h_{ie}. That is,

$$Z_{\text{in}} = h_{ie} = h_{ib}\,(1 + h_{fe})$$

$$= 30\ \Omega \times (1 + 50) = 1{,}530\ \Omega \qquad (9\text{-}7)$$

The designation h_{ie} is simply the hybrid parameter indicating that the input impedance is for a transistor in the common-emitter circuit.

In comparison to the input impedance of vacuum-tube circuitry, the value of 1,530 Ω is extremely low. It is therefore essential that it not be overlooked and that Ohm's law be used when working with current concepts such as these. The symbols will seem less strange if you think in terms of their meanings as you use them.

VOLTAGE GAIN AND INPUT IMPEDANCE

If we double the power-supply voltage to about 16 V and readjust the base current until V_{CE} is reestablished at one-half the power-supply voltage, or about 8 V, as suggested before, the new current I_C will be doubled. Returning to the solid-state input-impedance formula, we have

$$h_{ib} = \frac{kT/q}{I_E}$$

$$= \frac{30\ \text{mV}}{2\ \text{mA}} = 15\ \Omega \qquad (9\text{-}8)$$

$$h_{ie} = h_{ib}\,(1 + h_{fe})$$

$$= 15\,(1 + 50) = 765\ \Omega \qquad (9\text{-}9)$$

The input impedance has indeed changed, as would be expected.

This functional relation, the change in input impedance with every change in emitter current, is one to remember. Technically, the relation requires that V_{CE} remain constant, but the error introduced in this instance is negligible. It is the problem of the change that we are concerned about, because this change in input impedance causes the voltage gain of simple circuits to vary widely from one transistor to the next.

If, for example, the transistor h_{fe} were 100 instead of 50, the voltage gain of the circuit in Fig. 9-4 would be different. The setting of the base resistance for the first transistor, with $h_{fe} = 50$, results in an emitter current of about 1 mA. If a transistor with $h_{fe} = 100$ is substituted, the emitter current will be 2 mA. The input impedance for the first transistor is about 1.5 kΩ, and for the second transistor it will be about the same,

$$h_{ie} = 15\ (1 + 100) = 1{,}515\ \Omega \tag{9-10}$$

However, according to the voltage-gain formula,

$$\begin{aligned} A_v &= -A_i A_r \\ &= \frac{-100 \times 4.7\ \text{k}\Omega}{1.5\ \text{k}\Omega} \\ &= -100 \times 3.1 = -310 \end{aligned} \tag{9-11}$$

This value of voltage gain for the same circuit is just twice the value calculated for the transistor with $h_{fe} = 50$. This is a very large voltage gain, but it is almost uncontrollable in that the next transistor selected may have some other value of h_{fe}. Unless the emitter current is re-adjusted for each transistor, the voltage gains will vary tremendously.

To summarize this voltage-gain determination, let us look at the formulas once again:

$$A_v = -A_i A_r \tag{9-12}$$

$$\text{where } A_v = \frac{V_{ce}}{V_{be}}$$

$$A_i = \frac{I_c}{I_b}$$

$$A_r = \frac{R_L}{Z_{\text{in}}} \cong \frac{R_L}{h_{ie}} \cong \frac{R_L}{(1 + h_{fe})\, h_{ib}} \cong \frac{R_L}{h_{fe} h_{ib}}$$

If these equivalent values of A_i and A_r are substituted back into the equation, a very simple form results,

$$A_v = \frac{-I_c}{I_b} \frac{R_L}{(1 + h_{fe})\, h_{ib}} \cong \frac{h_{fe}}{1} \frac{R_L}{h_{fe} h_{ib}}$$

$$\cong \frac{-R_L}{h_{ib}} \cong \frac{-4.7\ \text{k}\Omega}{30\ \Omega} \cong -155 \qquad (9\text{-}13)$$

Because the factor h_{fe} appears in both the numerator and denominator, it cancels out, leaving the final very simple version of the voltage-gain formula. Although it is not exact, it is a close approximation and will serve well in all but the most exacting engineering applications.

All that remains in using Eq. (9-13) is to compute the value of h_{ib} for each emitter current. Simply dividing the number 30 by the emitter current, expressed in milliamperes, will give a close enough value of h_{ib} for most applications. For example, if the emitter current is 3 mA, then h_{ib} will be about 10 Ω,

$$h_{ib} = \frac{30\ \text{mV}}{I_E}$$

$$= \frac{30\ \text{mV}}{3\ \text{mA}} = 10\ \Omega \qquad (9\text{-}14)$$

Of course, as very large amounts of current are used, the value of h_{ib} becomes very small, and thus the input impedance also becomes very low.

This problem of low input impedance can be resolved very easily, as we shall see in later discussions of other types of common-emitter amplifiers.

GRAPHICAL ANALYSIS

Up to this point we have taken a standard form of common-emitter amplifier with the values already chosen and proceeded from there. There is no mystery about how these values of resistance are chosen. Inspection of the set of characteristic curves for the transistor should be a great aid in understanding the circuit design.

From the set of characteristic curves in Fig. 9-6 let us design a simple amplifier to operate with a supply voltage of 8 V dc using a collector current of $I_C = 1.5$ mA.

Basic design procedures indicate that V_{CE} should be set at one-half the supply voltage; therefore find $V_{CE} = 4$ V on the horizontal

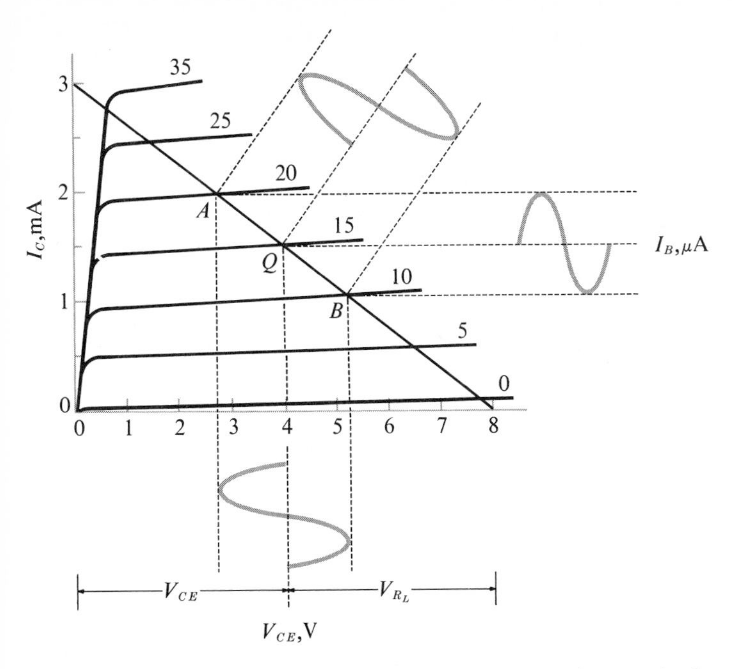

FIG. 9-6 The collector characteristics for a common-emitter amplifier, including input and output waveforms. The current gain $h_{FE} = h_{fe} = 100$.

axis of the graph. Now find $I_C = 1.5$ mA on the vertical axis of the graph. This is the quiescent point of operation, the Q point shown on the graph. A line has been drawn through point Q and 8 V and extended until it intersects the vertical axis at about 3 mA. This line is called the *load line* and represents the actual value of load resistance needed to produce the conditions desired for the circuit.

If the voltage across the load resistance is also 4 V, and a 1.5-mA collector current passes through it, the value of the load resistance must be 2.67 kΩ,

$$R_L = \frac{V_{R_L}}{I_C}$$

$$= \frac{4\text{ V}}{1.5\text{ mA}} = 2.67\text{ k}\Omega \qquad (9\text{-}15)$$

The value of standard or preferred resistor closest to 2.67 kΩ is 2.7 kΩ. The use of this standard value 2.7-kΩ resistance introduces very little error.

Now that our output circuit calculations are complete, let us turn to the input circuit. If the collector current is to be 1.5 mA, the base current needed to accomplish this is

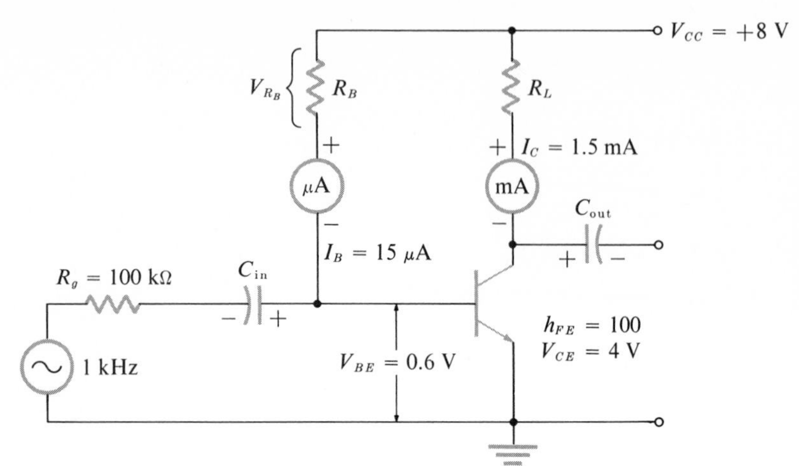

FIG. 9-7 The basic common-emitter amplifier circuit with values corresponding to the characteristics shown in Fig. 9-6.

$$I_B = \frac{I_C}{h_{FE}} = \frac{1.5 \text{ mA}}{100} = 15\,\mu\text{A} \tag{9-16}$$

This base current must pass through the series resistance of the base-emitter junction and the base resistance R_B. Since the base-emitter junction voltage is about 0.6 V, the voltage across R_B may be computed as

$$V_{R_B} = V_{CC} - V_{BE} = 8 \text{ V} - 0.6 \text{ V} = 7.4 \text{ V} \tag{9-17}$$

From the calculated value of 7.4 V for V_{R_L} and the known value of 15 μA for the base current I_B, the actual value of R_B may be determined as

$$R_B = \frac{V_{R_B}}{I_B} = \frac{7.4 \text{ V}}{15\ \mu\text{A}} = 493 \text{ k}\Omega \tag{9-18}$$

$V_{CC} = 8\text{ V} = V_{BE} + V_{R_B}$

V_{R_B} $R_B = \frac{V_{R_B}}{I_B}$

V_{BE}

(a)

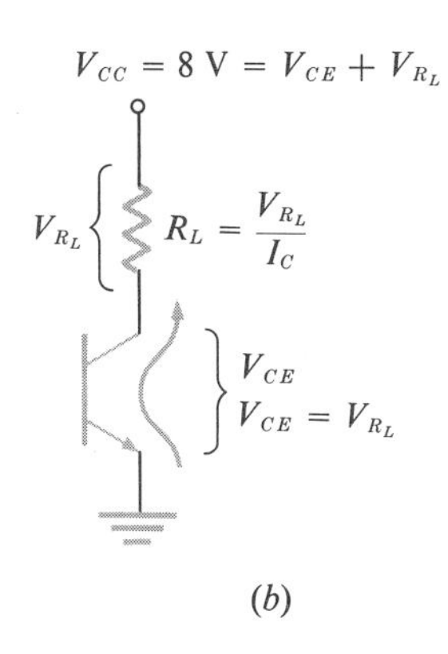

FIG. 9-8 (*a*) Base circuit and computation of R_B, the base-current-limiting resistor; (*b*) collector circuit and computation of R_L, the collector load resistance.

Since 493 kΩ is not a standard value, the choice must be between fixed resistances of 470 and 510 kΩ or a potentiometer set to 493 kΩ. All three should be tried, because the actual variation due to the approximately 20-kΩ error will be less than 5 percent. Most resistors that may be used have tolerances greater than 5 percent. Generally speaking, errors in actual component values as high as 10 percent can be tolerated without adverse effects on the circuit.

Computation of the capacitor sizes at this time will have little meaning, and so we shall defer this until later.

THE *Q* POINT AND THE LOAD LINE

The completion of calculations for the sizes of load and base resistors establishes the *Q* point or dc point of operation for the amplifier. If the signal generator is adjusted to give a peak-to-peak variation in base current of 10 μA (±5 μA), the collector current should vary above and below the *Q* point by 0.5 mA, or 1.0 mA peak to peak. This is indicated on the characteristic curves of Fig. 9-6.

The excursion of the collector current is shown on the load line from point *A* to point *B*, or $Q + A$ and $-B$. As the current increases to point *A*, note that the voltage V_{CE} drops from 4 V, to about 2.6 V. As the current decreases to point *B*, the voltage V_{CE} increases to about 5.4 V. In other words, as the input voltage and current increase, the output voltage decreases. They are said to be 180° out of phase. This is a vital concept to remember in working with the common-emitter amplifier.

In summary, graphical analysis of the characteristic curves of a transistor with h_{FE} value of 100 allows us to calculate the values of resistance needed to construct the common-emitter amplifier and to predict its behavior under dynamic, or ac, conditions. Knowing the voltage across the load resistor and arbitrarily choosing a collector current allows us to plot the load line and the value of the load resistance. From the value of collector current and the value of h_{FE}, the base current can be established. Once the base current is known and the voltage across the base resistance has been computed, the value of the base resistance may also be determined.

With the load line plotted on the characteristic curves, inspection of the distance between the curves should indicate the linear operating range of the transistor. If all curves are equally spaced, the transistor may be used over its whole range of operation—that is, anywhere along the load line from cutoff to saturation—without distortion. If the curves are not equally spaced, operation should be held to the small area where the curves are equidistant.

Variation of the base current from point A to point B on the load line, a variation of 10 μA *p-p*, produces a variation of collector current of 1.0 mA *p-p*, and a collector to emitter voltage variation of 2.8 V *p-p*.

The complete input and output characteristic curves are plotted in Fig. 9-9. In the first quadrant (upper right) the output collector curves are plotted or displayed. In the second quadrant (upper left) the *transfer* characteristics of input and output current are shown. In the third quadrant (lower left) are the input characteristics, indicating the relationship between input voltage and input current.

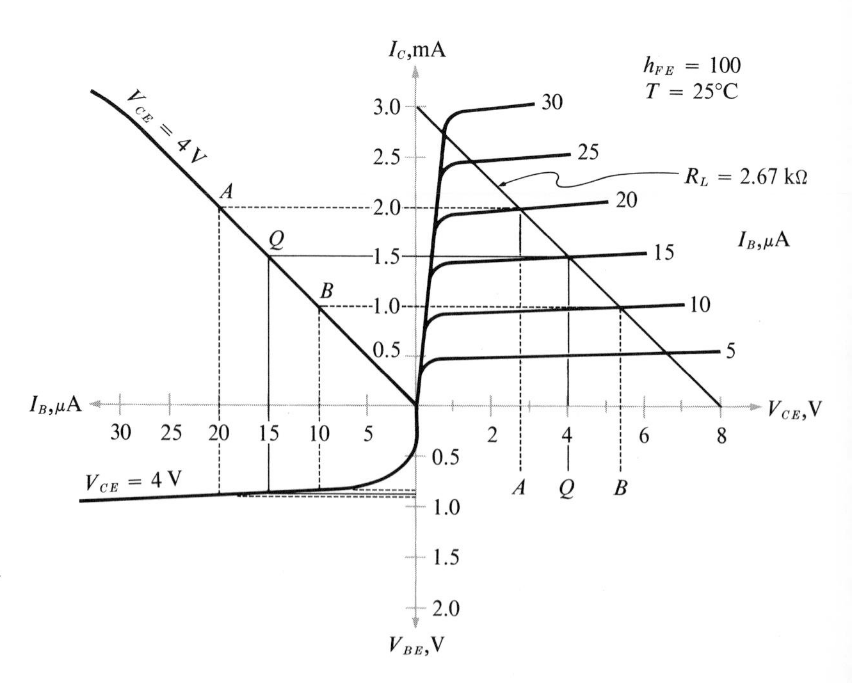

FIG. 9-9 A summary of common-emitter input and output characteristics for a transistor with $h_{FE} = 100$ at 25°C.

The Q point and the limits of signal excursion from point A to point B are indicated on each of the curves. Note the very small change in base-emitter voltage in comparison to the large change in collector-emitter voltage. Note also the linearity of the I_C-vs.-I_B transfer curve in comparison to the I_B-vs.-V_{BE} input curve.

SUMMARY OF CONCEPTS

1. A capacitor may be used to isolate two points of potential difference and prevent current flow between them.
2. Electrolytic capacitors must be placed in the circuit with the proper polarity direction observed.
3. The input to the common-emitter transistor amplifier must be from a constant current source, since the input voltage V_{be} is not linear with respect to the collector current.
4. The input circuit of a common-emitter amplifier may be discussed in terms of its Thévenin equivalent circuit.
5. Once the Q point of operation has been established and the collector current has been chosen, all resistances in the circuit may be computed from Ohm's law.
6. Voltage gain may be computed to a close approximation from the expression $A_v \cong -R_L/h_{ib}$.
7. The collector signal voltage V_{ce} is 180° out of phase with the base signal voltage V_{be} and with the input and output signal currents.
8. The input resistance, or impedance, of a common-emitter amplifier is $1 + h_{fe}$ times larger than h_{ib}, the input impedance to the transistor in the common-base circuit.
9. The value of h_{ib} is about 25 to 35 Ω at 1.0 mA of emitter current. To find the value of h_{ib} at currents other than 1.0 mA, simply divide this constant of 30 by the emitter current expressed in milliamperes.
10. From graphical analysis of the input and output curves of a transistor it is possible to determine the linearity of the transistor and how it will function in a circuit.
11. From the horizontal axis of the output collector characteristics it is possible to determine how much voltage is across the transistor and how much is across the load resistance.

GLOSSARY

Load line A line drawn on a set of characteristic curves, usually from the point of maximum current to the point of maximum voltage, with its slope representing the value of the load resistance.

Preamplifier An amplifier that is used to amplify very small signals to a level or size needed to drive a power amplifier.

Q-point The quiescent or operating point of an amplifier. As indicated on a set of characteristic curves, it represents the operating collector current, base current, and collector-emitter voltage with no signal applied.

Transfer characteristics The characteristic curves of a transistor relating the change in output or collector current to the change in base current for some value of collector voltage. They represent the ability of a device to transfer the input signal to the output.

Voltage gain The amount of voltage amplification possible from an amplifier, given by a ratio of the output voltage to the input voltage.

REFERENCES

Brazee, James G.: *Semiconductor and Tube Electronics: An Introduction*, Holt, Rinehart and Winston, Inc., New York, 1968, chap. 4.

Corning, John J.: *Transistor Circuit Analysis and Design*, Prentice-Hall, Inc., Englewood Cliffs, N.J., 1965, chap. 7.

Lenert, Louis H.: *Semiconductor Physics, Devices, and Circuits*, Charles E. Merrill Books, Inc., Columbus, Ohio, 1968, pp. 185–187, 196–272.

Malvino, Albert Paul: *Transistor Circuit Approximations*, McGraw-Hill Book Company, New York, 1968, chap. 6.

Surina, Tugomir, and Clyde Herrick: *Semiconductor Electronics*, Holt, Rinehart and Winston, Inc., New York, 1964, chap. 4.

REVIEW QUESTIONS

1. What is the basic difference between a dc amplifier and an ac, or small-signal, amplifier?

2. What is the purpose of the input capacitor to an amplifier?

3. What is the purpose of an output capacitor from an amplifier?

4. Why are electrolytic capacitors used for coupling capacitors?

5. What conditions must be met in using electrolytic capacitors that may be disregarded with ceramic or paper capacitors?

6. Why is it necessary to make a signal generator appear to deliver a constant-current signal instead of the voltage signal that it puts out?

7. Why is a transistor input considered to be base current rather than base-emitter voltage variation?

8. Describe the circuit action of R_g in Fig. 9-1.

9. What is the waveform indicated on an oscilloscope when the amplifier is being driven into cutoff?

10. What is the maximum peak-to-peak output voltage of an ac amplifier in the common-emitter configuration?

11. If the voltage gain of a particular amplifier is indicated as -45, what does this mean?

12. Why is the signal output voltage 180° out of phase with the input voltage in the common-emitter amplifier?

13. In what way may the voltage gain of an amplifier be increased?

14. Describe the input resistance to the common-emitter amplifier. In what way are h_{ib} and h_{ie} related to the input impedance?

15. Why is it not necessary to know the value of h_{ie} if the emitter current and the value of h_{fe} are known?

16. In designing an ac amplifier, why is it desirable to set the collector-emitter voltage to about one-half the supply voltage?

17. What are some reasons that graphical analysis is used in some cases of circuit design?

18. What is one disadvantage of a simple circuit such as Fig. 9-7 when a variety of transistors may be used?

19. What is a load line and what is its function?

20. How can one predict from viewing a set of characteristic curves what range of operation will produce low distortion?

21. From the graph in Fig. 9-9, why would the transfer curve I_B vs. I_C be preferable to V_{BE} vs. I_C for predicting circuit operation?

PROBLEMS

1. In a circuit such as Fig. 9-1, if the value of R_g is 10 kΩ and the input impedance is 2.0 kΩ, what percentage of the signal-generator output voltage will be delivered to the amplifier input?

2. If the signal-generator output voltage in Prob. 1 is 0.4 V *p-p,* what is the input voltage to the base of the amplifier?

3. With the conditions as stated in Probs. 1 and 2, what is the signal base current?

4. If the value of h_{ib} is known to be 28 when $I_E = 1.0$ mA, what is the corresponding value of h_{ie} if $h_{fe} = 75$?

5. In a particular common-emitter amplifier the input impedance h_{ie} has been measured at 2.2 kΩ when the emitter current is 1.0 mA, approximately how much is h_{ib} and what is the value of h_{fe}?

6. If the circuit current gain in a common-emitter amplifier is 90 and the resistance gain is 1.25, what is the voltage gain?

7. If an amplifier is biased so that $V_{CE} = 4$ V when the supply voltage is 10 V, what is the absolute maximum voltage peak to peak that an output signal may have without distortion? Why?

8. If the voltage gain of a common-emitter amplifier is 40 and the output signal voltage is 10 V *p-p,* what is the input signal voltage V_{be}?

9. If the value of $h_{ib} = 32$ Ω at $I_E = 1.0$ mA, what would be its approximate value at $I_E = 4$ mA?

10. From the graph in Fig. 9-6, if the Q point is at $I_B = 15$ μA and $I_C = 1.5$ mA, what is the swing in the collector-emitter voltage if the input base current signal is 10 μA *p-p*?

11. For the data in Prob. 10, what is the output current swing in peak-to-peak units? What is the signal current gain?

12. An amplifier such as that shown in Fig. 9-7 has an $h_{fe} = 80$ at $I_E = 1.0$ mA. What is the approximate voltage gain if the load resistance is 8.2 kΩ?

13. If good design was used in the circuit of Prob. 12, what supply voltage would you expect for the circuit?

14. In the circuit of Prob. 12, if the output signal voltage is found to be 8.2 V peak to peak, what is the output current? What is the input signal current?

15. From the graphs in Fig. 9-9, what would be the maximum current if the load resistance were 6.2 kΩ instead of the 2.67 kΩ given?

CHAPTER 10

OBJECTIVES

1. To consider leakage current as a potential problem in the common-emitter amplifier
2. To examine the nonlinearities of transistor parameters with changes in temperature
3. To explore the swamping action of emitter-bias stabilization and its effects
4. To describe the ease of operating with transistors when the circuit becomes insensitive to h_{FE}
5. To describe the control of voltage gain and input resistance

BIAS STABILIZATION AND THE COMMON-EMITTER AMPLIFIER

We have considered the operation of the ac amplifier, but with little attention to movement of the Q point of operation with changes in temperature. Only the majority current carriers (electrons) were discussed, since the effect of leakage current is minimal in the silicon transistor at normal temperatures.

In general there is no gain without some kind of cost. In order to stabilize the point of operation of an amplifier against variations of temperature and current gain, some of the voltage gain must usually be sacrificed. However, in a stabilized amplifier the input impedance may be significantly higher, allowing it to be used for many more applications.

LEAKAGE CURRENT

Remember from our study of diodes that when a diode is reverse biased there is a tiny leakage current. This leakage current, although small, seems to be independent of the applied voltage, but dependent on the temperature of the junction. The leakage at 25°C may be only 25 nA, but it about doubles with every 10°C rise in junction temperature. Thus if the temperature rises to 35°C, the leakage is 50 nA, at 45°C the leakage is 100 nA, and so on. Note that 45°C corresponds to 113°F, and the 100-nA leakage at this temperature is only 0.1 μA. This is an insignificant amount, but it can obviously become a problem in a common-emitter circuit operating at very high temperatures.

The leakage current diffuses through a reverse-biased junction, since minority carriers do not experience the high potential barrier that majority carriers do. In the operating transistor the only reverse-biased junction is the collector-base junction. Since the leakage current between the collector and the base is in the form of holes, this corresponds to additional electron movement into the collector.

The total current flow into the collector includes this minute amount of leakage current, as well as the current we have previously denoted as I_C. Since I_C in this case is

$$I_C = h_{FE} I_B \tag{10-1}$$

the total collector-current flow must include the current due to amplifier action and the leakage,

$$I'_C = h_{FE}\, I_B + I_{\text{leakage}} \tag{10-2}$$

Now any current changes in the base region may be amplified and will show up as much larger changes in collector current. This concept should help to clarify the leakage problem, since for every hole that enters the base from the collector, $1 + h_{FE}$ times that many electrons enter the collector from the emitter by amplifier action.

A third letter subscript in the parameter notation indicates the condition of the third transistor terminal. For example, I_{CEO} designates that the circuit is common emitter but that the base is open. The leakage current in the common-emitter circuit then is designated as I_{CEO}, which means the collector-emitter leakage current in the common-emitter circuit with none leaving at the base. It is $1 + h_{FE}$ times larger than the leakage current from the collector to the base with none affecting the emitter, as would be the case in a common-base amplifier. This latter current is usually written as I_{CBO}, or sometimes simply as I_{CO}.

In summary, the common-emitter circuit does have a small amount of leakage current, even when a silicon transistor is used. This amount is given by

$$I_{CEO} = I_{CBO}\,(1 + h_{FE}) \qquad (10\text{-}3)$$

Equation (10-2) may be restated as

$$\begin{aligned} I'_C &= h_{FE}I_B + I_{CEO} \\ &= h_{FE}I_B + I_{CBO}\,(1 + h_{FE}) \qquad (10\text{-}4) \end{aligned}$$

If the current gain h_{FE} is 100 and the leakage current I_{CBO} is 25 nA at 25°C, the actual collector current for a base current of 10 μA will be

$$\begin{aligned} I'_C &= (100 \times 10\ \mu\text{A}) + 25\ \text{nA} \times (1 + 100) \\ &= 1\ \text{mA} + 2{,}525\ \text{nA} = 1.002525\ \text{mA} \qquad (10\text{-}5) \end{aligned}$$

This 2.525 μA additional current is indeed negligible in comparison to 1 mA.

If we now consider this same hypothetical circuit operating at 45°C instead of 25°C, the situation changes somewhat. In this case we have

$$\begin{aligned} I'_C &= 100 \times 10\ \mu\text{A} + 100\ \text{nA} \times (1 + 100) \\ &= 1\ \text{mA} + 10{,}100\ \text{nA} = 1.010\ \text{mA} \qquad (10\text{-}6) \end{aligned}$$

This new value is still only about 1 percent of the main collector current. However, it may begin a spiral effect: an increase in current increases the power dissipated by the junction, which increases the temperature; this increases the leakage, which increases the temperature some more, and so on, until self-destruction occurs as a result of *thermal runaway*.

If by chance the temperature of the junction should reach 95°C, the value of I_{CBO} will have increased to about 3.2 μA. This, when amplified to I_{CEO}, would be a leakage of about 320 μA or 0.32 mA. If the main current is only 1 mA, the leakage has risen to 32 percent of the current caused by base current—a bad situation, since the Q point will be shifted considerably toward nonlinear operation.

Germanium transistors, although quite valuable for high-frequency applications, are very leaky. At 25°C typical leakage currents of $I_{CBO} = 5\ \mu A$ are common. When this leakage is amplified to I_{CEO} by means of an h_{FE} value of 100, it increases to 500 μA at only 25°C.

BIAS STABILIZATION

A very simple change in the basic circuit of Chap. 9 will add sufficient stabilization of the point of operation for most silicon-transistor circuits. This change is nothing more than the addition of a resistor between the emitter and ground, as indicated in Fig. 10-1. The addition of this resistor R_E changes the dc voltage arrangement somewhat, as shown. The emitter is no longer attached directly to ground, the negative terminal of the power supply. The emitter current consisting of 1 mA of collector current, 10 μA of base current, and 0.25 μA of I_{CEO} passes through this new emitter resistor. This is still very nearly 1 mA, actually 1.01025 mA. The voltage measured across this 1-kΩ resistor is about 1 V (1.01 mA $\times$ 1 kΩ = 1.01 V).

The dc voltage present at the emitter is then 1 V. Since V_{BE} is about 0.6 V when the emitter current is 1 mA, the base voltage must be about 0.6 V above the emitter voltage, or 1.6 V, as indicated in Fig. 10-1. Since V_{CE} should be equal to V_{R_L}, the voltage across the load, the voltage remaining after the emitter voltage has been subtracted from the supply voltage should be divided equally between V_{CE} and V_{R_L}. Thus

$$\begin{aligned} V_{CC} - V_E &= V_{CE} + V_{R_L} \\ 16 - 1 &= 15\ \text{V} = 7.5\ \text{V} + 7.5\ \text{V} \end{aligned} \qquad (10\text{-}7)$$

If V_{CE} is set at 7.5 V, the voltage at the collector with respect to ground will be

$$\begin{aligned} V_C &= V_E + V_{CE} \\ &= 1 + 7.5 = 8.5\ \text{V} \end{aligned} \qquad (10\text{-}8)$$

All voltage measurements should normally be made from ground in order to maintain the correct polarity and preserve the common point of the circuit. In some varieties of test equipment each piece has

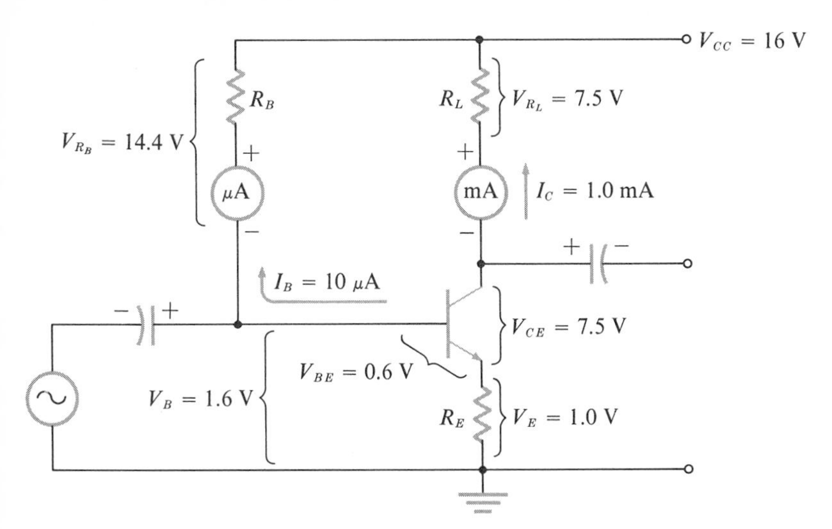

FIG. 10-1 Fixed emitter bias for stabilization of the common-emitter amplifier in a functional circuit.

a common ground. When several units are being used in the same experiment, the common terminal of the test equipment may short out some point in the circuit to ground and upset the circuit or even cause damage. For this reason each voltage should be measured from ground, and the value of some voltage in the middle of the string should be computed from Kirchhoff's law. For example,

$$V_{CE} = V_C - V_E$$
$$= 8.5 - 1.0 = 7.5 \text{ V} \tag{10-9}$$

Once the voltages and the collector current are established, application of Ohm's law will yield the values of the resistances:

$$R_E = \frac{V_E}{I_E} = \frac{1 \text{ V}}{1.01 \text{ mA}} = 1 \text{ k}\Omega \tag{10-10}$$

$$R_{CE} = \frac{V_{CE}}{I_C} = \frac{7.5 \text{ V}}{1.0 \text{ mA}} = 7.5 \text{ k}\Omega \tag{10-11}$$

$$R_L = \frac{V_{R_L}}{I_C} = \frac{7.5 \text{ V}}{1.0 \text{ mA}} = 7.5 \text{ k}\Omega \tag{10-12}$$

THE BASE-CURRENT-LIMITING RESISTANCE

The voltage across the base-current-limiting resistor R_B is determined in much the same way as V_{R_L}. From Kirchhoff's law,

$$V_{R_B} = V_{CC} - V_B$$

However,

$$V_B = V_E + V_{BE}$$

and therefore

$$\begin{aligned} V_{R_B} &= V_{CC} - (V_E + V_{BE}) \\ &= 16\text{ V} - (1\text{ V} + 0.6\text{ V}) = 16\text{ V} - 1.6\text{ V} \\ &= 14.4\text{ V} \end{aligned} \tag{10-13}$$

The exact value of R_B may be computed from Ohm's law, since we know that the base current is about 10 μA. Thus

$$\begin{aligned} R_B &= \frac{V_{R_B}}{I_B} \\ &= \frac{14.4\text{ V}}{10\ \mu\text{A}} = 1.44\text{ M}\Omega \end{aligned} \tag{10-14}$$

It is at this point that good judgment is required in the actual selection of a resistor for R_B. The standard-value resistor closest to 1.44 MΩ is 1.5 MΩ. This is the value that should be chosen to limit the base current. The actual base current will be slightly less than 10 μA, but the slight difference is unimportant in comparison to other factors, such as tolerance in resistance values. Remember that the value of R_B just selected has the job of limiting the base current to about 10 μA when the supply voltage is 16 V. If another supply voltage is desired at some time, the computation must be repeated if all the voltages in the amplifier are to maintain their design values. This is because the base current controls the collector current, and thus the collector voltage. Figure 10-2 indicates the current path of the base current.

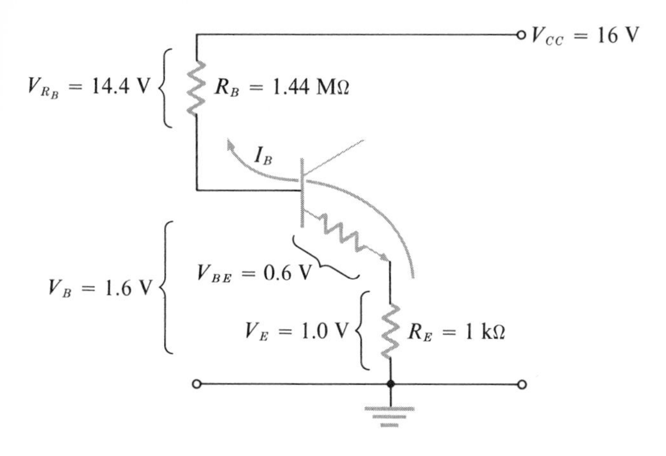

FIG. 10-2 Partial-emitter-bias common-emitter amplifier circuit showing the base-current path and its limiting resistances and associated voltages.

THE Q POINT

With the dc bias conditions set up to give an operating or Q point at $V_C = 8.5$ V, $I_C = 1$ mA, and $I_B = 10$ μA, let us now analyze the changes that will occur in these conditions as the temperature changes.

If the temperature of the junction rises enough to cause some significant amount of leakage current to flow, the voltage across the emitter resistance will increase proportionally. Since the base current is a function of the base-emitter voltage, the increase in the emitter voltage will be followed by a slight reduction in the base current. This results from the fact that

$$V_{BE} = V_B - V_E \tag{10-15}$$

Generally the base voltage does not increase as much as the emitter voltage, so that the difference voltage has the effect of reducing the amount of base current. As the base current is reduced, the collector current must decrease a much larger amount. Reducing the collector current should cool off the junction somewhat and at least slow down the shift in Q point as the junction temperature rises.

This is a simplification of the actual mechanism occurring in the transistor, but it is sufficiently accurate to be very useful. Of course, other factors do complicate the picture, however. As the temperature increases, so does the h_{FE} value of the transistor. This accounts for the slight increase in collector current in spite of this partial stabilization of the circuit for leakage current. However, as stated previously, this is not a problem with silicon transistors operated at normal temperature variations and low collector currents.

SWAMPING ACTION

One of the most important properties resulting from the addition of the emitter resistor is its *swamping* action. Recall from our previous discussion of the ac amplifier that voltage gain is in part a function of the input resistance of the transistor. Recall too that h_{ib} is 25 to 30 Ω at $I_E = 1$ mA. This value drops rapidly with an increase in current, carrying with it a drop in the common-emitter input impedance,

$$h_{ie} = (1 + h_{fe})h_{ib} \tag{10-16}$$

Observe from Fig. 10-3 that the emitter voltage is about 1 V and the base voltage is about 1.6 V owing to the placement of the emitter resistance. The base voltage *appears* to be a result of base current, even though it is primarily a function of the emitter current. It will not change much if significant changes in h_{ib} are disregarded completely, since h_{ib} is so small in comparison to 1,000 Ω. Any changes in input impedance due to changes in temperature and h_{ib} will tend to be swamped out by the fixed value of the emitter resistor.

Since the voltage V_B appears to be caused by the base current, which is $1 + h_{fe}$ times smaller than the emitter current actually causing most of it, the amplifier input impedance due to this emitter resistor appears to be $1 + h_{fe}$ times the value of the resistance,

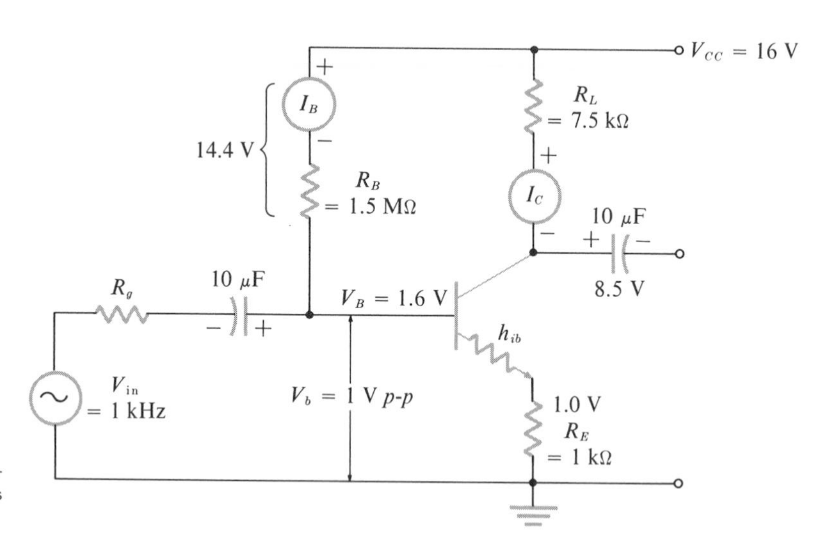

FIG. 10-3 Experimental common-emitter amplifier with emitter-bias stabilization.

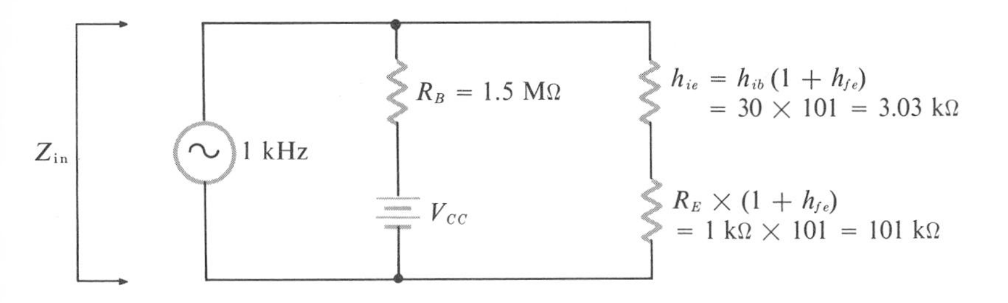

FIG. 10-4 An equivalent circuit for the approximate input impedance in the common-emitter emitter-bias circuit.

$$
\begin{aligned}
Z_{in} &= (1 + h_{fe})(R_E + h_{ib}) \\
&= (1 + 100)(1\ \text{k}\Omega + 30\ \Omega) \\
&= 101 \times 1{,}030 = 104{,}030\ \Omega
\end{aligned}
\tag{10-17}
$$

This value of input impedance for the amplifier in Fig. 10-3 is quite accurate even though it is really an approximation in terms of the total circuit. There are other factors that exert some influence on the actual value, such as the size of the load resistor, the voltage feedback ratio, the output conductance, and any resistance in parallel with the input circuit. Figure 10-4 illustrates how resistances appear in parallel with the input circuit of the amplifier of Fig. 10-3. Note that both R_B and R_L in the circuit return to ground through the power supply, which has a resistance of less than 1 Ω. For this reason no power supply has to be indicated in the equivalent circuit, although it is shown here for clarity.

The input impedance for the circuit of Fig. 10-4 may be computed from the equation for parallel resistances as

$$
\begin{aligned}
Z_{in} &\cong R_B \parallel [(h_{ib} + R_E)(1 + h_{fe})] \\
&\cong 1.5\ \text{M}\Omega \parallel 0.104\ \text{M}\Omega
\end{aligned}
\tag{10-18}
$$

$$
\begin{aligned}
Z_{in} &\cong \frac{R_1 R_2}{R_1 + R_2} \\
&= \frac{1.5\ \text{M}\Omega \times 0.104\ \text{M}\Omega}{1.5\ \text{M}\Omega + 0.104\ \text{M}\Omega} = \frac{0.156 \times 10^{12}}{1.604 + 10^{6}} \\
&\cong 0.0975\ \text{M}\Omega = 97.5\ \text{k}\Omega
\end{aligned}
\tag{10-19}
$$

The two parallel lines used to separate numbers in any equation are mathematical directions. They mean that the two numbers represent two components in parallel and that the value should be calculated from the formula for total resistance of two resistances in parallel. $R_1 \parallel R_2$ means, therefore, that R_1 is in parallel with R_2, and that this summation must be taken in solving the problem.

It is important to note that when resistances are in parallel, the total resistance is always smaller than the smallest resistance involved. For this reason the input impedance to the amplifier is always smaller than the impedance calculated as the value of $(h_{ib} + R_E)(1 + h_{fe})$. However, if the resistances in parallel include a base resistance which is at least 10 times greater than the other values, its effect on the total parallel value is negligible.

When lower values of supply voltage are used lower values of base resistance must be used, since it takes less resistance to limit the base current. In cases such as this the value of R_B may be so low that it does alter the total input impedance and must therefore be considered in any computations.

To test the principles of bias stabilization we have discussed, connect the circuit of Fig. 10-3. Place it in operation and note the values of base current and collector current. Compute the value of emitter current. Also note the voltages at the emitter, base, and collector with respect to ground.

In order to check the circuit stability bring a hot soldering iron or a hot light bulb near (*but not touching*) to the transistor and observe any changes in any voltages or current over a time period of at least 2 min.

Next set up the circuit of Fig. 9-1 without the emitter resistor and readjust the base current of this circuit to the previous Q point. Again heat the transistor and observe any differences in the actions of the two circuits.

If a germanium transistor with an h_{FE} value greater than 50 is available, set up a similar circuit for it and repeat the process. Note the more sudden increase in current with the germanium transistor. This should be expected, since the mobility of the carriers in germanium is about 1,000 times greater than in silicon.

As the leakage current increases, the total current passing through the load resistance increases, which in turn increases the voltage across the load resistance V_{R_L}. Since R_L is in series with the transistor, the voltage across the transistor decreases with the increase in leakage current. This in turn shifts the Q point toward saturation.

If a can of Freon or similar aerosol coolant is available, or some dry ice is obtainable, repeat the previous experiments, but this time cool the transistor. Cooling the transistor junction reduces the leakage as well as the mobility of the carriers. The net effect is that less current flows, and the Q point will begin to shift toward cutoff.

It should be pointed out that the temperature of the junction is always higher than the *ambient* temperature, the temperature of the air and material surrounding the transistor. Even if the ambient temperature T_A is only 25°C, the temperature of the junction T_J may rise to 100°C if too much current is allowed to pass through the transistor.

The use of a fixed bias by means of the emitter resistor is only one means of achieving bias stabilization. With germanium transistors this means of stabilization is generally not adequate protection, since the leakage is several orders of magnitude higher than in silicon transistors. However, before we consider other means of stabilizing the Q point, let us consider the remaining effects of adding the emitter resistor.

BIAS STABILIZATION AND DEGENERATION

When the circuit of Fig. 10-5 is placed in operation with a suitable input signal, it will be immediately noticed that the signal voltage gain is rather low in comparison to the previous experimental circuits. This apparent loss in signal voltage gain is one of the results of bias stabilization and increased input impedance. The new gain of 7 or 8 is quite respectable in comparison to vacuum-tube circuits, but not in comparison to the value of 250 obtainable with other transistor circuits. As noted earlier, any gain usually entails some sacrifice; in this case voltage gain is sacrificed for stability.

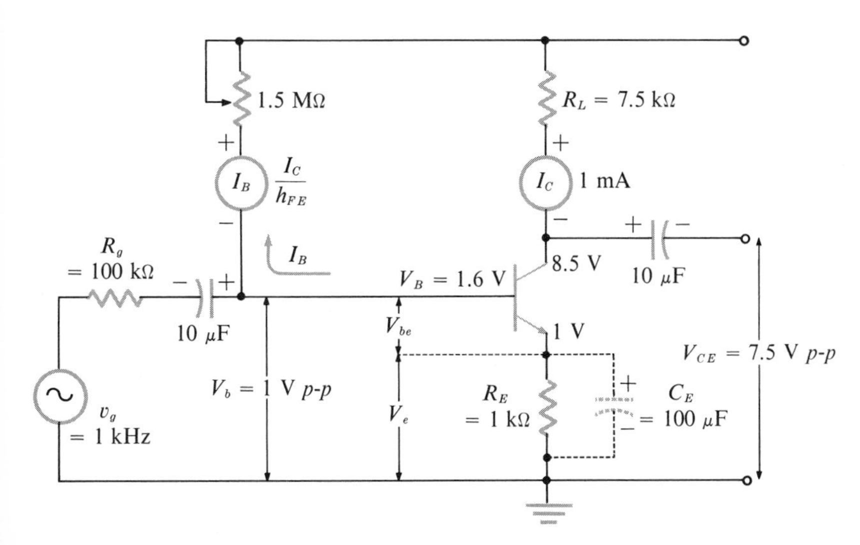

FIG. 10-5 Experimental common-emitter amplifier with emitter-bias stabilization and degeneration.

Observe the input circuit to the transistor in Fig. 10-5. Notice that the signal generator must now deliver most of its signal voltage to the emitter resistor and only a small portion of the total voltage to the base-emitter junction. It is this voltage V_{be} that the transistor will actually amplify.

Ohm's law can be used to show that only about 3 percent of the signal voltage delivered to the base is actually experienced by the base-emitter junction. If a sensitive oscilloscope is used, the signal voltage measured at the base of the transistor should show up as a larger signal than that measured at the emitter. Since h_{ie} from Fig. 10-4 is about 3 kΩ and R_E appears to be 101 kΩ, it is evident that if the same signal current passes through each resistance, the voltages must be proportional to the resistances. In other words, if the signal voltage applied to the base is 1.0 V *p-p*, V_{be} would be about 0.03 V *p-p*. When voltage gain is considered as V_{ce}/V_{be} rather than V_{out}/V_{in}, it should be observed that the transistor is amplifying just as before, but the circuit is degenerating or reducing the effect of the full input signal to a smaller value.

Recall from Chap. 9 that the approximate voltage gain A_v of a transistor amplifier circuit without stabilization is

$$A_v \cong \frac{-R_L}{h_{ib}} = \frac{-7.5 \text{ k}\Omega}{30} = -250 \qquad (10\text{-}20)$$

If the base-emitter signal voltage V_{be} is only 0.03 V *p-p*, the output signal voltage V_{ce} should be

$$\begin{aligned} V_{out} &= V_{ce} = A_v V_{be} \\ &= -250 \times 0.03 \text{ V } p\text{-}p = -7.5 \text{ V } p\text{-}p \qquad (10\text{-}21) \end{aligned}$$

A check of the actual circuit will show that this value is quite accurate. The minus sign in front of the output voltage only indicates that the signal voltage at the collector is 180° out of phase with the signal voltage at the base; that is, when the base signal goes positive the collector to ground signal goes negative.

Modification of Eq. (10-20) to include the value of the emitter resistor in series with h_{ib} yields

$$A_v \cong \frac{-R_L}{h_{ib} + R_E} = \frac{-7.5\ \text{k}\Omega}{30 + 1{,}000} = -7.5 \qquad (10\text{-}22)$$

There is good agreement among Eqs. (10-20) to (10-22), since a voltage input signal to the base of the transistor of 1.0 V *p-p* would yield an output voltage of 7.5 V *p-p*,

$$A_v = \frac{v_{\text{out}}}{v_{\text{in}}} = \frac{7.5\ \text{V}\ p\text{-}p}{1.0\ \text{V}\ p\text{-}p} = -7.5 \qquad (10\text{-}23)$$

In actual practice a close approximation to the voltage gain of a circuit using degeneration may be obtained from

$$A_v \cong \frac{-0.9R_L}{R_E} \qquad (10\text{-}24)$$

The full equation for very accurate calculation of this voltage gain is quite complex and requires a much greater understanding of the *h* parameters. Many engineers use this shorter rule-of-thumb method for quick calculations and find that its accuracy is generally within ±5 percent.

Note that h_{fe} does not appear at all in Eq. (10-24). Its effect on the voltage gain has been almost entirely eliminated by the size of the emitter resistor. This means that a circuit designed for a transistor with $h_{fe} = 50$ will also function with h_{fe} values of 20 or 80. Of course, the *Q* point is likely to move somewhat, but the voltage gain will remain essentially the same. This phenomenon will be discussed in more detail when we examine voltage-divider bias.

THE BYPASS CAPACITOR AND VOLTAGE GAIN

When a large capacitor is placed in parallel or in shunt across the emitter resistor, the voltage gain of the amplifier can be restored to its nondegenerated value. All capacitors display capacitive reactance when they are placed in a circuit where alternating current or pulsating direct current is present. This capacitive reactance, however, depends on the frequency of operation. As the frequency increases, X_C decreases, and as the frequency decreases, X_C increases. As the frequency is increased, a point is reached where the capacitor seems to act as a short circuit for the alternating current, since its reactance is very low in relation to the resistance it shunts. At the same time the capacitor's opposition to the direct current does not change.

When a sufficiently large bypass capacitor is chosen, it will act essentially as a short circuit even for very low audio frequencies, of the order of 50 Hz. Under this condition the bypass capacitor in parallel with the emitter resistor effectively reduces the ac impedance of the emitter circuit to a value approaching zero.

The emitter resistor is said to be effectively bypassed when the value of X_C is less than one-tenth the resistance R_E which it bypasses at that frequency. The actual mechanism of the bypassing action is similar to that of a shock absorber. With each increase or decrease in emitter current, the current starts to move into the capacitor to charge or discharge it, rather than to move through the resistor. The net result is that little or no ac voltage can be measured across the resistance R_E.

In Fig. 10-5, the 100-μF bypass capacitor C_E has only 80 Ω of reactance at 20 Hz, less than 10 percent of the 1,000-Ω emitter resistor it bypasses. The value of the capacitive reactance of this 100-μF capacitor is shown as a function of frequency in Fig. 10-6. Most speakers will not respond to frequencies as low as 20 Hz, and so audio amplifiers are not required to amplify voltages at that low frequency. At 100 Hz the 100-μF capacitor has only 15.9 Ω of reactance, and it indeed appears almost as a short circuit to the emitter resistor.

Although bypassing the emitter resistor avoids the negative feedback or degeneration that reduces the gain of the amplifier, note that it does not alter the dc bias stabilization. The dc emitter voltage has

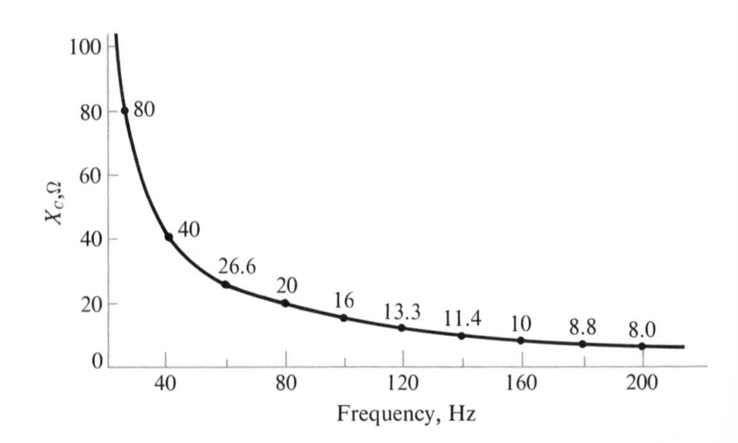

FIG. 10-6 Bypass capacitive reactance versus change in frequency for a 100-μF capacitor.

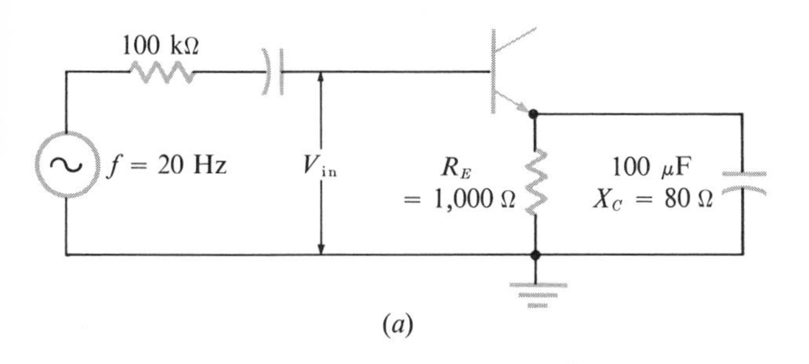

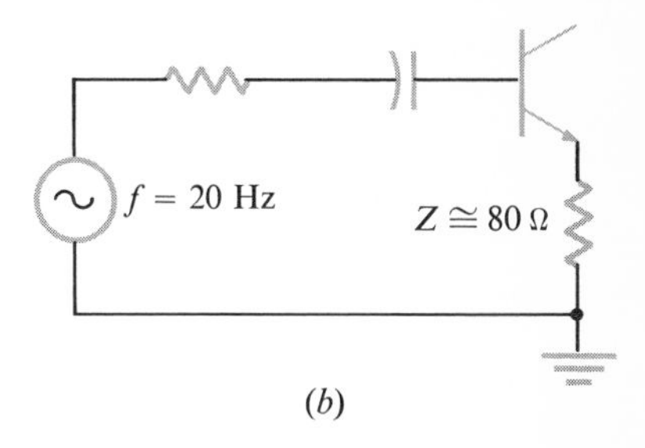

FIG. 10-7 (a) The input circuit with R_E bypassed at 20 Hz and (b) the equivalent circuit at 20 Hz.

not changed. What has changed, however, is the input resistance to the amplifier. The circuit of Fig. 10-5 has an input impedance of 101 kΩ with the emitter resistor unbypassed, but the introduction of a bypass capacitor changes this picture. Figure 10-6 indicates the change in reactance of the capacitor with frequency, and Fig. 10-7 indicates the ac effect on the emitter impedance at 20 Hz.

Remember that the input impedance for a common-emitter amplifier is an amplified version of the resistance or impedance between the base and ground; thus the change in input impedance is dramatic. If at 100 Hz the bypassed resistance is equivalent to about 16 Ω, the input impedance Z_{in} is reduced from 101 kΩ to less than 5,000 Ω,

$$Z_{in} \cong (1 + h_{fe})(30 + 16) = 101 \times 46 = 4{,}646\ \Omega \qquad (10\text{-}25)$$

This input impedance may be so low as to load down a previous stage of amplification, even though it gives a higher voltage gain,

$$A_v \cong \frac{-h_{fe} R_L}{(1 + h_{fe})(h_{ib} + R_{E,\,eq})} = \frac{-100 \times 7.5\ \text{k}\Omega}{4{,}646} = -162 \qquad (10\text{-}26)$$

INCREASING THE INPUT IMPEDANCE

The basic fixed-bias circuit may be slightly modified as indicated in Fig. 10-8 by replacing the emitter resistor with two resistors whose total value is approximately 1,000 Ω. Although 500-Ω resistors are not easy to obtain, the standard values of 470 or 510 or a combination of the two will suffice. Instead of bypassing the whole 1,000 Ω of emitter

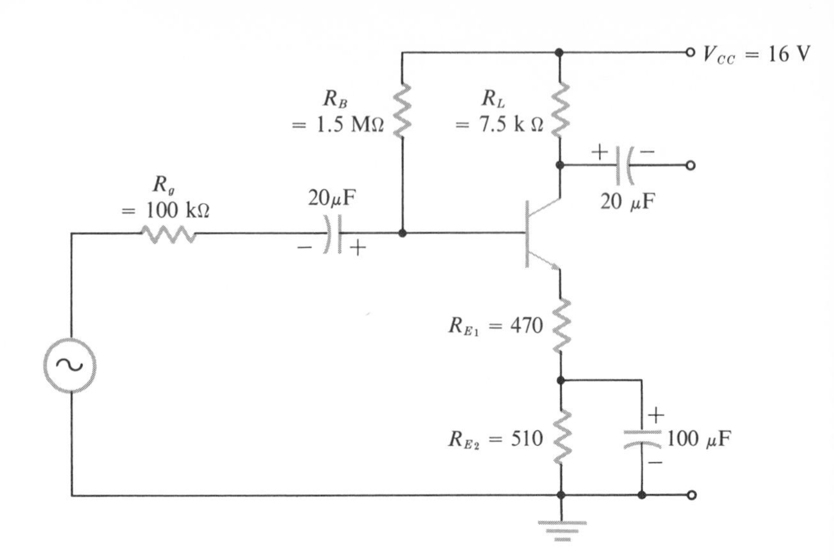

FIG. 10-8 A common-emitter amplifier with a partially bypassed emitter-bias resistance.

bias resistance, only one-half is bypassed. The net result of this modification is that the input impedance is restored to about 50 kΩ, while the voltage gain is increased from about 7.5 to about 15. This common circuit modification is found in various forms in many high-fidelity circuits. The actual amount of resistance left unbypassed, of course, determines what the input impedance and the voltage gain will be.

If the two resistors used in Fig. 10-8 are replaced by a potentiometer, as shown in Fig. 10-9, a very interesting circuit design results. Since the body of the potentiometer is in the main-flow circuit for the emitter current, the dc bias is maintained. However, since the bypass

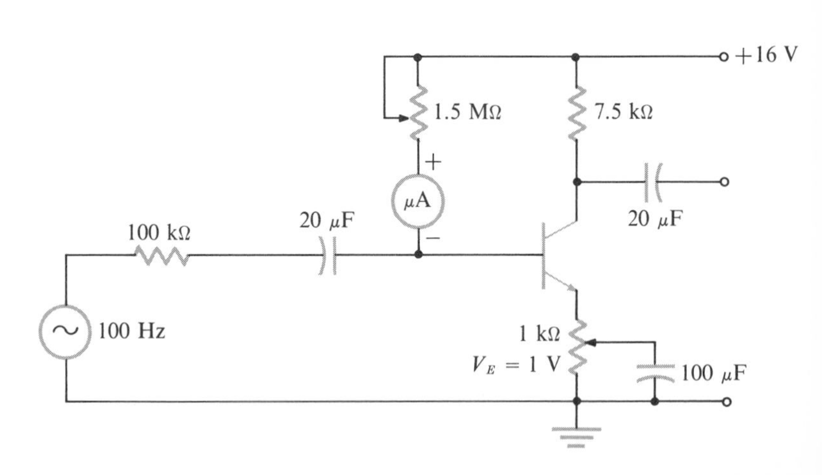

FIG. 10-9 Experimental emitter-bias common-emitter amplifier with a variable input impedance and voltage gain.

capacitor is attached to the wiper arm of the potentiometer, any amount of the total bias resistance may be biased at will. Virtually any input resistance from 2.5 to 100 kΩ is possible with this arrangement. The resultant change in voltage gain is from about 7.5 with no bypass to almost 200 when the potentiometer is fully bypassed at 100 Hz.

This experimental circuit should be built on a permanent circuit board or chassis. Accessible contacts or jacks should be attached to all three contacts of the potentiometer, so that when the correct input impedance and gain have been established the resistance values can be readily measured.

SUMMARY OF CONCEPTS

1. Leakage current I_{CBO} is a function of junction temperature and passes through the reverse-biased collector-base junction, since minority current carriers do not encounter a potential barrier at a reverse-biased junction.
2. In a common-emitter amplifier the leakage identified as I_{CEO} is the amplified form of I_{CBO} and is $1 + h_{fe}$ times as large.
3. Leakage current approximately doubles for every 10°C rise in temperature.
4. At high temperatures leakage currents will cause shifting of the Q point.
5. Bias stabilization is more necessary in circuits with germanium transistors than in those with silicon transistors.
6. An emitter resistor acts to swamp out the changes in h_{ib} that result from changes in temperature owing to its comparatively larger size.
7. Emitter bias increases the input impedance by a factor of $1 + h_{fe}$.
8. Emitter-bias stabilization makes a circuit voltage gain dependent on the load resistance and emitter resistance rather than on h_{fe} variation among transistors.
9. Input impedance depends on the size of the base-limiting resistor, which appears in parallel with the transistor input under signal conditions.
10. The voltage gain of the common-emitter amplifier is significantly reduced with emitter bias as a result of degeneration, or negative feedback, since the emitter signal voltage subtracts from the base signal voltage, allowing amplification of only the difference, V_{be}.
11. Bypassing the emitter resistance with a large capacitor has no effect on the dc resistance, but shorts out a significant amount of ac resistance.
12. Bypassing the emitter-bias resistor increases the voltage gain, but radically lowers the input impedance.

GLOSSARY

Degeneration A process whereby a signal is reduced in strength by application of out-of-phase or negative feedback, with the feedback subtracting from the size of the original input signal.

Leakage current A reverse-bias junction current due to minority carriers which encounter no potential barrier at the junction. It is a function of the absolute temperature rather than the applied voltage, and although it is only a few nanoamps for silicon devices at room temperature, it about doubles with every 10°C rise in temperature.

Nonlinear The inability of a component or circuit to amplify without introducing distortion, so that the output signal is not a faithful reproduction of the input signal.

Output conductance A measure of the change in the output current of a transistor with increases in collector-emitter voltage when the base current is held constant; the inverse of the transistor output impedance.

Stabilization The process by which a transistor circuit is made insensitive to changes in h_{FE} and leakage currents.

Swamping action The circuit action resulting from introduction of an emitter resistance several times larger than the transistor's change in h_{ib}, or leakage effects.

Voltage feedback ratio A transistor hybrid parameter h_{re} or h_{rb} referring to the change in V_{be} with changes in V_{ce}, a measure of the internal voltage feedback through a device.

REFERENCES

Corning, John J.: *Transistor Circuit Analysis and Design*, Prentice-Hall, Inc., Englewood Cliffs, N.J., 1965, chap. 6.

Cutler, Phillip: *Semiconductor Circuit Analysis*, McGraw-Hill Book Company, New York, 1964, chap. 4.

Gillie, Angelo C.: *Principles of Electron Devices*, McGraw-Hill Book Company, New York, 1962, pp. 124–127.

Horowitz, Mannie: *Practical Design with Transistors*, The Bobbs-Merrill Company, Inc., Indianapolis, 1968, chap. 3.

Malvino, Albert Paul: *Transistor Circuit Approximations*, McGraw-Hill Book Company, New York, 1968, chap. 6.

Surina, Tugomir, and Clyde Herrick: *Semiconductor Electronics*, Holt, Rinehart and Winston, New York, 1964, pp. 111–115.

Temes, Lloyd: *Electronic Circuits for Technicians*, McGraw-Hill Book Company, New York, 1970, chap. 8.

REVIEW QUESTIONS

1. What is meant by leakage?

2. What are the principle causes of leakage currents?

3. Discuss the problem of leakage in germanium transistors as compared with silicon transistors.

4. What is the difference between I_{CBO} and I_{CEO}?

5. What is a swamping resistor and why is it used?

6. In what way does the addition of a swamping resistor make a circuit more stable?

7. What happens to the input impedance of an amplifier when a swamping emitter resistor is added to the simple fixed-bias circuit?

8. What is meant by degeneration? How is it prevented in a circuit with emitter bias?

9. Does the use of a bypass capacitor to prevent degeneration change the swamping action of the emitter resistor? Why?

10. What happens to the input impedance of an emitter-bias transistor amplifier if the emitter resistor is bypassed?

11. How is the voltage gain of a common-emitter ampliner affected by the addition of an emitter resistor? Why?

12. At the lowest frequency to be amplified by the circuit, what is the maximum capacitive reactance that should be tolerated from a bypass capacitor?

13. How can a high input impedance be developed for an amplifier while still maintaining some relatively high voltage gain?

14. Why is the voltage gain of a common-emitter amplifier considered as negative?

15. How does the use of an emitter-bias resistor make the voltage gain of a common-emitter circuit essentially independent of the h_{fe} of the circuit?

PROBLEMS

1. If the value of I_{CBO} for a particular transistor is 50 nA at 300°K (27°C), what will be the leakage at 310°K (37°C)?

2. In Prob. 1, if $h_{FE} = 50$, what is the leakage current I_{CEO} at 300°K? At 310°K?

3. In a germanium transistor, if $I_{CBO} = 5\ \mu A$ at 300°K, what is I_{CEO} if $h_{FE} = 30$?

4. In a circuit like that of Fig. 10-3, if the emitter resistance is 750 Ω and the supply voltage is 20 V, what is the emitter voltage when the emitter current is 1.2 mA?

5. In Prob. 4, what is the input impedance of the circuit if $h_{fe} = 20$? (Hint: First determine R_B and h_{ib} at 1.2 mA, where $h_{ib} = 30\ \Omega$ for $I_E = 1.0$ mA.)

6. In Prob. 5, what is the input impedance if the emitter resistance is completely bypassed?

7. If the Q point of a circuit is $I_C = 2$ mA, $I_B = 15\ \mu A$, $V_C = 8.0$ V, and $V_{CC} = 15$ V, how much is V_{R_L}? If $V_E = 0.8$ V, how much is V_{CE}? How much is V_B?

8. From the data collected in solving Prob. 7, draw the circuit and compute the values of all resistances.

9. In Prob. 7, what is the value of h_{FE}? Assume $h_{fe} = h_{FE}$ and compute the input impedance to the circuit.

10. If an emitter-bias common-emitter amplifier has an emitter resistance of 470 Ω and a load resistance of 8.2 kΩ, what is the approximate voltage gain?

11. What is the voltage gain of an amplifier having a load resistance of 6.8 kΩ with an emitter resistance of 820 Ω, where 220 Ω is bypassed?

12. What is the approximate input impedance of an emitter-bias amplifier where $h_{fe} = 90$ and the unbypassed emitter resistance is 300 Ω with $I_E = 1.0$ mA?

13. If a signal generator delivers a signal voltage of 1.2 V *p-p* to an emitter-bias amplifier and the signal voltage measured across the emitter resistor is 1.14 V *p-p*, how much signal voltage is actually delivered to the transistor in the form of V_{be}? If the output voltage is 6.0 V *p-p*, what is the total circuit voltage gain?

14. If an amplifier must have a "flat" frequency response down to 40 Hz, how large must the bypass capacitor be for an emitter resistance of 1,000 Ω?

15. What size of load resistance would be needed to yield a voltage gain of 15 if a 360-Ω emitter-bias resistor is used?

16. In the circuit of Fig. 10-9, what resistance of the potentiometer should be left unbypassed to yield an input impedance of 50 kΩ if $h_{fe} = 110$?

17. Design a circuit similar to Fig. 10-9 having an input impedance of 15 kΩ and a voltage gain of 10 when $V_{CC} = 5$ V, $h_{fe} = h_{FE} = 100$, $I_C = 0.8$ mA, and $V_E = 1.0$ V. Find R_L, V_{R_L}, V_B, R_B, and R_E (both bypassed and unbypassed).

OBJECTIVES

1. To examine circuit bias and self-correction
2. To utilize Ohm's law in analyzing complex circuits
3. To describe all circuits as combinations of series and parallel resistance or impedance networks
4. To compare feedback at the collector to feedback at the emitter of an amplifier circuit
5. To examine equivalent circuits for complex networks of impedance
6. To compare the β-independent voltage-divider-biased circuit to other common-emitter circuit variations
7. To develop experimental techniques in measuring voltage gain

SELF-BIAS AND VOLTAGE-DIVIDER BIAS IN THE COMMON-EMITTER AMPLIFIER

The prevalence of inexpensive transistor radios is ample proof that the problems of thermal runaway and leakage in transistors have been dealt with effectively in manufacture. This was made possible by the use of two rather simple variations of the common-emitter circuit: the self-biased circuit and the voltage-divider-biased circuit. We shall examine these two circuits and consider some experiments that illustrate the ingenuity of manufacturers.

SELF-BIAS

Next to the basic common-emitter amplifier circuit, perhaps the simplest circuit in appearance is the self-bias arrangement of Fig. 11-1. Do not be misled by the simplicity of the circuit design, however; the operation of this circuit is more complex than it looks. The basic circuit functions are not difficult to understand. The difficulty lies in the fact that this circuit will seek its own bias level as a result of negative feedback from the collector voltage.

If the self-biased circuit is approached from the standpoint of Ohm's law and current flow, there is little difficulty in determining what values of resistance to use. It is simply a case of deciding what supply voltage must be used and what transistor is available. For example, in Fig. 11-1 a supply voltage of 10 V has been assumed for simplicity, and a transistor with $h_{FE} = 50$ has been selected because of its availability. If the usual safe 1.0 mA of collector current is used, and the usual design procedure of $V_{CE} = V_{CC}/2$ is followed, we have $V_{CE} = 5$ V. Since the remaining 5 V will appear across the load resistance, R_L will require a 5-kΩ resistance. In this case a standard value of 4.7 kΩ or 5.1 kΩ should be used. The choice of either standard value will have little effect on the overall performance of the circuit, but the 4.7-kΩ resistor is easier to obtain.

Notice that the path of the base current in Fig. 11-1 is no longer directly back to the positive terminal of the power supply. Instead it reenters the mainstream of current, rejoining the collector current at the entrance to the load resistance. It is obvious, then, that *the voltage across the load resistance must be a function of the collector current plus the base current.* This means that the load current is the same as the emitter current.

The voltage across the load resistance is found as follows:

$$I_{R_L} = I_C + I_B$$

$$= 1 \text{ mA} + 20\ \mu\text{A} = 1.02 \text{ mA} \qquad (11\text{-}1)$$

$$V_{R_L} = I_{R_L} R_L$$

$$= 1.02 \text{ mA} \times 4.7 \text{ k}\Omega = 4.8 \text{ V} \qquad (11\text{-}2)$$

$$V_{CE} = V_{CC} - V_{R_L}$$

$$= 10 \text{ V} - 4.8 \text{ V} = 5.2 \text{ V} \qquad (11\text{-}3)$$

With the collector voltage established at about 5.2 V, the size of the base-current-limiting resistance may be found. At 1 mA of emitter current the base-emitter voltage is about 0.6 V for a silicon transistor. The collector-base voltage is then, by subtraction,

$$V_{CB} = V_C - V_B$$

$$= 5.2 \text{ V} - 0.6 \text{ V} = 4.6 \text{ V} \qquad (11\text{-}4)$$

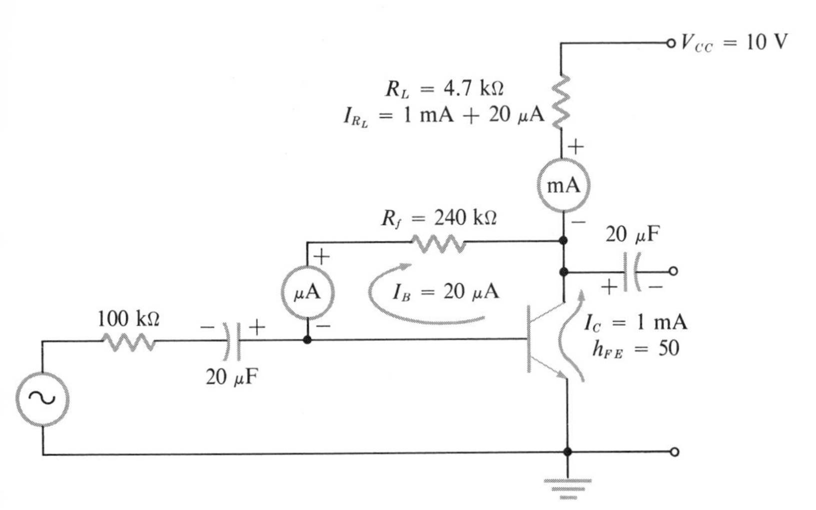

FIG. 11-1 The self-biased common-emitter circuit.

The value of the base-current-limiting resistor, also termed the *collector-base feedback resistance* R_f, is, from Ohm's law,

$$R_f = \frac{V_{R_B}}{I_B} = \frac{4.6\ \text{V}}{0.02\ \text{mA}} = 230\ \text{k}\Omega \qquad (11\text{-}5)$$

Even though this value is derived rather indirectly, it is realistic, *since R_f should normally be approximately h_{fe} times larger than the resistance of the collector-base junction which is reverse biased.* This value is not determined exactly, but it is known to be slightly smaller than R_{CE}, which in this case is 5.2 kΩ (5.2 V/1.0 mA). If R_{CB} is considered to be somewhat smaller, say, 4.7 kΩ, then

$$h_{fe}R_{CB} = 50 \times 4.7\ \text{k}\Omega = 235\ \text{k}\Omega$$

which is rather close to the value of 230 Ω calculated by reasonable application of Ohm's law. Since the standard resistance closest to 230 kΩ is 240 kΩ, this is what should be used in the circuit.

The self-biased circuit is interesting in that it does seek its own bias level. It also tends to correct itself with changes in temperature. For this reason it was one of the most popular circuit designs in the early days of germanium transistors.

If the collector current begins to increase, the collector voltage will begin to decrease. Since the base current is supplied by the collector terminal voltage, this drop in voltage is automatically fed back to the base terminal of the transistor. This lowered voltage at the base reduces the base current. Since the collector current is a function of the base current and the gain h_{fe}, the collector current is reduced. This results in a smaller load voltage V_{R_L} and a larger collector-emitter voltage V_{CE}. Of course, this process will restabilize the circuit rapidly.

DEGENERATION

As with other types of bias stabilization, however, this process results in a loss in voltage gain due to degeneration, or negative feedback from the collector. This circuit is therefore sometimes called the *constant-collector-voltage circuit*. The process is sometimes called *voltage-feedback stabilization*.

Part of the negative feedback may be cancelled for small-signal applications by the same process used with the emitter resistor. That is, the feedback resistance R_f may be made up of two resistors instead of one, with a bypass capacitor connected as shown in Fig. 11-2. This may be desirable, because the signal variation at the collector will feed back to the base, but 180° out of phase with the base signal, cancelling most of the input signal. All the resistance R_f should not be bypassed, however, since this would have the effect of shorting out the signal at the base directly to ground.

Another way of looking at the self-biased circuit is to consider the voltage changes that occur at the collector. These changes may be caused by any variation of current, whether they arise from tempera-

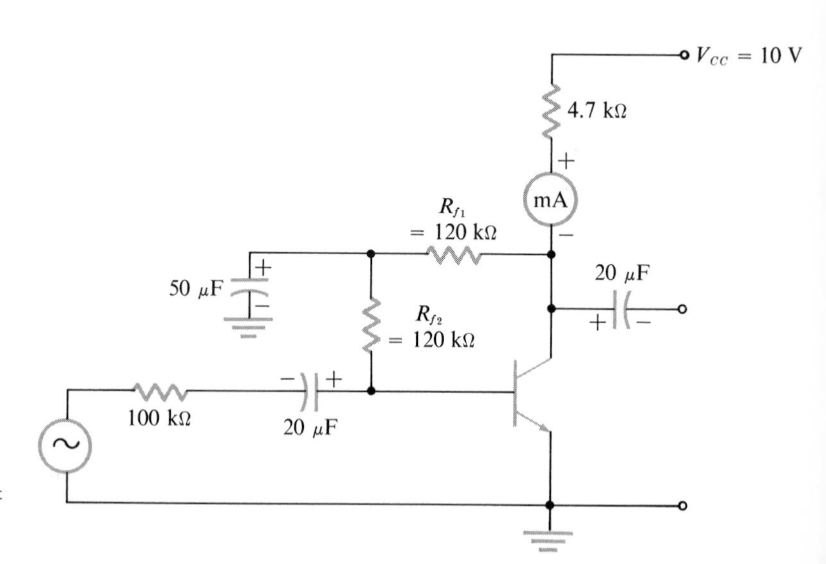

FIG. 11-2 The self-biased circuit with feedback control by using a bypass capacitor.

ture changes, increased base current, signal-current variations, or variations in supply voltage. When a voltage variation does occur, the base current must change a small amount as a consequence of its supply-voltage change. This apparent voltage feedback is in a direction that results in less current flow; it is negative feedback.

Whether this feedback is thought of as voltage or current feedback is of no serious concern in terms of the overall effect, since there can be no current without voltage. The net result is a correction of a situation that is a potential source of distortion or even device destruction. The side effect is a change in the input impedance of the circuit and a greater loss in voltage gain than in the fixed-bias common-emitter circuit.

The feedback current indicated above will always develop a change in voltage across the feedback resistance from the collector to the base. If the resistance is divided as indicated in Fig. 11-2, and a capacitor is attached to ground from this intersection, the resulting voltage variation across the resistor R_{f1} will be shunted to ground through the low impedance of the capacitor. Thus the dc bias can be stabilized without the enormous loss in voltage gain. As with emitter bias, the feedback resistance may be a potentiometer with the capacitor connected to the wiper arm, allowing variable feedback.

INPUT IMPEDANCE

Any circuit may be analyzed in terms of equivalent resistance. If the circuit of Fig. 11-1 is redrawn as a resistance equivalent as in Fig. 11-3, it is not difficult to see that the feedback resistance is in parallel with the collector-base junction resistance. The input to the amplifier is to the base, so that as far as the signal source is concerned, the resistance into which it is working consists of h_{ie}, which is in turn in parallel with the series-parallel combination $(R_f \parallel R_{CB})+R_L$. The power

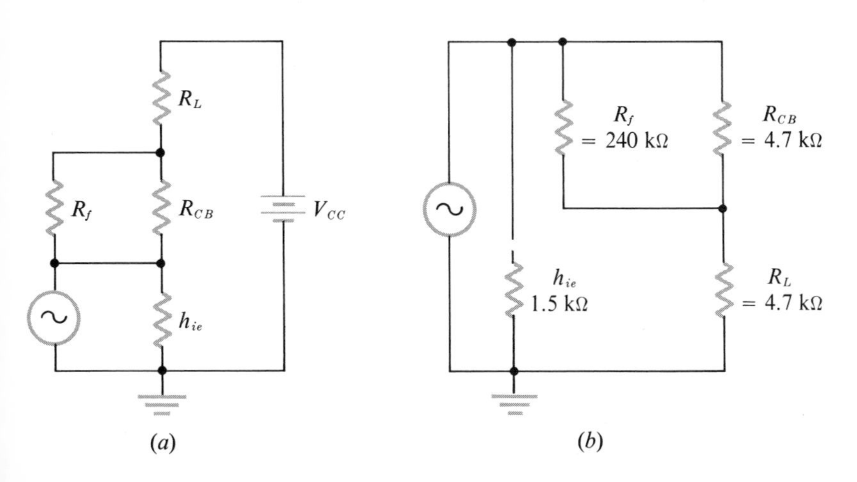

FIG. 11-3 (*a*) The resistive equivalent of a self-biased circuit; (*b*) equivalent circuit showing the series-parallel nature of the circuit as its input impedance.

supply, which has very little ac resistance, is not indicated in Fig. 11-3*b*. Note that the smallest resistance in the parallel network is h_{ie}, about 1,500 Ω for the transistor with an h_{fe} value of 50. The input impedance must therefore be less than 1,500 Ω,

$$
\begin{aligned}
Z_{\text{in}} &= h_{ie} \parallel (R_f \parallel R_{CB} + R_L) \\
&= 1.5 \parallel (240 \parallel 4.7 + 4.7) \\
&= 1.5 \parallel (4.6 + 4.7) = 1.5 \parallel 9.3 = 1.3 \text{ k}\Omega \qquad (11\text{-}6)
\end{aligned}
$$

Note that the parallel nature of the feedback resistance is such that low values will allow a great deal more feedback and will also lower the input impedance considerably. This is the opposite of the effect with emitter bias, where an increase in resistance appears to amplify the input resistance by a factor of $h_{fe} + 1$.

As previously indicated, this is an approximation of the actual input impedance as it would be developed from the more rigorous mathematical treatment it deserves. To determine a more exact value consult the references at the end of the chapter for the more exact formulas.

VOLTAGE GAIN

The voltage gain of the self-biased circuit, like the input resistance, is also developed by approximation. However, the circuit developed from Fig. 11-1, which used a load resistance of 4.7 kΩ and a feedback resistance of 240 kΩ, was computed to have an input resistance of approximately 1.3 kΩ. From these values it is possible to compute the current gain, the resistance gain, and hence the voltage gain. Remember that our circuit was developed with an approximation that gave a feedback resistance about h_{fe} times larger than the load resistance. We then chose a standard-size resistance which still gives a close approximation. This precaution is not absolutely necessary, but without it the computation is more difficult. Thus we find the voltage gain to be

$$
\begin{aligned}
A_v &= -A_i A_r \\
&= -A_i \frac{R_L}{R_{\text{in}}} \\
&= -50 \times \frac{4.7 \text{ k}\Omega}{1.3 \text{ k}\Omega} = -181 \qquad (11\text{-}7)
\end{aligned}
$$

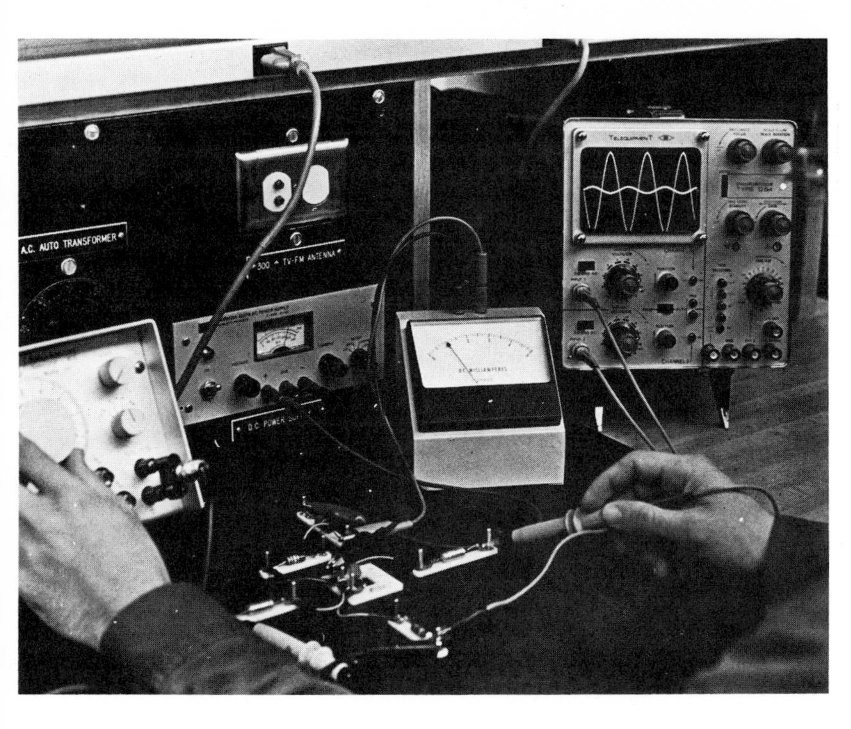

FIG. 11-4 The input and output waveforms for the self-biased circuit as viewed on a dual-trace oscilloscope.

The current gain of the circuit may not always be near to the value of h_{fe}. An approximation of A_i is given by

$$A_i \cong \frac{R_f}{R_L} \qquad (11\text{-}8)$$

This expression indicates that with lower values of R_f than those indicated above, the current gain of the circuit will be significantly lower.

SELF-BIAS WITH EMITTER RESISTANCE

Although the self-biased circuit has been popular because of its simple design, the addition of an emitter resistance, as indicated in Fig. 11-5, adds more bias stabilization. The same theory and principles that apply to the fixed-bias common-emitter amplifier also apply in general to this circuit. If the same conceptual approach is used to determine the resistance values, the additional complexity of the circuit should not pose difficulties.

For the sake of comparison, let us use the same transistor as in the previous experiment. Since $h_{FE} = 50$, if 1 mA of collector current is to be used, the base current must be 1 mA/50 = 0.02 mA = 20 μA.

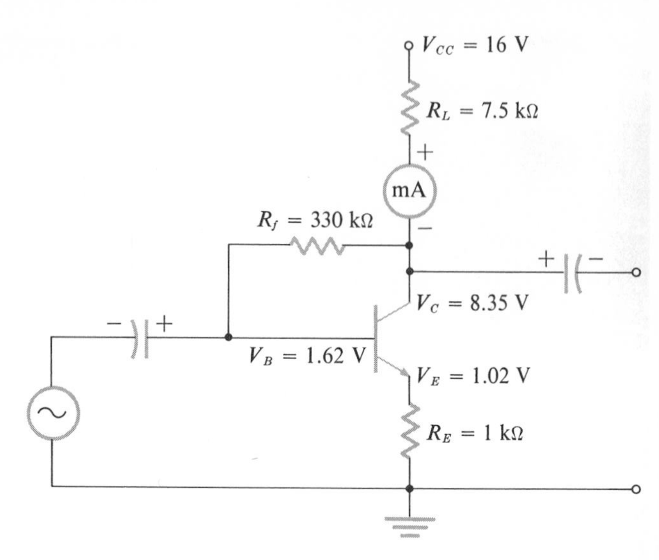

FIG. 11-5 The self-biased circuit with emitter-bias stabilization.

This means the resulting current through the load resistance will again be

$$I_{R_L} = I_C + I_B = 1.02 \text{ mA}$$

An emitter resistance of 1 kΩ will give an emitter voltage of 1.02 V, leaving about 15 V to be divided between R_L and R_{CE}. Theoretically this voltage should be divided equally, but in practice exactly equal division is not feasible because of the base current that passes through the load resistor and the fact that standard resistor sizes must be used.

With 1 mA of collector current and 15 V to be divided evenly if possible, the standard 7.5-kΩ resistor would be a good value to begin with, since 1 mA × 7.5 kΩ = 7.5 V = 15 V/2. If the 7.5-kΩ resistance is used for the collector load, the current passing through it will develop a voltage of 1.02 mA × 7.5 kΩ = 7.65 V. When this is added to the emitter voltage of 1.02 V the voltage V_{CE} may be computed as

$$\begin{aligned} V_{CE} &= V_{CC} - (V_{R_L} + V_E) \\ &= 16 - (7.65 + 1.02) \\ &= 16 - 8.67 = 7.33 \text{ V} \end{aligned} \qquad (11\text{-}9)$$

The dc voltage at the base is about 0.6 V above the voltage at the emitter, or about 1.62 V. Since $V_C = V_E + V_{CE} = 8.35$ V, the collector-base voltage is computed as

$$V_{CB} = V_C - V_B$$
$$= 8.35 - 1.62 = 6.73 \text{ V} \qquad (11\text{-}10)$$

To compute the value of the feedback resistance R_f it is only necessary to compute the resistance needed to limit the base current to 20 μA with 6.73 V across it, or

$$R_f = \frac{6.73 \text{ V}}{0.02 \text{ mA}} = 337 \text{ k}\Omega \qquad (11\text{-}11)$$

The closest standard value in this case is 330 kΩ.

INPUT IMPEDANCE AND VOLTAGE GAIN

As with the common-emitter fixed-bias circuit, the addition of an emitter resistance swamps out the changes in h_{ib} that occur with changes in temperature and leakage current. At the same time the input impedance increases by a factor influenced by the current gain of the circuit. The resulting increase in input impedance introduces degeneration, or negative feedback, into the circuit at the emitter as well as at the collector. The result is additional loss in voltage gain.

To verify this construct the circuit of Fig. 11-5. If a transistor with $h_{FE} = 50$ is not available, replace the feedback resistor R_f with a 500-kΩ potentiometer hooked up as a rheostat by connecting two of the terminals as shown in Fig. 11-6. If the base current is adjusted so that the voltages measured at the terminals are as shown, the circuit should function well. Be sure to start the base-current adjustment with the potentiometer-rheostat turned all the way up to its highest resistance. This will protect the circuit against a surge of base current when the power is first turned on. Gradually reduce the resistance by turning the potentiometer counterclockwise until the correct emitter current flows and establishes the correct voltages. When this dc point of operation, the Q point, has been set, the amplifier is ready to operate as a small-signal amplifier.

Attach the output capacitor of the amplifier to the vertical input terminals of the oscilloscope. Connect the signal generator to the input capacitor of the amplifier and set the frequency of operation to some frequency in the middle of the audio range, between 1 and 10 kHz. Observe the output signal on the oscilloscope screen and adjust the input signal until the output shows no distortion. With the amplifier so operating, compare the output signal voltage (peak to peak) to the

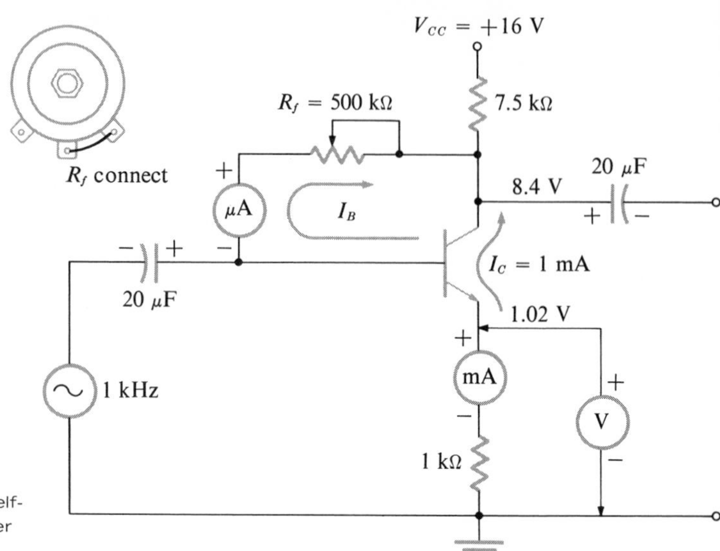

FIG. 11-6 An experimental setup for a self-biased common-emitter circuit with emitter resistance for stabilization.

input signal voltage at the signal input to the amplifier. Record all dc and ac voltages and compute the voltage gain. (For a review of oscilloscope measurement techniques see the discussion of voltage measurement in Chap. 4.)

After you have taken and recorded all voltage and current measurements, turn off the power and, without changing its setting, measure the resistance of the feedback resistor. Record this as the actual value of the feedback resistance. Measure the actual value of the load resistance and record this as well. These two measurements will be used to compute the approximate circuit current gain from Eq. (11-8),

$$A_i \cong \frac{R_f}{R_L}$$

With the voltage gain and current gain computed from these measured values we can now determine the input impedance to the amplifier. Rearranging Eq. (11-7), we have

$$A_v = -A_i A_r = -A_i \frac{R_L}{R_{\text{in}}} \qquad (11\text{-}12)$$

$$R_{\text{in}} = \frac{-A_i R_L}{A_v} \qquad (11\text{-}13)$$

THE THÉVENIN EQUIVALENT CIRCUIT

The value of input impedance may also be confirmed, if very careful oscilloscope measurement is used, by means of a simple Thévenin equivalent circuit. The series resistance represents the Thévenin equivalent resistance, with the resistor R_{in} denoting the actual but unknown value of input impedance. Reset the amplifier for operation as before. When a small signal is being amplified without distortion, disconnect the signal generator and insert a 100-kΩ potentiometer in series with the signal as shown in Fig. 11-7. Connect the potentiometer to act as a rheostat, just as for the feedback resistance. Adjust the signal generator until it puts out a signal that is two spaces high on the oscilloscope screen. Next move the scope to view the signal at the base of the transistor and adjust the series potentiometer until the signal at the base is just one space high. The signal may be very noisy, but adjust it as carefully as possible.

When the signal is just one-half the output voltage of the signal generator, the other half will be found across the series potentiometer. If the same voltage is found across both these resistances, the resistances themselves must be equal. Remove the potentiometer from the circuit and measure its value with an ohmmeter. How close is the value to that computed from Eq. (11-13)? What are possible causes for errors?

SELF-BIAS WITH THE EMITTER RESISTOR BYPASSED

If the circuit of Fig. 11-6 is modified by placing a bypass capacitor around the emitter resistor, its negative feedback will be cancelled. This, of course, will increase the voltage gain and reduce the input resistance to the level of the circuit without the emitter resistor. The dc bias stabilization will be preserved, however.

Repeat the experiment outlined for Fig. 11-6, but with the modifications indicated in Fig. 11-8. Note the addition of a large series resistor, which again makes the signal generator appear to be a constant-current source. Without this resistance severe distortion may result unless the signal generator has a very low output impedance and is capable of a great deal of signal attenuation without distorting the signal.

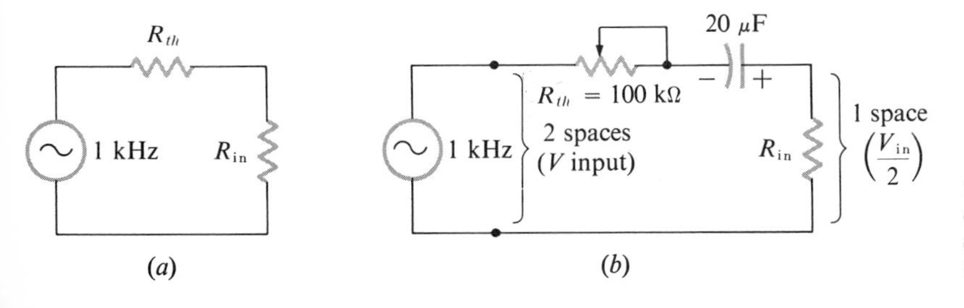

FIG. 11-7 (a) The Thévenin equivalent circuit, and (b) its counterpart in determining circuit input impedance.

VOLTAGE-DIVIDER BIAS: THE β-INDEPENDENT CIRCUIT

Perhaps the most widely used means of stabilizing a common-emitter amplifier circuit is voltage-divider bias. This type of circuit is exceptionally stable in terms of leakage and is relatively unaffected by changes in h_{fe}. Hence it is sometimes called the *β-independent* circuit. Almost any transistor with h_{fe} values from 20 to 200 can be used without too much difference in voltage gain or input impedance. Although the voltage gain is very low in comparison to other arrangements of the common-emitter amplifier circuit, voltage-divider bias is almost universally dependable under extreme conditions. It may also be used as a common-base or a common-collector circuit simply by changing the input and the output connections. The whole circuit is a function of its resistance values, and is not dependent on the transistor parameters. The only exception is that the breakdown voltage of the transistor must be greater than the supply voltage.

The circuit of Fig. 11-9 is similar to that of Fig. 11-8 in that the same values of load resistance and emitter resistance are used in conjunction with the same power supply voltage. Of course, with a different collector current—say, 2 mA—the resistances would have to be changed accordingly to reestablish the correct voltages for optimum design.

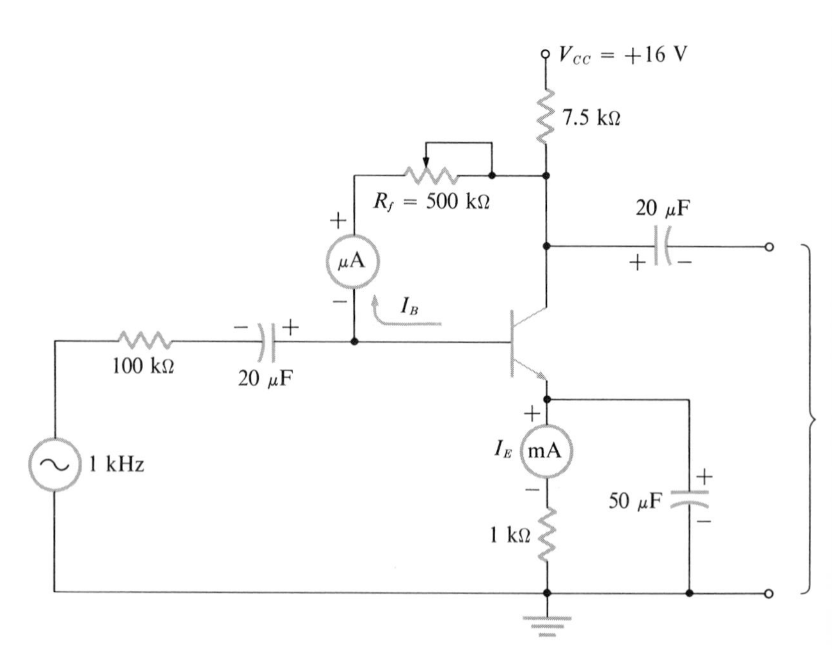

FIG. 11-8 An experimental self-biased amplifier with the emitter resistance bypassed.

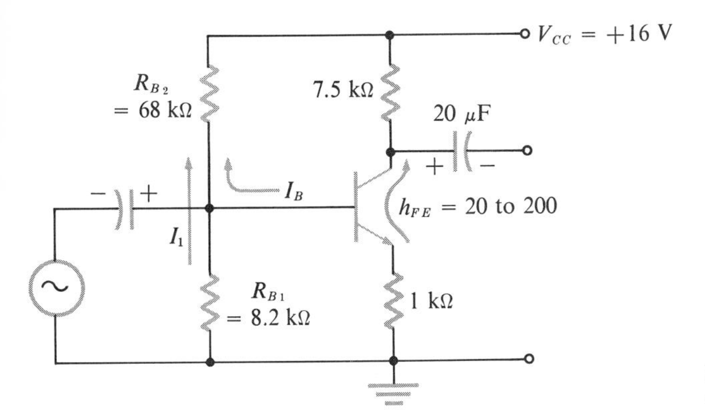

FIG. 11-9 A common-emitter amplifier with voltage-divider bias.

The voltage stability of this circuit is almost unique. The base-emitter voltage, which is the key voltage in all the circuits, has a tendency to vary slightly with changes in temperature. As we have seen, because of the increase in leakage current I_{CEO}, the base voltage creeps up slightly even with bias stabilization. The voltage-divider circuit, however, tends to keep the base-emitter voltage at a present level and to decrease it as the amount of leakage current increases. This decrease in V_{BE} is followed by a decrease in base current, which prevents thermal runaway.

Observe in Fig. 11-9 that current I_1 is allowed to pass through resistances R_{B1} and R_{B2} independently of the transistor currents. In addition to a *bleeder current*, I_1, the base current also passes through R_{B2}. If the bleeder current is at least 10 times greater than the base current, the base current will have almost no effect on the voltage at the junction of the two resistors, the base voltage. If the expected leakage is very low, the amount of bleeder current may be decreased accordingly—say, by a factor of 5. This will also result in a higher input impedance, a desirable outcome.

First consider the voltages that must be found across the resistors R_{B1} and R_{B2}. A base voltage of about 1.6 V must be allowed for R_{B1}. This leaves about 14.4 V across R_{B2}, since this is the difference between the supply voltage of 16 V and the base voltage of 1.6 V.

After the desired voltages have been established, the appropriate amount of bleeder current must be determined. If, for the sake of discussion, the transistor has a gain of $h_{fe} = 50$, then a collector current of 1 mA would require a base current of 20 μA. This base current, as well as the bleeder current I_1, will pass through R_{B2}. If the rule-of-thumb requirement of a safety factor of 10 is followed, 20 μA of base current would call for a bleeder current of $10 \times 20\ \mu\text{A} = 0.2$ mA.

With the bleeder current of 0.2 mA determined, R_{B1} may be computed from Ohm's law as

$$R_{B1} = \frac{V_B}{I_1} = \frac{1.6\ \text{V}}{0.2\ \text{mA}} = 8\ \text{k}\Omega \qquad (11\text{-}14)$$

As before, we shall use the closest standard-value resistor, which is 8.2 kΩ. With this value determined, we may compute R_{B2} by similar application of Ohm's law. Since the resistance computed for R_{B1} was slightly larger than 8 kΩ, strictly speaking, a bleeder current of 0.195 mA instead of 0.2 mA would be needed to fix the base voltage at 1.6 V. Since the circuit is self-correcting, this difference is not significant. The bleeder current will be sufficiently large to handle any thermal problems. However, in the interests of accuracy we shall consider the bleeder current as 0.195 mA.

In order to compute the size of R_{B2} we must estimate the currents actually passing through R_2. The bleeder current of about 0.195 mA added to the base current of 0.02 mA yields a total current of 0.215 mA. From Ohm's law,

$$\begin{aligned} R_{B2} &= \frac{V_{CC} - V_B}{I_1 + I_B} \\ &= \frac{16\ \text{V} - 1.6\ \text{V}}{0.195\ \text{mA} + 0.02\ \text{mA}} = \frac{14.4\ \text{V}}{0.215\ \text{mA}} \\ &= 67\ \text{k}\Omega \end{aligned} \qquad (11\text{-}15)$$

The nearest standard-value resistor here is 68 kΩ.

In review, a bleeder current passing through R_{B1} in series with R_{B2} will develop the base voltage. This bleeder current is many times larger than the base current which is to pass through R_{B2}, and so changes in base current should have little effect, if any, on the base voltage. Great increases in base current, however, will increase the voltage across R_{B2} slightly, thus decreasing the base voltage slightly.

If extremes of temperature cause some excess leakage current, this additional current passing through the emitter resistance will cause the emitter voltage to rise. The net effect of the rise in emitter voltage and the drop in base voltage is a rapid drop in emitter-base voltage V_{BE}, with a consequent drop in base current. Thus, the voltage-divider-biased common-emitter amplifier circuit exerts a great amount of bias stabilization by offsetting potential thermal-runaway conditions as soon as they arise.

INPUT IMPEDANCE

In the common-emitter amplifier circuit with voltage-divider bias and an unbypassed emitter resistance the input impedance is somewhat more difficult to derive than in other circuits. The easiest approach is the equivalent circuit of Fig. 11-10, which shows how the signal generator sees the input circuit as a group of parallel resistances. The output circuit is not included, since it offers such a small correction factor to the circuit that it is of little significance.

In any parallel grouping of resistances it is the smallest resistance that determines the highest value of the total resistance. In this case the smallest resistance is R_{B1} which is 8.2 kΩ. The emitter resistor and h_{ie} are in parallel with the other resistances, but in series with each other. R_{B2} returns to ground through the power supply. However, the combined resistance of $h_{ib} + R_E$ is amplified by a factor of $1 + h_{fe}$. For $h_{fe} = 50$ this yields a value of 52.5 kΩ, much larger than the 8.2-kΩ parallel resistance. It is this fact that makes the input impedance relatively independent of the h_{fe} value of the transistor. If h_{fe} is larger—say, 100—the effective impedance of $h_{ib} + R_E$ would be over 100 kΩ, more than 10 times larger than the 8.2-kΩ resistance, and hence of little effect.

The total input impedance may be computed from the formula for parallel resistances. For $h_{fe} = 50$

$$Z_{\text{in}} = R_{B1} \parallel R_{B2} \parallel (h_{ib} + R_E)(1 + h_{fe})$$

$$= \frac{1}{\dfrac{1}{R_{B1}} + \dfrac{1}{R_{B2}} + \dfrac{1}{(\mathrm{h}_{ib} + R_E)(1 + h_{fe})}}$$

$$= \frac{1}{(0.122 + 0.0147 + 0.0192) \times 10^{-3}}$$

$$= \frac{1}{0.156 \times 10^{-3}} = 6.45 \text{ k}\Omega \qquad (11\text{-}16)$$

This value differs little from the 6.65 kΩ found for $h_{fe} = 100$, or the 7 kΩ found for $h_{fe} = 200$.

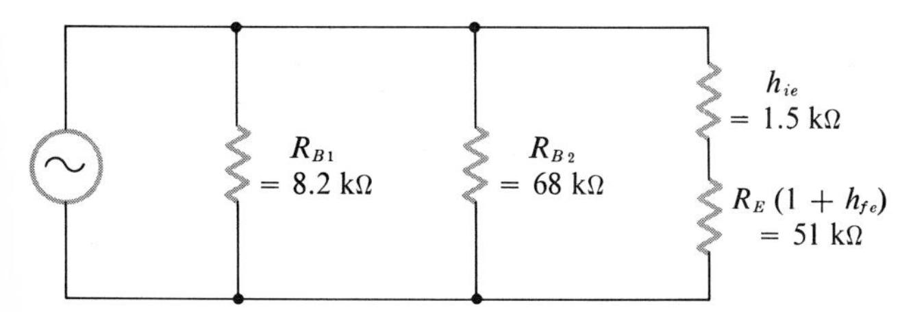

FIG. 11-10 The equivalent input circuit for the common-emitter amplifier with voltage-divider bias.

VOLTAGE GAIN

As would be expected, when the emitter resistance is left unbypassed the voltage gain is lower. In the amplifier circuit of Fig. 11-9 the gain is comparable to that of the simple fixed-bias common-emitter circuit with an emitter swamping resistor. The general formula for voltage gain yields

$$A_v = -A_i A_r \cong \frac{-R_L}{h_{ib} + R_E}$$

$$= \frac{-7.5\ \text{k}\Omega}{1.03\ \text{k}\Omega} = -7.4 \qquad (11\text{-}17)$$

It should be recalled from previous discussion that the current gain appeared in both the numerator and denominator in the derivation and was therefore cancelled. Thus the circuit is *essentially independent of the current gain.* In the lengthy mathematical solution of voltage-gain equations every possible resistance branch must be considered as drawing current from the signal source, leaving less signal to be amplified. However, the end result of such calculations will be within 10 percent of the value indicated above, and usually much closer. Recall, too, that *the voltage gain of the common-emitter circuit is negative, because the output signal is 180° out of phase with the input voltage signal.*

VOLTAGE-DIVIDER BIAS AND THE EMITTER BYPASS CAPACITOR

As with the fixed-bias circuits, bypassing the emitter resistor either fully or in part considerably improves the voltage gain of the circuit. At the same time the input impedance is reduced significantly. Figure 11-11 illustrates just such a situation. This circuit should be constructed in order to investigate the resistance and voltage-gain functions we have discussed. Remember that it is the unbypassed portion of the emitter resistor that enters into any of the calculations. The bypassed portion is, for all practical purposes, a short circuit to the signal current, but maintains its value for dc bias conditions.

After the circuit is constructed and the power supply is connected, measure all dc voltages and currents for the transistor used and record the values. Check the value of h_{FE} and record it with the data. Connect the signal generator and oscilloscope and inject a signal at 1 kHz without distortion. Adjust the input signal to a value that will

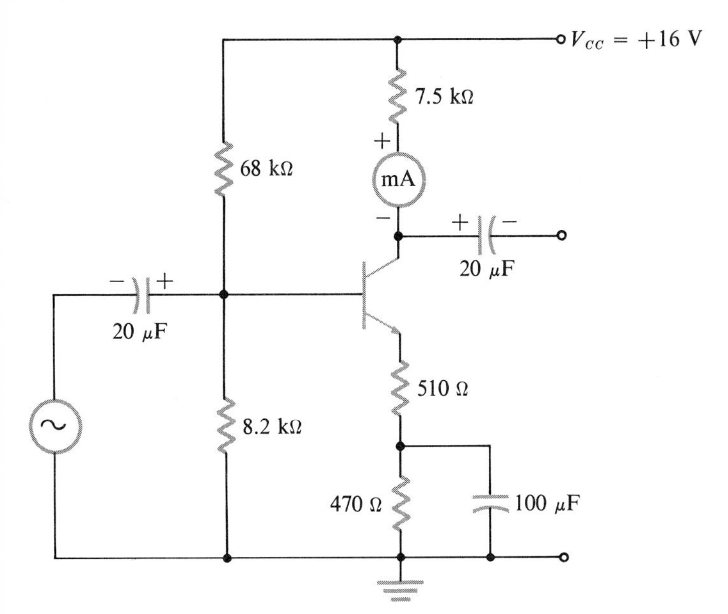

FIG. 11-11 The common-emitter amplifier with voltage-divider bias and partially bypassed emitter resistance for added voltage gain.

allow easy voltage-gain measurements. Record the input and output voltages, either as voltages or in terms of the number of spaces used by both signals with the same vertical-gain setting.

When this set of measurements has been completed, replace the transistor with one having some other h_{FE}. Make the same measurements for this transistor and compare the two sets of measurements.

If possible, repeat this experiment for several silicon and germanium *NPN* transistors with h_{FE} values ranging from 50 to 200. Remove the bypass capacitor and again repeat all the measurements; note that there is less variation in voltage gain when the bypass capacitor is removed.

Refer to the input-impedance experiment of Figs. 11-6 and 11-7. To verify the effect of voltage-divider bias on input resistance, repeat this experiment for the voltage-divider-biased circuit. Measure the circuit both with partial bypass and with the capacitor removed. Remember that the input resistance will always be less than the smallest parallel resistance, and so the potentiometer used in the experiment may be 20 kΩ, or even 10 kΩ, as long as it is larger than the actual input impedance.

SUMMARY OF CONCEPTS

1. When a supply voltage varies, the variation will be fed back through a series arrangement of resistances.
2. Input impedance to an amplifier is influenced by all the impedances that appear in parallel with it.
3. The input impedance to a common-emitter amplifier cannot be larger than any resistance appearing directly across the input from base to ground.
4. Standard values of resistance within 5 percent of the computed value may be used without downgrading the circuit action.
5. Degeneration, or negative feedback, may be used to control voltage gain and input impedance.
6. The self-biased circuit may be used to improve temperature stability and to reduce distortion by means of dc and ac feedback from the collector.
7. The Thévenin equivalent circuit is helpful in determining input impedance.
8. The current gain of a self-biased circuit is a function of the feedback resistance and the load resistance (R_f/R_L).
9. The common-emitter amplifier with voltage-divider bias operates independently of transistor current gain and may be used with a wide variety of transistors.
10. The voltage gain of a common-emitter amplifier is negative because the voltage signal at the collector or output is 180° out of phase with the output current and the input signal voltage.

GLOSSARY

β independent Generally a transistor circuit design where the circuit stability and voltage gain are not dependent on the current gain h_{FE} of the transistor.

Bleeder current A current that is allowed to flow in order to correct some situation or fault; e.g., the current through the voltage divider in a voltage-divider-biased amplifier whose function is to hold the base voltage to some value independent of the transistor.

Constant-current generator A simulated high-resistance current-generator circuit in which variation in the circuit imput impedance does not affect the amount of current delivered by the generator.

Self-bias A biasing scheme for a transistor circuit whereby a voltage feedback from the collector to the base automatically seeks its proper level.

Thermal runaway A situation in which an increase in temperature generates increased leakage current, which in turn causes further temperature rise, and hence more leakage current.

Voltage-divider bias A biasing scheme whereby the base voltage for the transistor is held constant by the voltage-divider action of two resistors connected in series across the power supply.

REFERENCES

Cutler, Phillip: *Semiconductor Circuit Analysis*, McGraw-Hill Book Company, New York, 1964, chap. 4.

Horowitz, Mannie: *Practical Design with Transistors*, The Bobbs-Merrill Company, Inc., Indianapolis, 1968, chaps. 3–4.

Malvino, Albert Paul: *Transistor Circuit Approximations*, McGraw-Hill Book Company, New York, 1968, chap. 10.

Seidman, A. H., and S. L. Marshall: *Semiconductor Fundamentals, Devices and Circuits*, John Wiley & Sons, Inc., New York, 1963, chap. 9.

Temes, Lloyd: *Electronic Circuits for Technicians*, McGraw-Hill Book Company, New York, 1970, chap. 8.

Veatch, Henry C.: *Transistor Circuit Action*, McGraw-Hill Book Company, New York, 1968, chap. 6.

REVIEW QUESTIONS

1. What is meant by self-bias in a common-emitter amplifier circuit?

2. In what way is the self-biased circuit different from the emitter-bias circuit?

3. In a self-biased circuit what is the effect on base current and base voltage when the collector-emitter voltage starts to drop?

4. Why is the resistor between the collector and the base in a self-biased circuit called a feedback resistance?

5. Why is the current through the load resistance of a self-biased circuit the same as the emitter current?

6. Approximately how large should the feedback resistance in a self-biased circuit be in relation to the collector-base junction resistance?

7. Why is it safe to say that the collector-base resistance is in general about the same as the collector-emitter resistance?

8. How can some of the effects of degeneration in a self-biased circuit be overcome?

9. Does the degeneration of the collector feedback of the self-biased circuit increase or decrease the input impedance? Why?

10. What is the effect on the input impedance of adding an emitter resistor to a self-biased circuit?

11. Describe an experiment to measure the input impedance to an emitter-bias self-biased circuit.

12. What are some of the advantages of voltage-divider-biased circuit designs?

13. What are some possible disadvantages of voltage-divider bias?

14. What circuit component limits the input impedance to a voltage-divider-biased common-emitter amplifier?

15. What is the function of the bleeder current in voltage-divider bias?

16. Why is a voltage-divider-biased circuit β independent?

17. What is the effect on voltage gain of bypassing the emitter resistance in a voltage-divider-biased circuit?

18. What is the effect on the input impedance of bypassing the emitter resistance of a voltage-divider circuit?

19. What would happen in a voltage-divider circuit designed for a silicon transistor if a germanium transistor were inserted instead?

20. In a voltage-divider-biased circuit designed for a voltage gain of 20 with $h_{fe} = 100$, what would the gain be if a transistor with $h_{fe} = 50$ were substituted?

PROBLEMS

1. In a circuit such as that of Fig. 11-1, if the supply voltage is 16 V and V_C is 10 V, how much is the resistance R_f if $h_{FE} = 100$ and $I_C = 1.0$ mA?

2. In Prob. 1, how large is R_L?

3. In Fig. 11-3, if $h_{ie} = 2.2$ kΩ, $R_f = 330$ kΩ, $R_{CB} = 4.5$ kΩ, and $R_L = 4.7$ kΩ, what is the approximate input impedance to the circuit?

4. What is the approximate voltage gain of a self-biased circuit if the current gain is 90, the load resistance is 5.6 kΩ, and the input resistance is 1.4 kΩ?

5. Since the feedback resistance of a self-biased circuit is important in determining current gain, what is the current gain if the load resistance is 6.8 kΩ and the feedback resistance is 470 kΩ?

6. Compute the values of R_E, R_L, and R_f for the circuit below.

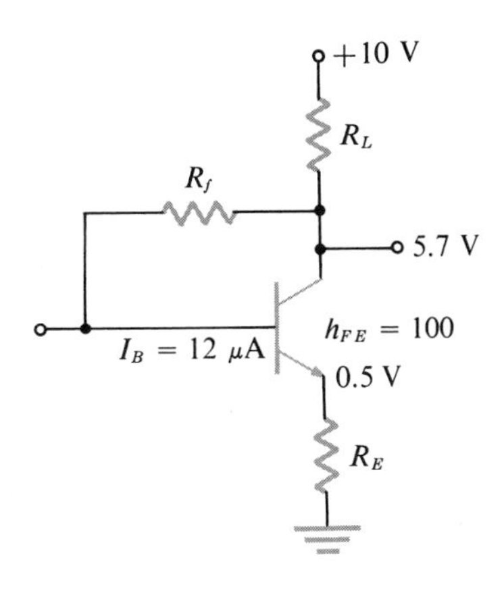

7. In Prob. 6, what is the current passing through R_L?

8. In Prob. 6, what is the approximate resistance of the collector-base junction?

9. Compute the values of R_E, R_L, R_1, and R_2 for the circuit below.

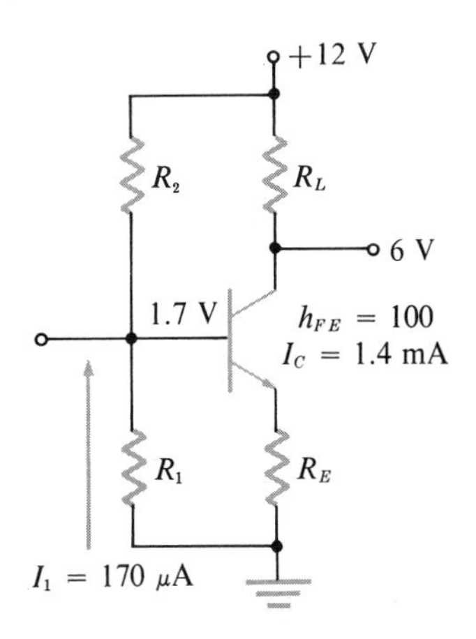

10. From the resistance values determined in Prob. 9, what is the approximate voltage gain of the circuit?

11. In the circuit of Prob. 9, what is the approximate input impedance?

12. In the circuit of Prob. 9, if the emitter resistance is bypassed, what will be the new input impedance?

13. In Fig. 11-11, if $h_{fe} = 75$ and the emitter current is 1 mA, what is the input impedance?

14. Compute the approximate voltage gain for the circuit of Fig. 11-11. (Hint: Do not forget to include h_{ib} as part of the emitter resistance.)

15. In Prob. 6, if the input signal is 1 V *p-p*, what voltage should be measured across R_E?

16. In the circuit of Prob. 9, if the voltage gain is 12 when the load resistance is 10 kΩ, what is the input signal voltage for an output voltage of 3 V *p-p*? What is the output signal current?

CHAPTER 12

OBJECTIVES

1. To discuss the structure and use of the common-collector and common-base amplifiers
2. To examine the significance of impedance matching between stages of amplification
3. To note the effect of emitter current on the input impedance of the common-collector amplifier
4. To investigate the possible circuit arrangements and uses of the common-collector and common-base amplifiers
5. To examine the effect of current gain on the input impedance of the common-base amplifier
6. To consider frequency response of various circuit arrangements
7. To discuss basic circuit-design concepts for the common-collector and common-base amplifiers

THE COMMON-COLLECTOR AND COMMON-BASE AMPLIFIERS

We have examined the common-emitter amplifier in depth because of its wide range of operation. It has good voltage gain, current gain, and power gain. However, in many cases it is necessary to have a very high input impedance in order to handle or to match the impedance of a voltage source. In other cases a low impedance, on the order of 25 to 50 Ω, is required. Some cases require that there be no phase inversion. In these instances the common-emitter circuit must give way to the common-collector or the common-base circuit arrangements.

THE COMMON-COLLECTOR AMPLIFIER

Since the transistor is a three-terminal device, any one of the three terminals may be used as the common point in the circuit. However, regardless of which terminal is used as the common, or ac, ground, the dc currents remain the same. Although there are several ways of drawing each circuit, the emitter current always divides into base and collector currents. Figure 12-1 shows two common circuit arrangements for the common-collector amplifier. Note that in each case the direct current enters the transistor at the emitter and leaves at the base and the collector.

As noted earlier, the common-collector amplifier is often termed an *emitter-follower* circuit, because the signal present at the emitter of the transistor follows, or is in phase with, the signal present at the base. For this reason the circuit is generally drawn so that it looks like a common-emitter amplifier, but with the signal leaving at the emitter rather than the collector terminal. Note that in this case the load resistor also appears in the emitter circuit rather than in the collector circuit.

DC BIASING

As in the common-emitter amplifier, the load resistance plays an important role. Its function is to provide a means of voltage division. Since the largest variation of signal voltage output is possible when the voltage across the load resistance is one-half the supply voltage, this is the Q point from which a design begins. In the case of the emitter-follower circuit the load current is the emitter current; it is this current that determines the actual load voltage when the circuit is in operation.

If the emitter current is selected as 1 mA, all that is needed to complete the design for dc biasing is the value of h_{FE} for the transistor to be used. From the known emitter current and h_{FE}, the base current may be found from the relationship

$$I_E = (1 + h_{FE})\, I_B \qquad (12\text{-}1)$$

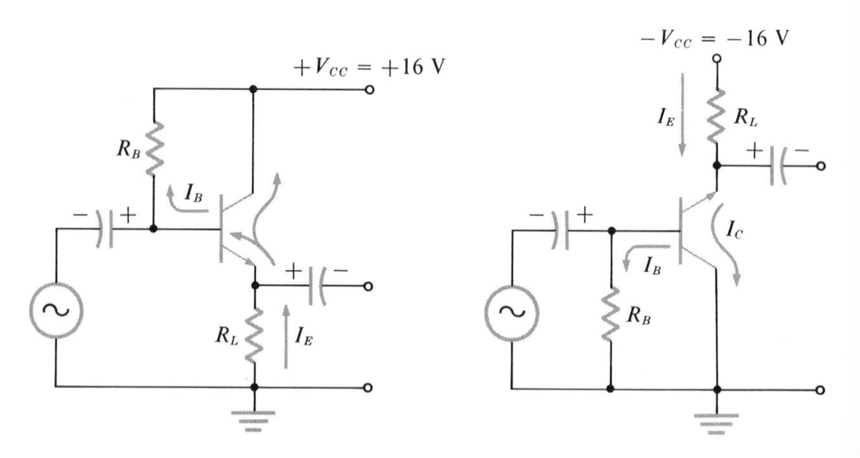

FIG. 12-1 Two versions of the common-collector (emitter-follower) amplifier circuit.

Simple rearrangement yields

$$I_B = \frac{I_E}{1 + h_{FE}} \qquad (12\text{-}2)$$

These equations result from Eqs. (7-12) and (7-13), which were derived in Chap. 7 and should be reviewed at this point.

In the circuit of Fig. 12-1, with an emitter current of 1 mA and $h_{FE} = 100$, the base current needed for the circuit must be

$$\frac{1\ \text{mA}}{1 + 100} = 9.9\ \mu\text{A}$$

The supply voltage is 16 V; therefore good design would place the emitter voltage, which is the load voltage, at $16/2 = 8$ V. The base-emitter voltage is about 0.62 V, and so the base voltage should be about

$$\begin{aligned} V_B &= V_E + V_{BE} \\ &= 8\ \text{V} + 0.62\ \text{V} = 8.62\ \text{V} \end{aligned} \qquad (12\text{-}3)$$

Hence the voltage across R_B is

$$\begin{aligned} V_{R_B} &= V_{CC} - V_B \\ &= 16\ \text{V} - 8.62\ \text{V} = 7.38\ \text{V} \end{aligned} \qquad (12\text{-}4)$$

Once the circuit voltages have been established, the resistances may be computed from Ohm's law:

$$\begin{aligned} R_L = R_E &= \frac{V_E}{I_E} \\ &= \frac{8\ \text{V}}{1\ \text{mA}} = 8\ \text{k}\Omega \end{aligned} \qquad (12\text{-}5)$$

$$R_B = \frac{V_{R_B}}{I_B}$$

$$= \frac{7.38 \text{ V}}{9.9\ \mu\text{A}} = 745 \text{ k}\Omega \qquad (12\text{-}6)$$

These values for R_L and R_B are based on theoretically available parameters. In reality, the nearest standard or preferred value of resistance would be used: 8.2 kΩ for the 8 kΩ in Eq. (12-5) and 750 kΩ for the 745 kΩ in Eq. (12-6). Even though a small difference is introduced into the operation of the circuit, the actual values of voltage and current that result will be very close to the predicted values.

When the dc bias for the amplifier has been established it may be put into operation as a signal amplifier. The emitter-follower circuit has certain limitations, as well as certain unique properties.

First of all, the output voltage from the emitter is less than the input signal voltage at the base. This means that the voltage gain A_V is less than 1. At first glance this would indicate that the circuit is of no value. However, the current gain is greater than h_{fe}; that is,

$$h_{fc} = 1 + h_{fe} \qquad (12\text{-}7)$$

This means that the power gain is significant, even though it is not as high as in the common-emitter circuit. Despite the fact that the voltage gain is less than 1, the tradeoff is for a greater value, since the properties gained are a *higher input impedance, a lower output impedance,* and *no phase reversal.*

INPUT IMPEDANCE

Let us first consider the increased input impedance. As in the common-emitter circuit, the current delivered to the base of the amplifier by the signal source is limited by the input impedance. The source sees the input impedance as an amplified version of the base-emitter resistance in series with the emitter load resistance. The forward-current-gain parameter h_{fc} is the equivalent of the familiar $1 + h_{fe}$. The parameter h_{fc} may be determined by dividing the emitter signal current by the base signal current. Since the base signal current is $1 + h_{fe}$ times smaller than the emitter current, it can be seen that the apparent impedance between the base and the bottom of the load resistance is $1 + h_{fe}$ times larger than the actual impedance. This must be so for the base signal current to be limited to its small value.

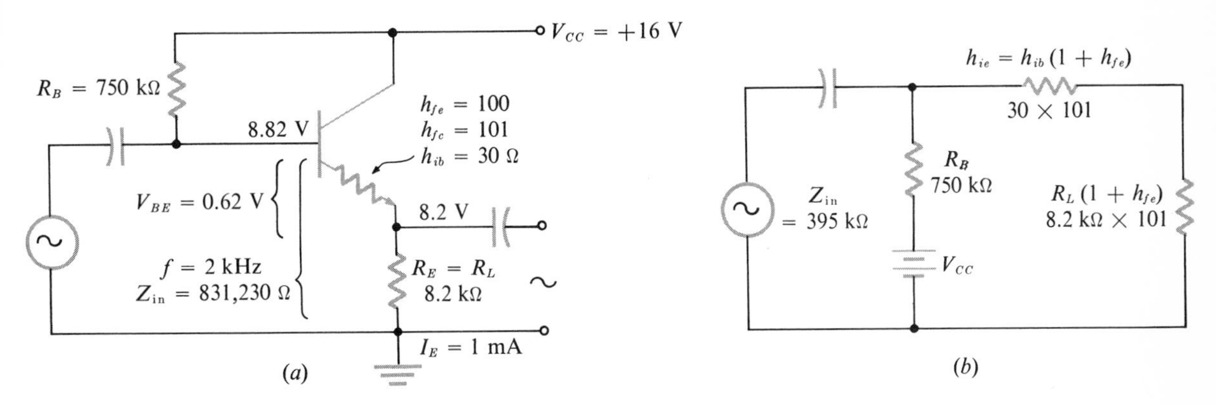

FIG. 12-2 (a) The emitter-follower circuit with factors affecting the determination of the circuit input impedance, and (b) the equivalent circuit for determining the input impedance for the emitter-follower amplifier.

Note in Fig. 12-2 that R_B actually appears in parallel with the input to the transistor and is therefore the resistance that limits the input impedance. Thus the input impedance is computed as

$$
\begin{aligned}
Z_{in} &= (h_{ib} + R_L)h_{fc} \parallel R_B \\
&= (30 + 8.2\ \text{k}\Omega)\ 101 \parallel 750\ \text{k}\Omega \\
&= 831{,}230 \times 750{,}000 = 395\ \text{k}\Omega \qquad (12\text{-}8)
\end{aligned}
$$

As would be expected from the results of an emitter-bias resistance, the input impedance is very high. In fact, it can be made even higher by using higher-gain transistors, larger load resistances, or higher voltages. If a larger load resistance is used, however, a lower emitter current will result, and other circuit values will also have to change. Of course, this is no problem in such a simple circuit.

The high input impedance available from the emitter follower makes it very useful as an impedance-matching circuit. There are, for example, many signal sources with high output impedances; that is, they can deliver only tiny signal currents. These signal sources must be loaded by very high impedances, or the signal source will be loaded down and may not function. The emitter-follower circuit has a voltage gain of less than 1, but it does fulfill a specific need.

VOLTAGE GAIN

It may not be apparent at first glance that the voltage gain of an emitter-follower circuit is less than 1. This face it evident, however, from a consideration of basic circuit concepts.

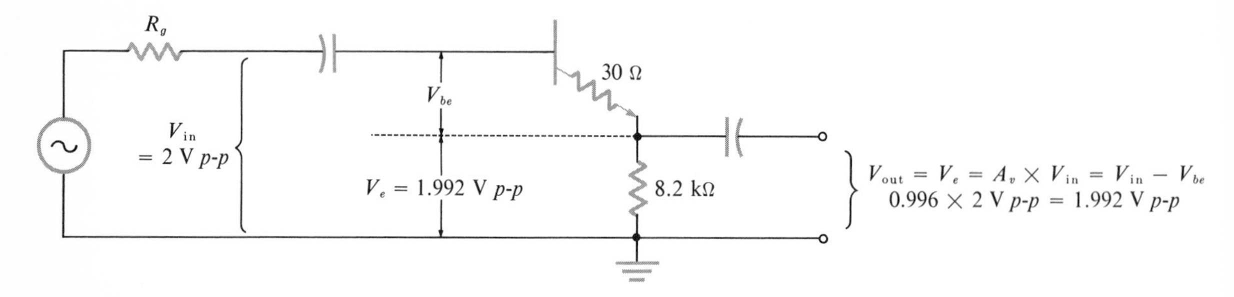

FIG. 12-3 The input and output signal indicating the degeneration and voltage relationships.

Recall from our discussion of degeneration in Chap. 10 that the input signal voltage is applied to the base, but that the transistor experiences only that part of the voltage that is impressed between the base and the emitter. The input voltage to the transistor is between the emitter and the base, but the input voltage to the circuit is between the base and ground. As a result, most of the input signal voltage is applied directly to the emitter resistor and is never experienced by the transistor. The input signal voltage, then, is degenerated by the amount found on the emitter resistor. If this value is subtracted from the input signal, the amount remaining is, for all practical purposes, the emitter-base voltage. There are other factors involved in the actual value of V_{be} when the voltage feedback ratio h_{re} is considered, but in most cases the amount of difference is insignificant.

In the circuit of Fig 12-3 note that because of the load resistance in the emitter circuit, the input signal has indeed been degenerated. Note also that the output signal voltage is taken from the same resistor that acted as *part* of the input impedance. If the same resistor is common to both input and output circuits, and degeneration has occurred, the output voltage must be about the same as the amount of degeneration. That is to say, the output signal voltage must be smaller than the input signal voltage.

From the voltage-divider concepts of Kirchhoff's law the actual voltage gain of the circuit can be determined. Remember that since the same current passes through each resistance of a series circuit, the voltage ratio must be the same as the ratio of the load resistance to the whole series circuit. In this case the load resistance is 8,200 Ω and h_{ib} is about 30 Ω. With these values we find

$$A_v = \frac{R_L}{R_L + h_{ib}}$$

$$= \frac{8{,}200}{8{,}200 + 30} = \frac{8{,}200}{8{,}230} = 0.996 \qquad (12\text{-}9)$$

For all practical purposes this is a voltage gain of 1, but the actual value is always less than 1. The larger the load resistance or the emitter current, the closer the voltage gain is to 1.

OUTPUT IMPEDANCE

One of the most interesting characteristics of the common-collector, or emitter-follower, circuit is its extremely low output impedance. The output impedance generally ranges between 20 and 600 Ω even when the load resistance is several thousand Ohms. An examination of Fig. 12-4 should help clarify this property. This is the equivalent circuit of Fig. 12-2*a*. Note that the load resistance as seen from the output capacitor appears to be in parallel with several other resistances. Remember too that the output signal current is emitter current, which is $1 + h_{fe}$ times larger than the base signal current. This is why the resistances on the input side of the circuit appear to be $1 + h_{fe}$ times smaller than their dc values—the counterpart of their larger appearance when viewed from the input side of the circuit.

Since the generator resistance and R_B are in parallel, their combined resistance will be smaller than the generator resistance, and when this value is divided by $1 + h_{fe}$ it is very small. In terms of the total output impedance, the input resistance values indicated are in parallel with the load resistance; therefore the output impedance must be a very low value. In fact, it depends largely on how small the generator resistance is. If the generator is an external source, its output impedance will be between 50 and 600 Ω in most cases. If it is a previous stage of common-emitter amplification, the output impedance may be several thousand Ohms, approximately equal to the load resistance. Even in this case, then, the output impedance of the emitter-follower circuit is quite low.

FIG. 12-4 (*a*) A simplified approximate equivalent circuit for Fig. 12-2*a*, used to determine the output impedance, and (*b*) a rearranged equivalent series-parallel form indicating the output impedance of the emitter follower.

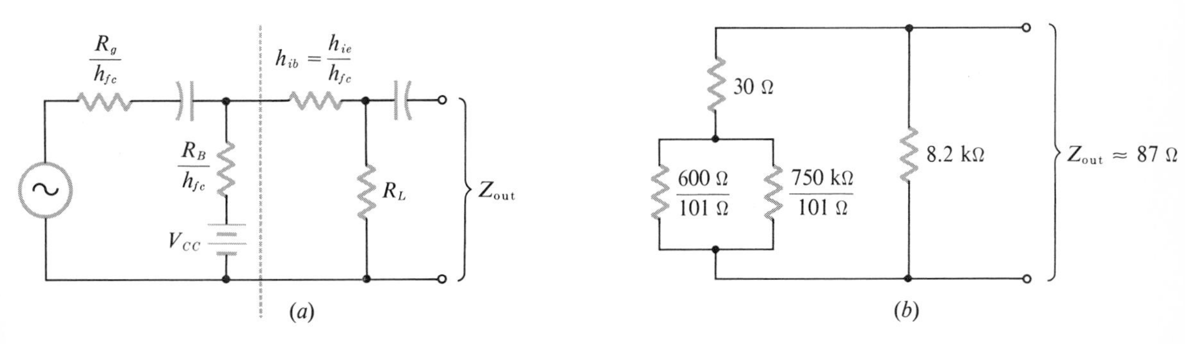

It is not difficult to compute output impedance if the parallel-resistance formulas and an equivalent circuit are used as guides. For example, from Fig. 12-4 or Fig. 12-5,

$$\begin{aligned} Z_{\text{out}} &\cong \left(\frac{R_G \parallel R_B}{1+h_{fe}} + h_{ib}\right) \parallel R_L \\ &= \left(\frac{600 \parallel 750\ \text{k}\Omega}{101} + 30\right) \parallel 8.2\ \text{k}\Omega \\ &\cong \left(\frac{600}{101} + 30\right) \parallel 8.2\ \text{k}\Omega \\ &= (5.94 + 30) \parallel 8.2\ \text{k}\Omega \cong 35.9\ \Omega \end{aligned} \tag{12-10}$$

The two parallel lines in the equations mean that the parallel-resistance formula must be used. In this case, however, it is obvious that 600 Ω in parallel with 750 kΩ is about 600 Ω. Since $h_{fe} = 100$, the effective resistance on the input side of the amplifier is less than 6 Ω. When this equivalent resistance is added to h_{ib}, the sum of 35.94 Ω is so small when it is placed in parallel with the 8.2-kΩ load resistance that the parallel combination results in an output impedance of less than 36 Ω.

As a second example, if the generator output impedance were 6,000 Ω instead of 600 Ω, the results would be

$$\begin{aligned} Z_{\text{out}} &\cong \frac{6{,}000}{101} + 30 \parallel 8.2\ \text{k}\Omega \cong 59.4 + 30 \parallel 8.2\ \text{k}\Omega \\ &\cong 89.4 \parallel 8.2\ \text{k}\Omega \cong 87\ \Omega \end{aligned} \tag{12-11}$$

EMITTER-FOLLOWER OPERATION

The circuit of Fig. 12-6 will serve to verify each of the points we have just discussed. The actual resistance values of the circuit will depend to a great extent on the h_{fe} value of the transistor used. In every case the values used for computation should be the measured values, and not the color-coded values. The input and output impedance of the circuit will also depend on the value of h_{fe} and the emitter current used.

Construct the circuit of Fig. 12-6. If a microammeter is available, place it in the base-current circuit to accurately determine the value of h_{fe}. Measure the value of the load resistance R_E as accurately as possible and record it. Connect the circuit to the 16-V supply and adjust the base resistance R_B as you measure V_E. Set R_B at a value that yields $V_E = 8.2$ V for $R_E = 8.2$ kΩ. If R_E is not 8.2 kΩ, adjust R_B until 1 mA of

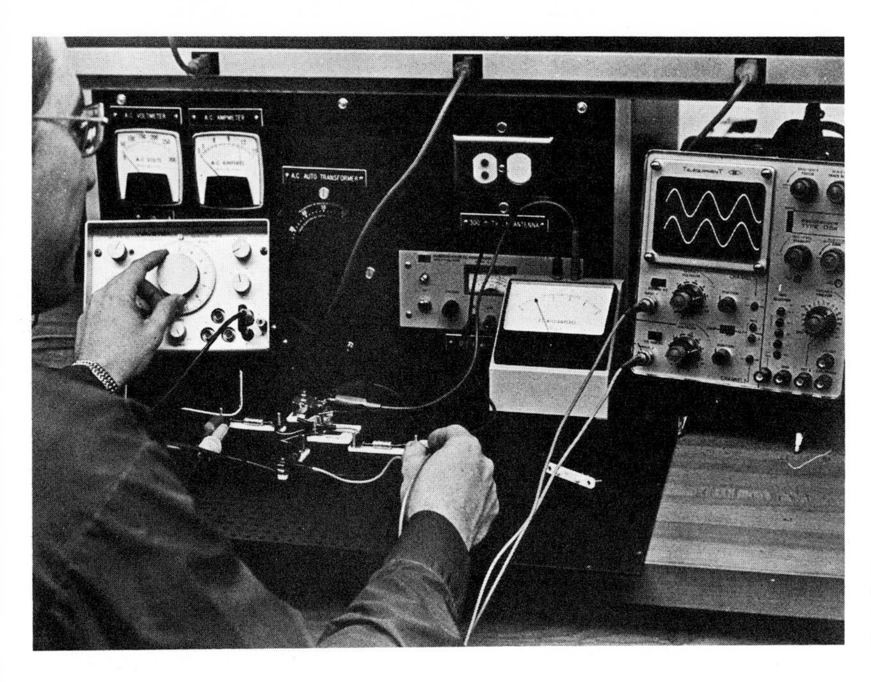

FIG. 12-5 Measurement of the input and output voltages for the emitter follower of Fig. 12-6 with a dual-trace scope.

emitter current is flowing. For example, if $R_E = 8.4\ \text{k}\Omega$, then 1 mA of emitter current will yield

$$V_E = I_E R_E = 1\ \text{mA} \times 8.4\ \text{k}\Omega = 8.4\ \text{V}$$

Record this voltage and current. Also record the base current that produced the 1 mA of emitter current, as well as the base voltage.

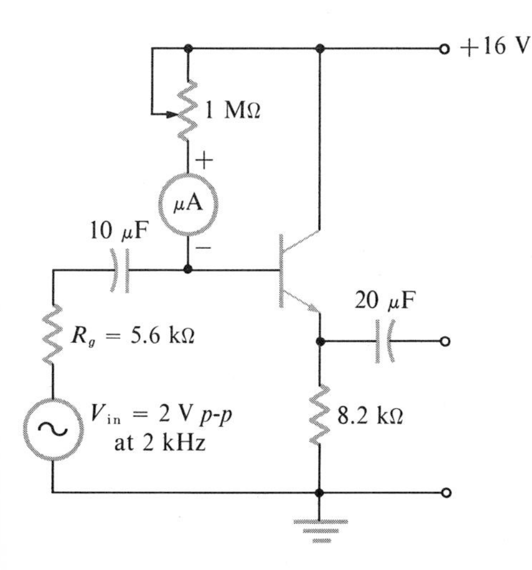

	Trial 1	Trial 2
R_B	725 kΩ	249 kΩ
R_E	8.2 kΩ	8.2 kΩ
I_B	9.9 μA	14.9 μA
V_B	8.82 V	12.92 V
I_E	1.0 mA	1.5 mA
V_E	8.2 V	12.3 V

FIG. 12-6 Experimental emitter-follower amplifier for determination of h_{FC}, h_{fe}, voltage gain, and input and output impedance.

Next adjust the base-current-limiting resistor R_B to increase the emitter current. Increase the base current by 5 μA and record the new values of base current, base voltage, and emitter voltage. Compute the new value of emitter current.

From the values of the base and the emitter current, the h_{fe} value of the transistor can be determined as

$$h_{fc} = \frac{\Delta I_E}{\Delta I_B} = 1 + h_{fe}$$

$$= \frac{1.5 \text{ mA} - 1.0 \text{ mA}}{14.9\ \mu\text{A} - 9.9\ \mu\text{A}} = \frac{0.5 \text{ mA}}{5.0\ \mu\text{A}} = 100 \qquad (12\text{-}12)$$

$$h_{fe} = h_{fc} - 1 = 100 - 1 = 99 \qquad (12\text{-}13)$$

This check is necessary because in some cases the value of h_{fe} may be very different from h_{FE}. Although most transistor checkers do give some value of h_{FE}, it is not usually constant over much of a current range. Since the variation in current due to the signal is close to the 1-mA dc level of emitter current, this short experiment should give a very close approximation of the true ac gain h_{fe}. As indicated in Eqs. (12-12) and (12-13), either h_{fc} or $1 + h_{fe}$ may be used in the equations for the common-collector circuit.

Readjust the base-resistance R_B back to the 1-mA emitter-current setting. Attach the signal generator to the circuit with the 5.6-kΩ resistance in series as shown. Some other resistance may be used as long as its value is known. The purpose of this resistance is to mask the output impedance of the signal generator and give a known value instead. If the output impedance of the generator is known, its value may be substituted into the equations later, and the generator may be connected directly to the amplifier *without* the 5.6-kΩ resistor.

Set the output frequency of the signal generator to around 2 kHz. The frequency is not critical, but the reactance of the input and output capacitors at 2 kHz is very low, and the impedance measurements will be more accurate. Adjust the output of the signal generator to give a signal at the input to the amplifier of 2.0 V *p-p*. If the series 5.6-kΩ loading resistor is used, measure the signal voltage on both sides. The difference is the voltage across the resistance. Ohm's law may then be used to compute the input signal current from the actual value of the resistance and the value of the voltage just measured.

Adjust the oscilloscope to give a full-scale reading for the 2-V *p-p* input signal and then connect it to the output. Measure the output signal voltage very accurately, record both input and output readings, and compute the gain.

MEASURING INPUT IMPEDANCE

The procedure for measuring input impedance was discussed in Chap. 11 and should be reviewed at this point. A Thévenin equivalent circuit will be used (see Fig. 12-7). Connect the 1-MΩ potentiometer into the circuit as a rheostat, but turn the wiper arm to the point of zero resistance. Reconnect the scope to point *A*, just before the rheostat, and again adjust the input signal to 2 V *p-p*. Adjust the scope so that the signal is either two or four spaces high. With the scope and the signal generator so adjusted, disconnect the scope and connect it to the amplifier input at point *B*. The waveform should be the same as before.

Next slowly increase the resistance of the rheostat until the waveform on the scope is only one-half as high as in the previous measurement. One-half the signal voltage is now across the rheostat and one-half is across the input resistance to the amplifier. This means that the resistance of the rheostat is equal to the input impedance to the amplifier and may be measured with an ohmmeter. With certain kinds of test equipment the signal in this part of the experiment may become very noisy and ragged. If this happens, acquire several measurements and take their average as the best value.

Use Eq. (12-8) to compute the input impedance. This requires that both R_L and R_B be measured very accurately. Before measuring, remove the power from the circuit, and if possible, remove the resistances from the circuit as well.

Compare the computed value of input impedance with the experimental value. Can you account for any discrepancies? With a crude

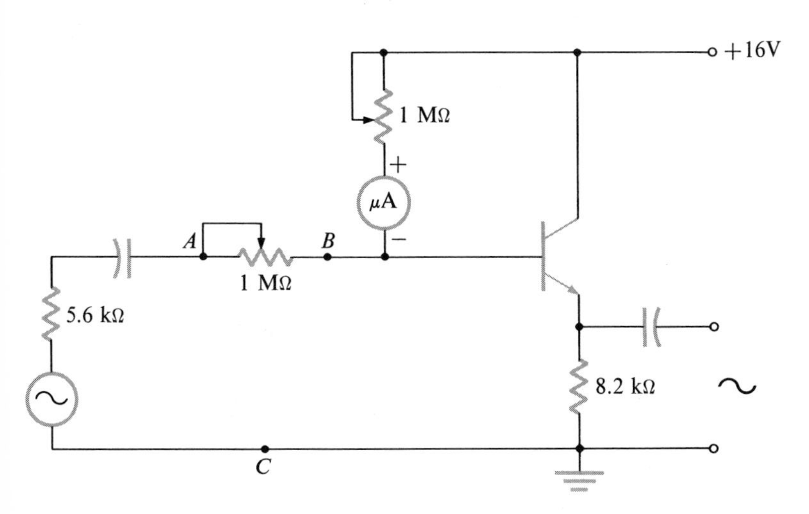

FIG. 12-7 Experimental emitter-follower circuit setup for measuring the input impedance. Resistance R_{AB} is to be adjusted until it equals Z_{in}.

experimental setup, the values should be within 10 percent of each other.

MEASURING OUTPUT IMPEDANCE

Measuring the output impedance of the common-collector amplifier is a little more difficult than it looks. The theory, based on Norton's theorem, is simple enough; it is the techniques that make the difference between success and failure.

Figure 12-8 illustrates the procedure to be used when measuring the output impedance. As indicated above, if the output impedance of the signal generator is not known, a 5.6-kΩ resistor will simulate a generator impedance of about 6,000 Ω. This approximation results from the series combination of the 5.6-kΩ resistor and the generator's actual output impedance, which is between 50 and 600 Ω.

Norton's theorem states that any two-terminal linear network may be replaced with a current generator in parallel with an impedance. In this experiment the common-collector circuit acts as a current generator, causing current to pass through the load impedance. If another impedance of exactly the same value is placed in parallel with the load impedance, half the current delivered by the circuit will pass through the normal load R_L and the other half will pass through the newly attached external load. Since only half the previous current is allowed to pass through the normal load under these conditions, the output voltage will be only half its previous value.

Connect the circuit of Fig. 12-8, but temporarily leave the external load rheostat R_{L2} disconnected. Connect the scope to the output capacitor and adjust the signal generator until a small 2-kHz signal can be measured easily. The signal must be very small, because as the external rheostat is adjusted, large signals will be clipped. Use the most sensitive range of the scope and adjust the input signal until the signal

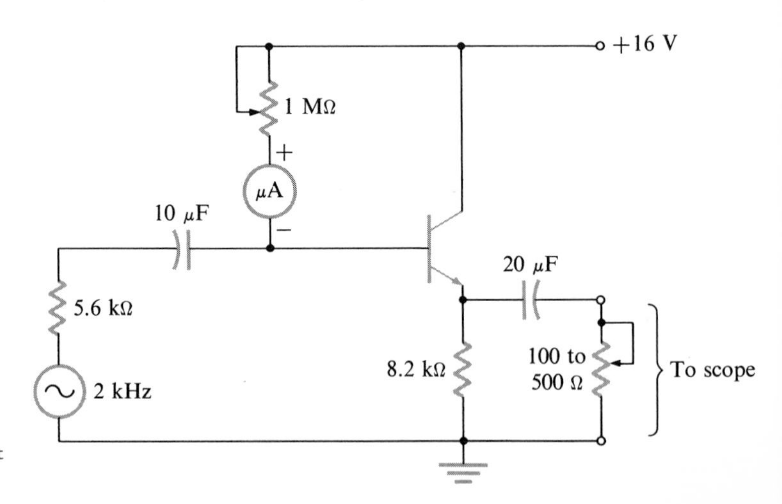

FIG. 12-8 An experiment to measure output impedance of the emitter-follower circuit.

is two spaces high. Now connect the rheostat R_{L2} to the output and slowly reduce the resistance until the signal voltage is only one space high on the scope. If severe distortion develops, reduce the size of the input signal and try again. When the output signal is only one-half its unloaded value, the rheostat resistance is equal to the output impedance of the circuit. Disconnect the rheostat from the circuit, measure its value, and record.

The computation for this output impedance is similar to that in Eqs. (12-10) and (12-11). Depending on the actual circuit values, the experimental value should be between 35 and 100 Ω. The most critical effect on the impedance will come from the h_{fe} value of the transistor used.

How did your experimental results compare with the computed results? Were you able to pinpoint possible errors that might be avoided in the future?

THE EMITTER FOLLOWER AS AN IMPEDANCE MATCHER

In most cases where a preamplifier is used before a signal is put into a power amplifier, as in most modern stereo amplifiers, an emitter-follower amplifier is used as the last stage. Its function is to reduce the output impedance of the preamplifier to a very low value, because the input impedance to the power amplifier is generally less than 2,000 Ω. The power amplifier's input impedance would severely load down the output from a common-emitter amplifier. For this reason the emitter follower is able to match the low input impedance of the power amplifier to the high impedance of the common-emitter amplifier.

As an example, consider Fig. 12-9. The output impedance of the common-emitter stage is 7.5 kΩ. This means that the input impedance to the next stage of amplification must be at least 10 times this value, or it will load down the amplifier. It is obvious from previous discussion

Fig. 12-9 A system diagram showing how the common-collector circuit is used to match the output impedance of the voltage amplifier to the low input impedance of the power amplifier.

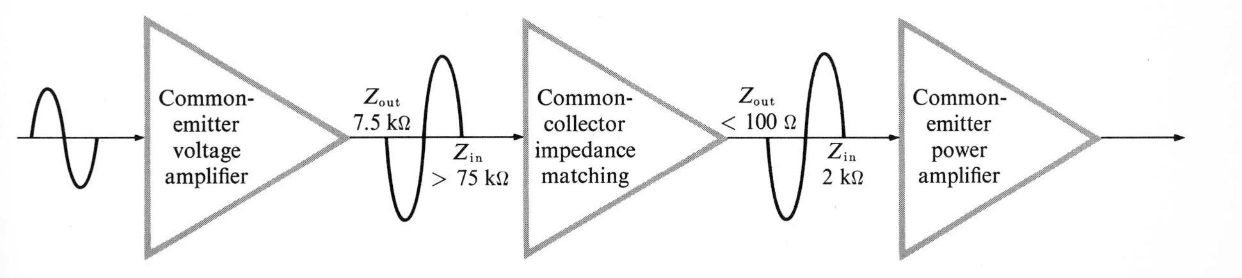

that the emitter follower can develop input impedances greater than 10 times 7.5 kΩ.

If the input to the power amplifier is 2,000 Ω, then the output impedance of the emitter follower must be less than one-tenth of 2,000 Ω. Since the output of the emitter follower just constructed is less than 100 Ω, it again passes the test. It is able to match the high impedance to the lower impedance.

In addition to being able to match impedances, the emitter-follower circuit provides a nice means of level control. If the load resistor is replaced by a potentiometer, the amount of output signal from the emitter follower can easily be varied. A resistor in series with the potentiometer and the input to the power amplifier prevents the base of the power amplifier from being connected directly to ground at low signal levels.

Another valuable property of the emitter-follower circuit is that it does not introduce a phase reversal. There are times, particularly in logic circuitry, where this is important. The reason no phase inversion is introduced is quite straightforward. An increase in base current is caused by an increase in base voltage. This is accompanied by an increase in emitter current, which causes an increase in the emitter voltage. Since the emitter signal voltage is the output signal voltage, and since it goes up as the input voltage goes up, the two signals are in phase.

Another obvious reason is that the input and output signal voltages are developed across the same resistor. High fidelity must then be present.

THE COMMON-BASE AMPLIFIER

The common-base amplifier circuit is perhaps the least understood transistor circuit arrangement, even though it is usually the first type of circuit discussed. Let us consider it now in light of what we know about circuit current flow. One reason that the common-base circuit appears to be difficult is that it is usually presented as a two-power-supply circuit. Of course, any of the circuit arrangements can use two power supplies, but they normally do not. Again it should be emphasized that regardless of which terminal is used as the ground or circuit common, the current still enters at the emitter and divides, with some going out at the base and the remainder going on to the collector.

DC BIASING

Figure 12-10 shows several methods of biasing the common-base amplifier. Note the current-flow directions in each case. Since the

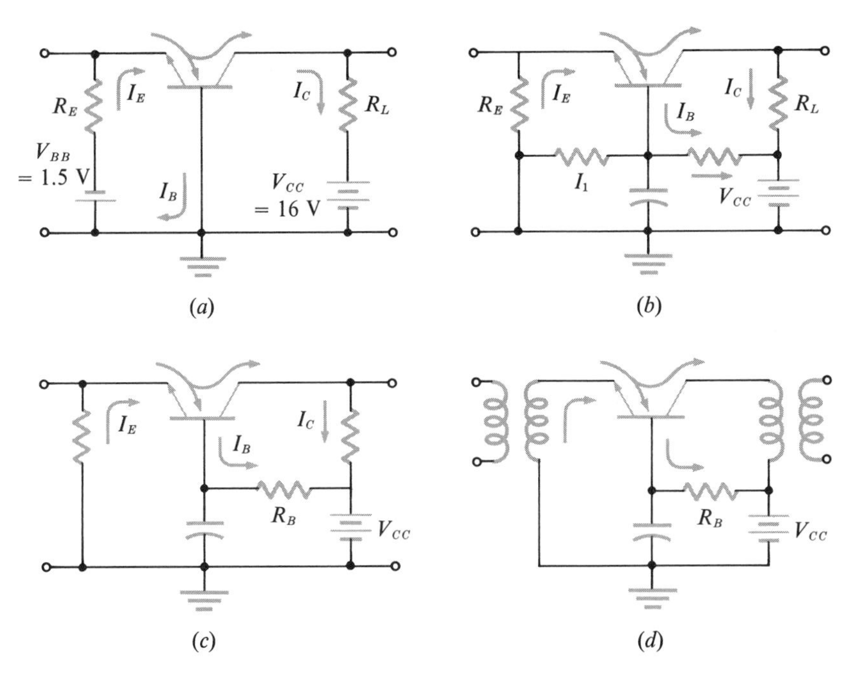

FIG. 12-10 (*a*) A two-supply common-base circuit; (*b*) a common-base circuit with voltage-divider bias; (*c*) a common-base circuit with emitter-bias stabilization; (*d*) an RF common-base amplifier.

circuit is used as an impedance-matching device and as a high-frequency amplifier, it deserves some close attention. Let us begin with the first circuit.

The emitter-base junction is forward biased by means of supply voltage V_{BB}. The collector-base junction is reverse biased by means of supply voltage V_{CC}. The emitter resistor R_E limits the amount of emitter current, and R_L both limits the collector current and acts as the load resistance. As we saw in Chap. 7, the function of the supply in the emitter-base circuit is simply to give the current carriers sufficient energy to get over the potential hill into the base region. From there the recombination that occurs in the base allows some base current, but most of the carriers are able to go on through the base into the collector, to be taken out of the transistor by the influence of V_{CC}.

The forward current gain of the common-base circuit is always less than 1 for a junction transistor, since not all the current carriers get through to the collector. That is,

$$I_E = I_C + I_B \qquad (12\text{-}14)$$
$$I_C = I_E - I_B$$

Therefore the forward current gain may be stated as

$$h_{FB} = \frac{I_{\text{out}}}{I_{\text{in}}} = \frac{I_C}{I_E} \tag{12-15}$$

The ac or signal current gain is derived in a similar manner as

$$h_{fb} = \frac{\Delta I_{\text{out}}}{\Delta I_{\text{in}}} = \frac{\Delta I_C}{\Delta I_E} = \frac{I_c}{I_e} \tag{12-16}$$

If, for example, the collector current in Fig. 12-10 is 1 mA and the base current is 15 μA, the emitter current will be 1.015 mA. From these data the current gain α, or h_{FB}, is computed as

$$h_{FB} = \frac{I_C}{I_E} = \frac{1.0 \text{ mA}}{1.015 \text{ mA}} = 0.985 \tag{12-17}$$

In order to relate the value of h_{FB} to the more familiar h_{FE}, recall from Eq. (7-11) that

$$h_{FB} = \frac{h_{FE}}{1 + h_{FE}}$$

That is, if h_{FE} is known, h_{FB} may be computed easily. If h_{FE} is not known, it may be computed from h_{FB},

$$h_{FE} = \frac{h_{FB}}{1 - h_{FB}}$$

$$= \frac{0.985}{1 - 0.985} = \frac{0.985}{0.015} = 65.6 \tag{12-18}$$

The actual numerical value of h_{FB} is deceptive in that a small deviation may produce a large variation in h_{FE}. For example, when $h_{FB} = 0.98$, then $h_{FE} = 49$; however, when $h_{FB} = 0.99$, then $h_{FE} = 99.0$.

In order to complete the circuit design for the simple common-base circuit, let us assume that our transistor has an h_{FB} value of 0.985, and that we have only one power supply, $V_{CC} = 16$ V. Since only about 0.62 V is needed to overcome the junction potential of the emitter-base junction, we may use a common 1.5-V D cell. Any other voltage would also work.

The function of the emitter resistor in Fig. 12-10*a* is to limit the emitter current to about 1.015 mA. In order to compute its value, we must determine the approximate voltage across it. Since the junction potential will be about 0.62 V, the difference between 1.5 and 0.62 V must be across the emitter resistor. Thus the voltage across R_E is

$$V_{R_E} = V_{BB} - V_{EB}$$
$$= 1.5 \text{ V} - 0.62 \text{ V} = 0.88 \text{ V} \qquad (12\text{-}19)$$

The value of R_E, computed from Ohm's law, is

$$R_E = \frac{V_{R_E}}{I_E}$$
$$= \frac{0.88 \text{ V}}{1.015 \text{ mA}} = 876\ \Omega \qquad (12\text{-}20)$$

The value of resistance needed to give just 1.015 mA of I_E is 876 Ω. This is not a standard or preferred value of resistance, and the only way to get it is to connect a potentiometer as a rheostat and set it to 876 Ω. Of course, there is nothing sacred about the 1.015-mA value for the emitter current. Whether it is a little larger or a little smaller is of no great consequence. For that matter, the normal D cell only puts out 1.5 V when it is new, and so the whole computed system would soon be off value just from deterioration of the supply voltage. In fact, if an 820-Ω resistance is used for R_E, the circuit will function quite well. However, all measurements should be recorded as they are, and not as they are supposed to be.

The value for R_L is selected in much the same way as for other circuit arrangements. One-half the supply voltage should appear across the load resistance. If V_{CC} is set at 16 V, the load voltage V_{R_L} should be about 8 V. A collector current of about 1 mA would require a load resistance of about 8 kΩ for $V_{R_L} = 8$ V. The nearest standard-value resistance, as previously indicated, is 8.2 kΩ.

VOLTAGE GAIN

Even though the current gain is less than 1, the voltage-gain possibilities for the common-base amplifier are as large as or larger than for the common-emitter circuit. The chief reason for the high voltage gain is the high resistance gain. Recall that the input impedance to the common-base circuit is very low, about 30 Ω at 1 mA of emitter current. However, the output impedance is quite high, since the collector-base junction is reverse biased. The voltage gain is the product of the current gain and the resistance gain. The current gain is approximately 1, and so the voltage gain is essentially equal to the resistance gain. For the circuit of Fig. 12-10*a* the input impedance is about 30 Ω and the output impedance is about equal to the load resistance, or 8.2 kΩ. The voltage gain should therefore be about

$$A_v = A_i A_z \cong 0.985 \times \frac{8{,}200}{30} \cong 270 \qquad (12\text{-}21)$$

Even though the voltage gain is very high, the circuit has such a low input impedance that it can be used only with very-low-impedance sources. This simply means that the signal source must be capable of delivering signal currents in the milliamp range rather than in the microamp range. Certain dynamic microphones have very low output impedances, ranging from 25 to 250 Ω. The common-base circuit can be used effectively to match the output impedance of the microphone to the higher input impedance of the common-emitter circuit. Note that it is a large *current* signal that is needed, and not a large voltage signal. In fact, only a few millivolts of signal may overdrive the circuit.

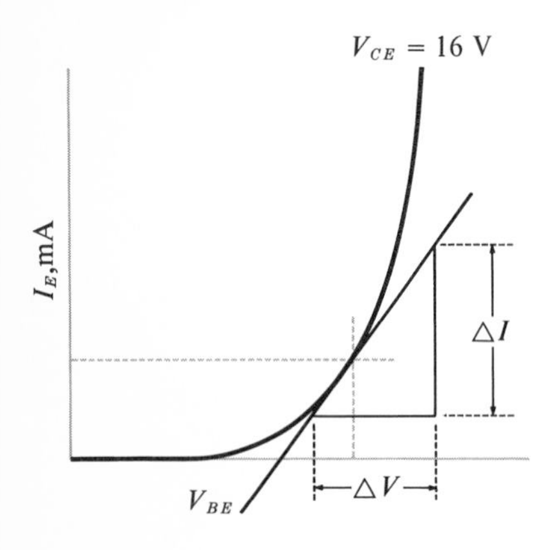

FIG. 12-11 Derivation of the transistor input impedance h_{ib}, where h_{ib} is the slope of the line that is tangent to the point of operation on the curve.

INPUT IMPEDANCE

We have made frequent reference to the hybrid parameter h_{ib}. Literally this is the *input* impedance to a transistor in a common-*base* circuit. Figure 12-11 indicates the source of h_{ib}. Mathematically it is the slope of the line tangent to the point of operation on the input characteristic curve, as is indicated. At an emitter current of 1 mA and a junction temperature of 25°C, the value of h_{ib} is approximately 30 Ω. Data sheets indicate that the actual value may vary from about 25 to 35 Ω.

The variation of h_{ib} with changes in emitter current is essentially linear. Since

$$V_t = \frac{kT}{q} = 26 \text{ mV}$$

where V_t = temperature voltage of the junction
k = Boltzmann's constant
T = absolute temperature, °K
q = charge on the electron

If this temperature voltage is divided by the emitter current, the answer is the input impedance,

$$h_{ib} = \frac{kT/q}{I_E} = \frac{26 \text{ mV}}{1 \text{ mA}} = 26\ \Omega \qquad (12\text{-}22)$$

As discussed earlier, the actual temperature voltage is closer to 30 mV than to the theoretical value of 26 mV.

If the emitter current is doubled to 2 mA, the new value of h_{ib} is about $30/2 = 15\ \Omega$. If the current is 0.5 mA, h_{ib} is about $30/0.5 = 60\ \Omega$.

The input impedance to the transistor is not the only impedance to be reckoned with. However, the other resistances appear in parallel with h_{ib}, and so h_{ib} becomes the limiting resistance. Figure 12-12 shows an equivalent circuit of the input resistances in Fig. 12-10*a*. Note that the 820-Ω resistance is in parallel with h_{ib}. The emitter resistance is so large in comparison that the input impedance is still 30 Ω.

OUTPUT IMPEDANCE

The output impedance for the common-base amplifier is determined by essentially the same process as for the common-emitter amplifier. The natural reverse-biased junction between the collector and the base in itself gives a high impedance. This parameter is referred to as the *output conductance* h_{ob} to the signal. Since conductance, measured in *microsiemens* (μS), is the reciprocal of resistance, we could take the reciprocal of h_{ob} and get a fair idea of the output impedance of the transistor. For example, suppose the output conductance is 0.5 μS at

FIG. 12-12 (*a*) A common-base amplifier showing the approximate source of h_{ib}, and (*b*) the equivalent circuit for approximating the input impedance shown.

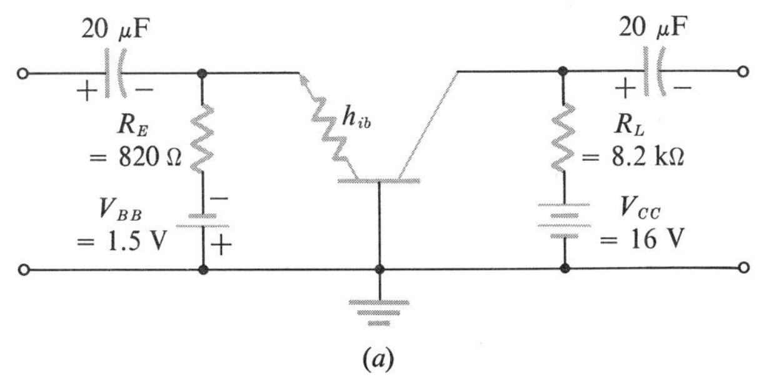

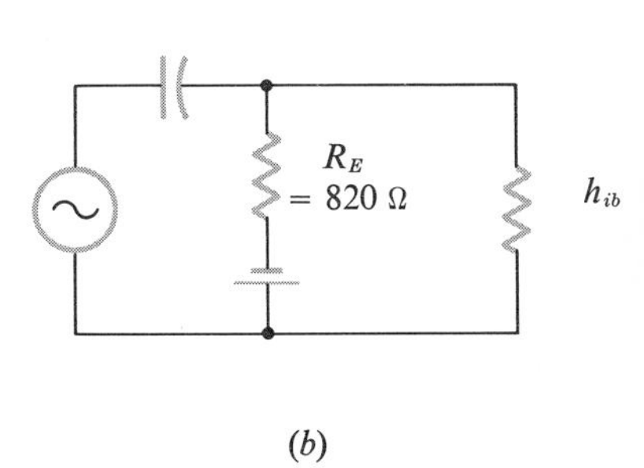

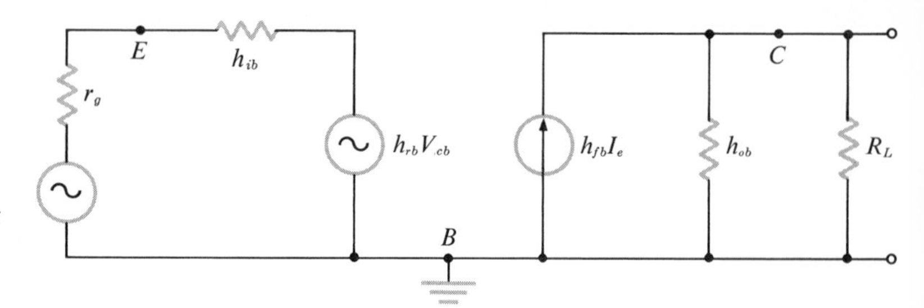

FIG. 12-13 An equivalent circuit for a common-base amplifier used to determine the circuit output impedance.

1 mA of emitter current at 25°C. This would correspond to an output resistance of 2 MΩ. The specific value is not so important as the resistance range. A value of 2 MΩ is extremely large in comparison to the size of the load resistance.

The output conductance is not the only factor involved in the output impedance in the common-base amplifier circuit. The signal-source impedance, the current gain, the reverse-voltage transfer ratio, and the load resistance are also considerations. For our purposes, however, only the output conductance and the load resistance are of consequence. Note in Fig. 12-13 that the Norton equivalent circuit shows the load resistance in parallel with the output conductance and a current generator. The current generator in this case is simply an indication that the signal-current output is the product of the current gain and the input signal current. It delivers this signal to both the output conductance and the load resistance. It should be apparent that the load resistance is extremely small in comparison to the reciprocal of the output conductance; hence the load resistance would get almost all the signal current.

We can conclude, then, that for a low output conductance h_{ob} the output impedance of the common-base amplifier circuit is essentially equal to the load resistance. This approximation holds as long as the emitter current is relatively low. As the emitter current increases, however, so does the output conductance.

VOLTAGE-DIVIDER BIAS

Frequently voltage-divider biasing—the so-called *universal circuit*—is used for a common-base amplifier. This arrangement has many advantages. Remember that in the common-emitter circuit of Fig. 11-9 the biasing resistors only set the proper voltages and limit the current. However, the circuit type depends on where the signal goes into the circuit and where it comes out. Thus, with the modification shown in Fig. 12-14, the circuit of Fig. 11-9 is converted to a common-base amplifier. The advantage of this circuit form over the common-base circuit discussed above is that it uses only one power supply.

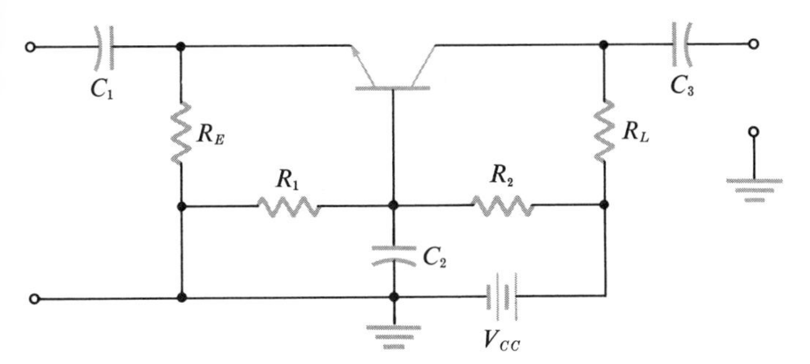

FIG. 12-14 The common-base amplifier with voltage-divider bias.

Voltage-divider bias, you will recall, ensures stability by preventing thermal runaway. It serves this function regardless of which of the transistor terminals is common to both the input and output signals. One disadvantage is that even when the transistor is not in the circuit current will pass through the series voltage divider, and this will cause a drain on the battery used to power the circuit. However, if an ac power supply is used, this circuit is superior in many ways to the simpler two-supply common-base circuit.

For all practical purposes, the input and output impedance analysis discussed for the previous common-base circuit also holds for the voltage-divider-biasing arrangement. Note, however, that a bypass capacitor (C_2 in Fig. 12-14) is needed, since the circuit ground is no longer in actual contact with the base. The ground is now at the end of the emitter-bias resistor, and so it is necessary to bypass R_1 from base to ground. With R_1 bypassed, the input impedance is essentially the same as for the previous case except at very low frequencies; that is, it depends on the size of the bypass capacitor.

COMPARISON OF COMMON-BASE AMPLIFIERS

Some experiments with one- and two-supply common-base circuits will help illustrate their similarities and differences. Construct the circuit of Fig. 12-15. Be certain that the two power supplies are connected with the proper polarity. It is pointless to try to measure both the emitter and the collector currents. If two meters are used, the error in one of them is likely to be at least 3 percent, and the difference between the emitter and collector current is less than 2 percent. If the measurements are made by shifting one meter back and forth between the emitter and collector circuits, the resistance of the meter itself must be accounted for in each measurement. Unless leakage is significant, the collector current can be determined much more easily, and with good accuracy, by measuring the emitter current and the base current and taking the difference between them as the collector current.

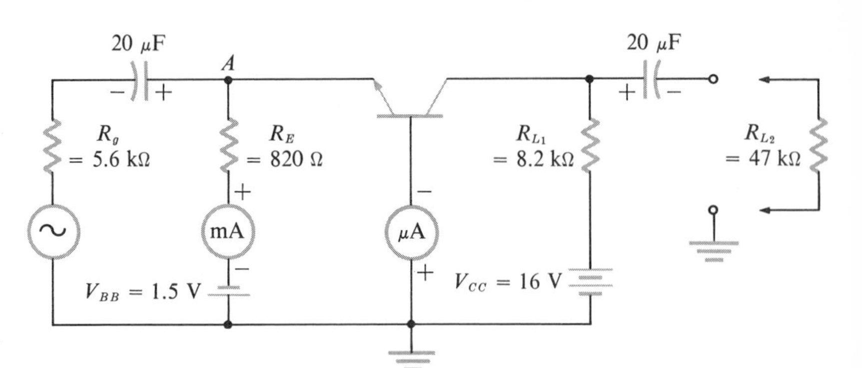

FIG. 12-15 Experimental common-base amplifier.

If the same transistor is used as for the common-collector experiment, h_{fe} should be known and the value of h_{fb} may be computed from Eq. (12-18). Record the values of emitter current, the collector current, and h_{fb}. Then, using the equation $h_{ib} = 30\ \text{mV}/I_E$, compute and record the value of h_{ib}.

After making certain that the resistance R_g is in the circuit, connect the signal generator. The value of R_g is not critical. Its purpose is to load down the signal generator so that it acts like a constant-current source. Otherwise the voltage signal may be rectified or distorted.

Connect the scope to the output of the amplifier and adjust the signal generator for an undistorted signal at about 2 kHz. Adjust the vertical gain of the scope and the output of the signal generator until the signal displays a full-scale reading on the scope. Measure it very accurately. Next move the scope to point A at the amplifier input and, without changing the setting of the signal generator or the scope, measure the input signal. It may be necessary to change the attenuation control on the scope to make this measurement accurately, since the voltage gain of the circuit is about 270. Compute the voltage gain and record its value.

After completing the voltage-gain measurements, open the circuit at point A. Insert a low-range potentiometer, 100 to 250 Ω maximum resistance, connected as a rheostat. Adjust its resistance to zero. After reconnecting the signal generator and oscilloscope, adjust the generator for a signal that is two spaces high on the scope. Without changing the scope settings, slowly increase the resistance of the rheostat until the signal display is only one space high. The resistance of the rheostat is now equal to the input impedance of the amplifier. Disconnect the rheostat and measure and record its resistance. Repeat the input-impedance measurement several times to check its accuracy (if you get different measurements, use the average value). Compare the measured value of input impedance with the value previously computed. How do you account for any differences?

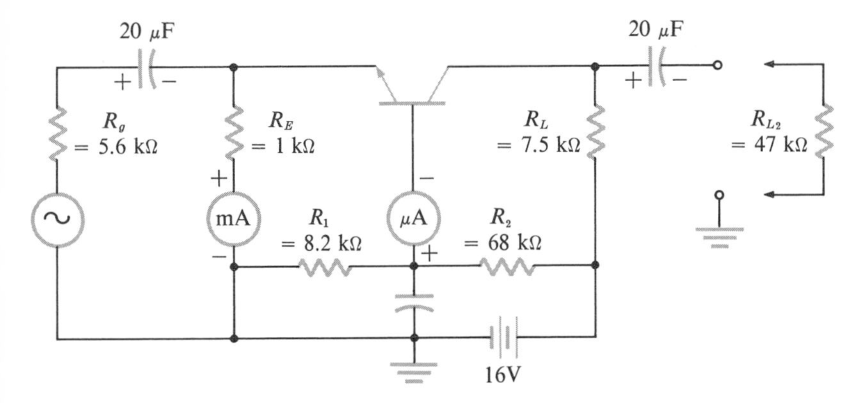

FIG. 12-16 Experimental common-base amplifier with voltage-divider bias.

To check for any output loading effects on the input impedance, connect a 47-kΩ resistance between the output capacitor and ground. This will simulate a second stage of amplification with an input impedance of 47 kΩ. Repeat the input-impedance measurement. Did the loading cause any changes? Change the simulated load from 47 to 33 kΩ and measure again. Does a greater amount of loading have a noticeable effect?

Repeat the above procedure for the common-base amplifier of Fig. 12-16 and compare the results. Are there any major differences?

Measure and record all dc voltages. Connect the signal generator and monitor the output signal on the oscilloscope. Bring a hot soldering iron or a hot light bulb close to the transistor and observe the output waveform as the transistor is heated. Are there any changes in the dc or ac measurements? If so, how do you account for them?

THE COMMON-BASE RADIO-FREQUENCY AMPLIFIER

A detailed study of the radio-frequency (RF) amplifier is not within our present scope, but very-high-frequency operation is one of the most important applications of the common-base circuit. Until now we have been concerned with audio frequencies, where the common-emitter amplifier has a decided advantage over other circuit arrangements. Certain limitations of the transistor make it less useful for high-frequency amplification in the common-emitter circuit than in the common-base circuit.

Recall that electrons do not travel through an *NPN* transistor in zero time. It takes some definite amount of time for any electron to cross the base region, for example. Moreover, not all electrons make the trip across the base in the same amount of time, since they follow different paths and have different energies. At signal frequencies that are in the hundreds of mega hertz range, this can cause difficulties. The input signal will be finished before most of the electrons get through the base region of the transistor. If some get through sooner than others, part of the signal will undergo a shift in phase and will therefore be cancelled. As the signal frequency increases, so does the phase shift and the resulting signal cancellation. Thus the *transit time* necessary for electrons to cross the base region places an upper frequency limit on transistors.

Figure 12-17 shows an equivalent circuit for a common-base amplifier circuit at very high frequencies. Note the junction capacitances. Recall that as the frequency of the signal current increases, the capacitive reactance decreases. At some high frequency the reactance of the tiny junction capacitance will drop to a value equal to the load resistance. At this point half the signal current will be lost through the shunting action of the capacitance. Of course, as the frequency increases further, so does the *loss* in output signal. Thus at very high frequencies the amount of signal output becomes insignificant.

When a common-base circuit is designed for RF amplification the combined effects of transit time and junction capacitance must be taken into consideration. These data are sometime given by the transistor manufacturer in terms of an *α cutoff frequency*. This is the frequency at which the forward current gain of the transistor has dropped to 0.707 times its midfrequency value. For example, if the transistor has an α, or h_{fb}, value of 0.98, at the cutoff frequency h_{fb} would be 0.693.

The common-base amplifier will operate at a much higher frequency and be subject to less instability and oscillation than the common-emitter amplifier because of its capacitance properties. The emitter-base junction capacitance is relatively high—hundreds of picofarads in many cases—because the junction is forward biased. This makes the depleted region very narrow, which means less distance between the capacitor plates and hence larger capacitance. In contrast, the collector-base junction is reverse biased, so that it has a wide depleted region and hence low capacitance. In older transistors the collector-base capacitance may be 50 pF or more, but some of the

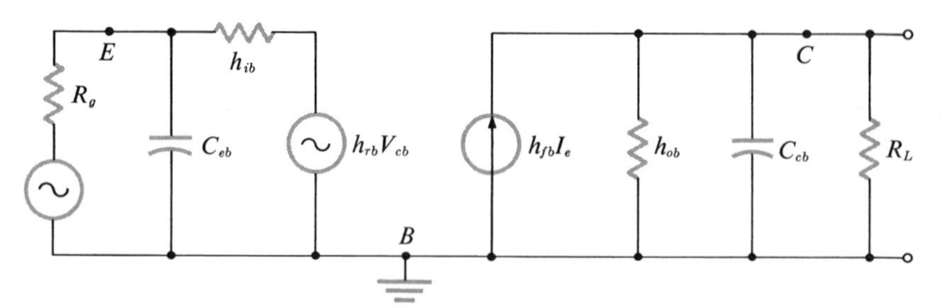

FIG. 12-17 The high-frequency equivalent circuit of a common-base amplifier.

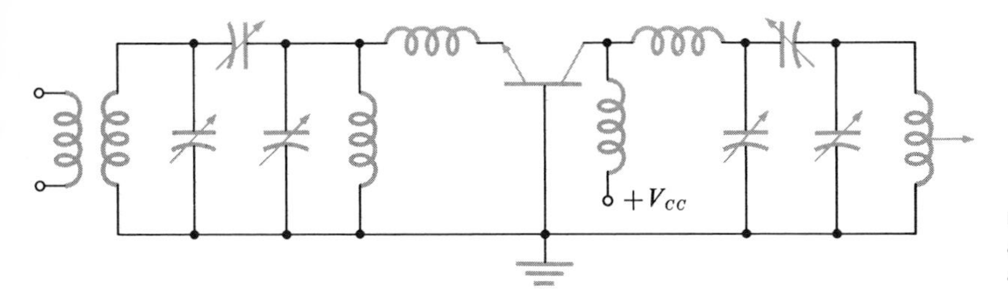

FIG. 12-18 A common-base circuit arrangement that may be used to amplify signals in the UHF range.

newer ones have values as low as a fraction of a picofarad. In general the lower the collector-base capacitance, the higher the possible frequency of operation.

For common-emitter circuits the output capacitance is the collector-emitter capacitance. This value may be 2 to 10 times higher than the collector-base value, depending on the method of transistor construction. For this reason the common-emitter circuit is limited in operating frequency.

Figure 12-18 shows one possible circuit arrangement for use at very high radio frequencies. Transformer coupling rather than capacitive coupling is used to match impedances. The transformers have only a few turns and are generally air-core structures. Circuits of this general type now operate at frequencies in excess of 1,000 MHz, whereas a few years ago the transistor was limited to an operating range of only a few kilohertz.

SUMMARY OF CONCEPTS

1. Direct current in a transistor is independent of the common-terminal arrangement of the circuit.
2. The current gain in the common-collector amplifier circuit is larger than in the common-emitter circuit.
3. The voltage gain of the common-collector circuit is always slightly less than 1.
4. The input impedance to the common-collector amplifier is much higher than for the common-emitter circuit.
5. The output impedance of the common-collector circuit is dependent on the signal-generator internal resistance, but it is generally less than 200 Ω.
6. The common-collector, or emitter-follower, circuit is often used for impedance matching—that is, to match a high-impedance source to a lower-impedance circuit to prevent source loading.
7. The common-base circuit amplifier generally has an input impedance less than 50 Ω.
8. The voltage gain of the common-base amplifier is about the same as the resistance gain.
9. The output impedance of the common-base amplifier is approximately equal to the load resistance as long as a small emitter current is used.
10. In both the common-base and the common-collector amplifier there is no phase reversal from input to output.
11. Voltage-divider biasing allows the use of one power source instead of two for the common-base circuit.
12. The current gain in the common-base amplifier is always less than 1.
13. If the value of h_{fe} is known, h_{fc} and h_{fb} may be computed.
14. Junction capacitance restricts the high-frequency response of transistor amplifiers.
15. The common-base amplifier is able to give significant voltage gains at higher frequencies than the common-emitter amplifier because of its lower output capacitance.
16. The value of h_{ib} is almost inversely proportional to the emitter current, with $h_{ib} = 30\ \Omega$ at 1 mA.
17. The amount of loading attached to a circuit has some effect on the input impedance.
18. The dc design procedure for the common-base and common-collector amplifiers is essentially the same as for the common-emitter amplifier.

GLOSSARY

α cutoff The upper frequency limit of transistor operation, where the h_{fb} value of the transistor has dropped to 0.707 times its midfrequency value.

Frequency response Gain as a function of frequency, usually plotted on semilog graph paper. The range of frequencies that fall between the low- and high-frequency half-power response points, that is, the −3 dB or 0.707 points on the graph.

h_{FC} The hybrid parameter defining the forward current gain in a common-collector amplifier ($h_{FC} = I_E/I_B = 1 + h_{FE}$).

h_{ob} *output conductance* The output conductance of a transistor in a common-base amplifier, a ratio of the change in collector current to the change in collector-base voltage for a constant value of emitter current ($h_{ob} = \Delta I_c/\Delta V_{\mathrm{CB}}|I_{\mathrm{E}}$), expressed in microsiemens.

Impedance matching The output impedance of a circuit must match the input impedance of the circuit it is driving for maximum transfer of power from one circuit to the next.

Output impedance The dynamic resistance of a transistor, equal to the summation of all resistances that appear in parallel with the load resistance. It is approximately equal to the reciprocal of the output conductance.

Transit time The time it takes a carrier to pass through the base region of a transistor, the limiting factor in high-frequency response for the device.

REFERENCES

Brazee, James G.: *Semiconductor and Tube Electronics: An Introduction,* Holt, Rinehart and Winston, Inc., New York, 1968, chaps. 3, 5.

Corning, John J.: *Transistor Circuit Analysis and Design*, Prentice-Hall, Inc., Englewood Cliffs, N.J., 1965, chap. 7.

Horowitz, Mannie: *Practical Design with Transistors*, The Bobbs-Merrill Company, Inc., Indianapolis, 1968, chap. 5.

Leach, Donald P.: *Transistor Circuit Measurements*, McGraw-Hill Book Company, New York, 1968, experiments 12, 14.

Lenert, Louis H.: *Semiconductor Physics, Devices, and Circuits*, Charles E. Merrill Books, Inc., Columbus, Ohio, 1968, pp. 181–250.

Malvino, Albert Paul: *Transistor Circuit Approximations*, McGraw-Hill Book Company, New York, 1968, chaps. 5, 7.

Veatch, Henry C.: *Transistor Circuit Action*, McGraw-Hill Book Company, New York, 1968, chaps. 7–8.

REVIEW QUESTIONS

1. What are the distinguishing characteristics of the common-collector, or emitter-follower, circuit?

2. Why is the input impedance to the common-collector amplifier so high?

3. Why is the voltage gain of the common-collector amplifier less than 1?

4. How do you account for the lack of phase shift in the signal transfer through an emitter follower?

5. Why is the voltage-divider variation of the common-collector circuit sometimes called a universal circuit?

6. What factors control input impedance to a common-collector amplifier?

7. Why is the output impedance of the common-collector amplifier generally less than 100 Ω?

8. What circuit-design technique is used to allow the voltage gain of the common-collector circuit to approach 1?

9. In what way is the generator resistance related to the output impedance of the common-collector amplifier?

10. What is the designation for forward current gain in a common-collector amplifier?

11. Describe how the input impedance of an amplifier may be determined experimentally.

12. Outline the method used to determine the output impedance of a common-collector amplifier.

13. Why should very small signals be used in measuring the output impedance of a circuit?

14. What are some of the advantages of a common-collector circuit?

15. In what way is the common-base amplifier different from the common-emitter circuit?

16. When a common-base amplifier is biased by means of one power supply, like the common-emitter circuit, what is considered as the input signal current? The output signal current?

17. Why is the voltage gain of the common-base circuit large when the current gain is less than 1?

18. Why is the common-base current gain h_{FB} always less than 1?

19. Why can the common-base amplifier generally operate at a higher frequency than a common-emitter circuit?

20. Why is the resistance gain of a common-base amplifier considered large?

21. What circuit element determines the approximate output impedance of the common-base amplifier?

22. Why is the input impedance of a common-base amplifier operating with an emitter current of 1.0 mA about 30 Ω?

23. What is meant by h_{ob}? What is its relationship to the circuit output impedance?

24. Why is it difficult to measure both the emitter and the collector current accurately?

25. What are some advantages of voltage-divider bias for the common-base amplifier? What are some disadvantages?

26. In what way does electron transit time become a factor in high-frequency response?

27. Discuss the effects of junction capacitance on the high-frequency response of an amplifier.

28. What is meant by the α cutoff frequency?

29. Why is the common-base amplifier circuit preferable to the common-emitter circuit for very-high-frequency applications?

PROBLEMS

1. If the emitter current of an amplifier is 2.0 mA, the base current is 20 μA, and the collector current is 1.98 mA, how much is h_{FE}? How much is h_{FC}?

2. Draw an emitter-follower amplifier stage in which the transistor has an input impedance $h_{ie} = 2$ kΩ at $I_E = 1.0$ mA. Determine the h_{fc} value of the transistor and compute the input impedance if $R_E = 10$ kΩ. (Hint: $h_{ib} = 30$ Ω at $I_E = 1.0$ mA.)

3. If an emitter follower has a load resistance $R_E = 10$ kΩ and $h_{ib} = 60$ Ω, what is the actual output voltage for an input voltage of 1.5 V *p-p*? What is the voltage gain of the circuit?

4. Compute the output impedance of the circuit of Fig. 12-2*a* for a generator output resistance of 1,200 Ω.

5. In Fig. 12-2 what is the input impedance if R_B is 330 kΩ?

6. Design an emitter-follower circuit to operate on 15 V at $I_E = 1.0$ mA, using a transistor with $h_{FC} = 100$ and a signal generator that has an output impedance of 600 Ω. Compute the input and output impedances of the circuit.

7. In a common-base amplifier circuit with voltage-divider bias, a supply voltage of 20 V, and a collector current of 1.0 mA, if the collector voltage is 11.8 V, how large is the load resistance? Approximately how large is the circuit output impedance?

8. In a circuit like that of Fig. 12-12*a*, if the load resistance is 15 kΩ, the emitter resistance is 1.5 kΩ, and $I_E = 0.5$ mA, what are the approximate input and output impedances?

9. In the circuit of Prob. 8, what is the approximate voltage gain?

10. How much is the approximate output impedance of a transistor whose output conductance h_{ob} is 0.25 μS at $I_E = 1.0$ mA?

11. If the effective output impedance of a transistor is 80 kΩ and the load resistance in a common-base circuit is 10 kΩ, what is the approximate total output impedance?

12. If $h_{fb} = 0.99$, what is the gain at the α cutoff frequency?

CHAPTER 13

OBJECTIVES

1. To compare the conventional bipolar transistor and the FET
2. To set up basic FET circuit design and analysis criteria
3. To consider the limitations and capabilities of the FET
4. To develop and test basic FET circuits

THE FIELD-EFFECT TRANSISTOR

To the person who has a background in vacuum tubes, transistors have been a foreign world. However, the field-effect transistor (FET), whose basic design is not really new, actually behaves much like a triode tube with pentode characteristics. Its development in the 1960s started slowly with a few companies such as Amelco Semiconductor and Siliconix. However, at this point, as a result of FET design processes combined with integrated-circuitry techniques, thousands of metal oxide silicon field-effect transistors (MOS-FETs) can be produced on a single chip for computer memories.

As with other devices, several varieties of the FET are now in production. The *N*-channel and *P*-channel junction FET, the depletion-mode MOS-FET, and the enhancement-mode MOS-FET are currently the most important.

THE JUNCTION FET

The basic concepts underlying junction phenomena were discussed in Chap. 5. Recall that whenever an *NP* junction is formed the region of this junction

is depleted region depends on the amount of reverse bias across the junction. The higher the voltage bias, the wider the depleted region. The FET makes use of this depleted region in an interesting way.

Current movement in semiconductor material depends on the resistivity of the material. The higher the resistivity, the lower the current—but the larger the cross-sectional area, the more current will pass through.

With these concepts in mind, look at the FET structure in Fig. 13-1, which shows two layers of *P*-type material diffused into a basic chip of *N*-type silicon, one on each side. (This process is no longer used, but it serves nicely to explain FET operation.) When a voltage supply is connected between the two ends of the chip a current moves down the channel between the two *P*-type layers. Since the channel is made from *N*-type material, the FET so constructed is referred to as an *N-channel FET*.

PINCHOFF VOLTAGE

Because of the resistance of a channel of *N*-type silicon, as current passes the length of the channel a voltage gradient is set up. That is, at one end the voltage is zero and at the other end it is equal to the supply voltage. From this bit of dc theory we can conclude that for the FET in Fig. 13-1 the voltage midway between the two ends of the channel must be one-half the supply voltage.

If we now connect the two *P*-type layers to the ground connection, the negative terminal, each *P*-type layer is at 0 V with respect to ground. As current proceeds through the channel from ground to the current drain end, the voltage gradient is such that an increasing reverse bias builds up between the *P*-type layer and the channel across

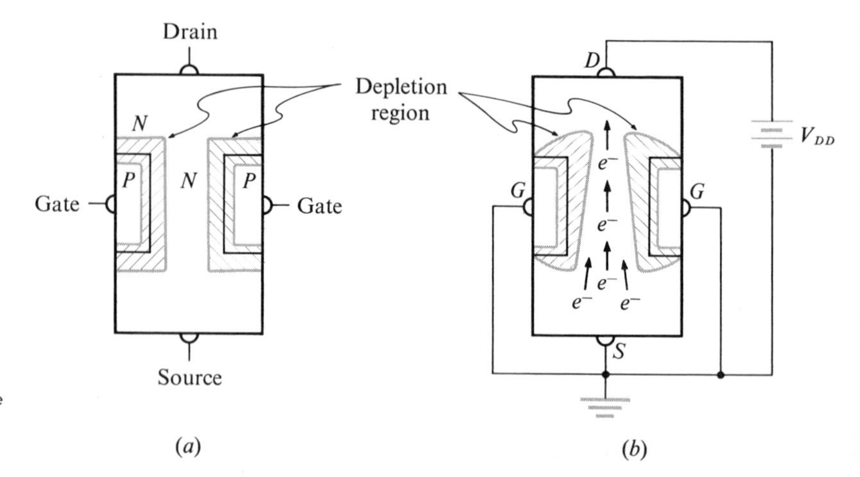

FIG. 13-1 (*a*) The field-effect structure without applied bias and (*b*) the FET with a gate-source voltage of 0 V and a supply voltage properly connected between the source and drain.

the junctions. As would be expected, the depleted regions are wider at the more positive end owing to the higher reverse-bias voltage.

Because the reverse-biased depleted regions extend so deeply into the current channel as to almost close it off, the *P*-type layers are referred to as *gates*. The gates are normally connected together and operate to gate, or pinch off, the channel.

As would also be expected, as the supply voltage is increased, the current increases, and so does the width of the depleted regions. The velocity of the electrons in the channel also increases until the channel is virtually saturated with rushing electrons. That is to say, beyond a certain supply voltage further increase in voltage will not result in further increase in the amount of current flow in the channel. Let us see why.

The depleted region can expand only so far before the channel becomes so constricted that only high-velocity electrons can push through; it has been pinched off. In fact, the channel is not actually closed, but the electrons have reached their approximate final velocity and cannot accelerate any further. That is, the rate of *change* in their velocity is zero. Their velocity, of course, is not zero. This pinchoff point, then, is the point at which the *rate of change* in current approaches zero, not the point at which the current is zero and ceases to flow.

The point of the FET where current enters the channel is called the *source* and the point where the current leaves is called the *drain*. The voltage between the drain and the source is called *drain-source voltage* V_{DS}. Figure 13-2 shows the schematic symbol of this circuit as well as the pictorial representation and a graph of the current change as a function of drain-source voltage. Note the point at which the rate of current change approaches zero. This is the *pinchoff point*, and the current at this point is denoted as I_{DSS}.

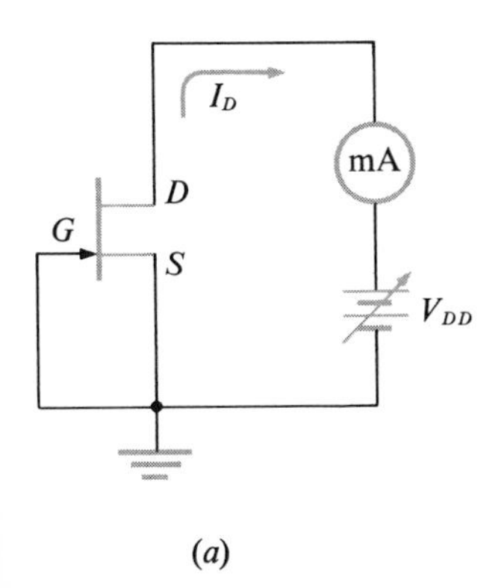

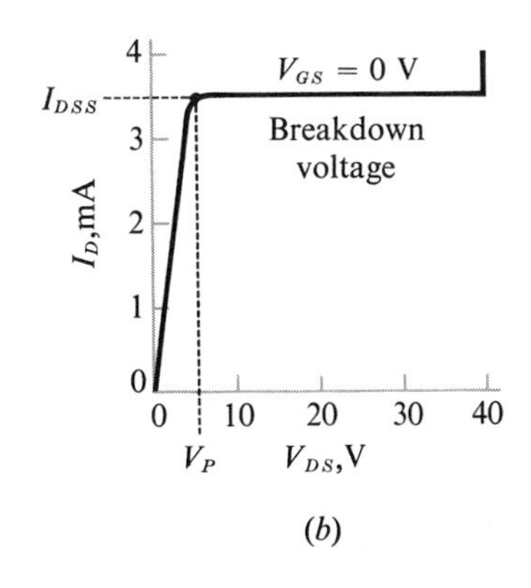

FIG. 13-2 (*a*) The schematic symbol of an *N*-channel FET in a circuit to measure V_P and I_{DSS} and (*b*) the drain characteristic curve showing I_{DSS} and V_P.

When the drain-source voltage V_{DS} increases beyond the pinchoff voltage V_P, the drain current I_D increases only slightly beyond I_{DSS}. If V_{DS} is increased sufficiently, the crystalline silicon channel begins to avalanche, and the current rises very sharply. Operation in this region for even a short period of time will destroy the FET.

The FET is generally operated in the constant-current region between V_P and the avalanche, or breakdown, voltage. Remember that this pinchoff voltage is *that drain-source voltage at which the rate of change of drain current approaches zero*, but with *the gates shorted to ground*; that is,

$$V_P = V_{DS} \qquad \text{at} \qquad I_D = I_{DSS},\ \Delta I_D \to 0 \tag{13-1}$$

GATE-SOURCE VOLTAGE

When the connection shorting the gate to the source of the FET is removed and the gate is biased negatively with respect to the source, another parameter of the FET comes into view. If the circuit of Fig. 13-3 uses a supply voltage greater than V_P, the drain current will be I_{DSS}, or maximum, when the gate-source voltage V_{GS} is zero, just as before. Now if V_{GS} is made gradually more negative, the drain current will begin to decrease. This is just as expected, since the increase in reverse bias on the gate pinches off the channel still further.

If the reverse bias is allowed to increase to an amount equal in magnitude to the pinchoff voltage, the gate-source voltage will eventually equal the drain-source voltage in magnitude, but will be opposite in polarity. The depleted regions should meet at the center of the channel, so that no current can pass; the drain current will have been truly pinched off. That is, when $V_{GS} = -V_P$, then $I_D = 0$. Although it is easier to determine the magnitude of the pinchoff voltage V_P in this manner, Eq. (13-1) is still valid and is also the accepted definition.

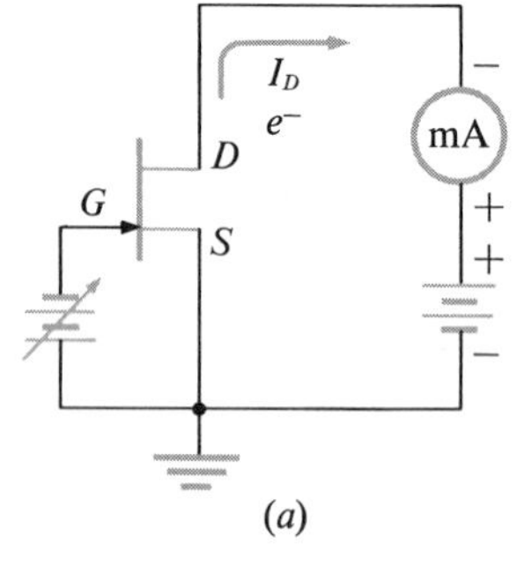

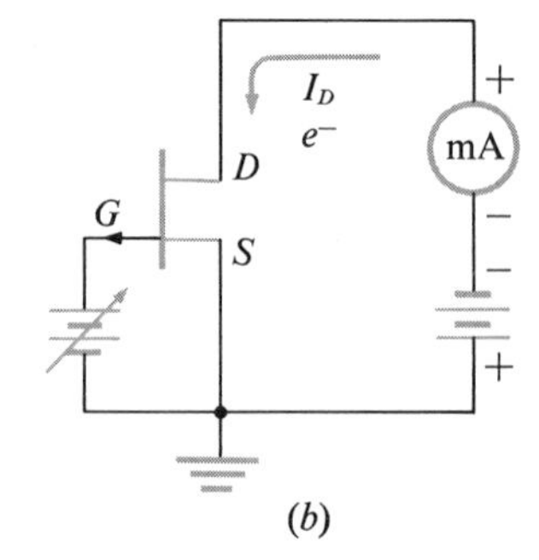

FIG. 13-3 (*a*) An *N*-channel FET circuit showing the normal gate biasing and (*b*) a *P*-channel circuit showing normal supply polarity and gate biasing.

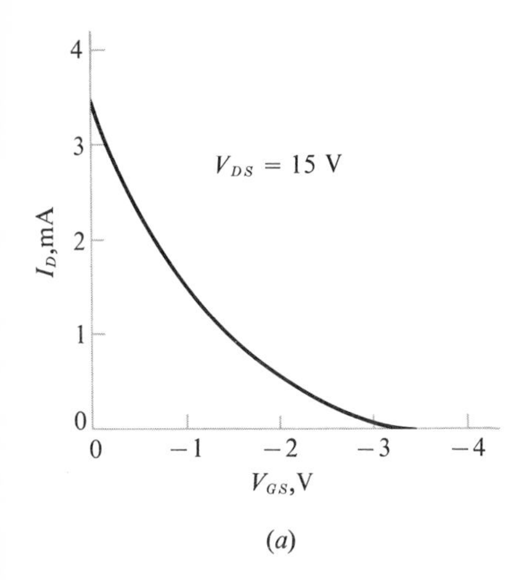

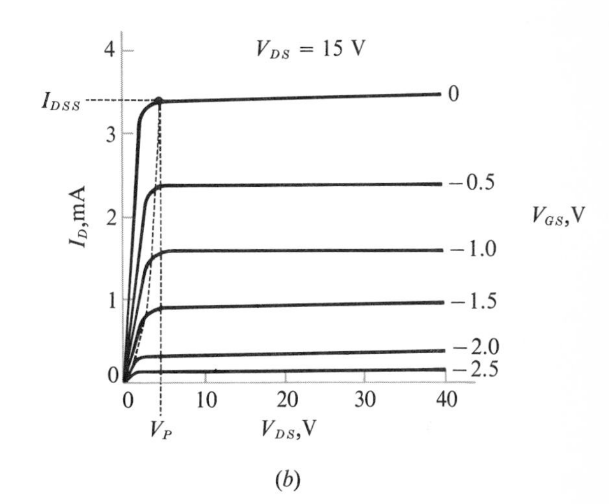

FIG. 13-4 (a) The transfer curve of I_D vs. V_{GS} and (b) the FET drain characteristics of I_D vs. V_{DS} for V_{GS} values from 0 to −2.5 V.

The graphs in Fig. 13-4 show the variation of drain current with changes in gate-source voltage while the drain-source voltage is held constant. Figure 13-4*b* is obtained by setting V_{GS} to some value and then increasing V_{DS} from zero to some voltage greater than V_P. This set of curves resembles the pentode characteristics. They are called the *drain characteristics* and are used in designing amplifiers. Note that the top curve of Fig. 13-4*b* is the same as that in Fig. 13-2*b*, and that equal increments of gate bias produce curves that are not exactly equally spaced. The importance of this point will become quite apparent if you build an amplifier and bias it incorrectly.

INPUT IMPEDANCE

The circuit of Fig. 13-5 differs in only a few respects from the conventional common-emitter amplifier, but the differences are significant. The drain current does not pass through a junction. It is varied by means of a bias voltage at the gate, and not by gate current. The FET is a *voltage-controlled device*. The reason is that the voltage control is a reverse bias on the gate-to-channel diode, and so this diode has a very high impedance and will pass only a tiny leakage current.

The typical leakage current of an *N*-channel FET is generally less than 5 nA and may be as low as 100 pA. The *P*-channel FET may have a leakage of 10 nA, but even this much leakage corresponds to 100 million Ω. The MOS-FET has a thin layer of glass between the gate and the channel and has input impedances ranging up to 100 trillion Ω.

As a result of the very high junction input impedance, the FET draws no current from any signal source. However, many signal sources require a specific loading; that is, they are designed to deliver

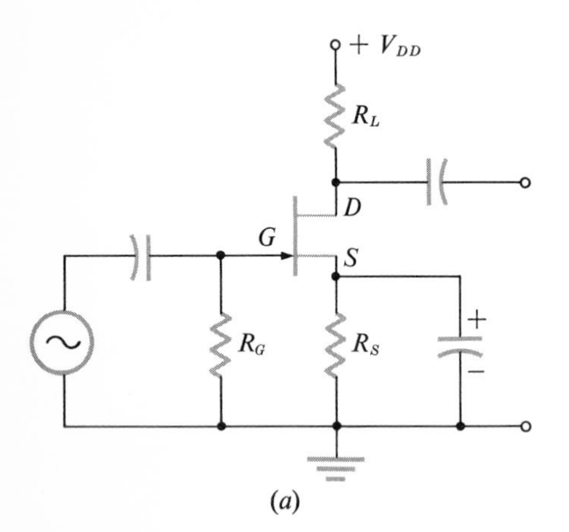

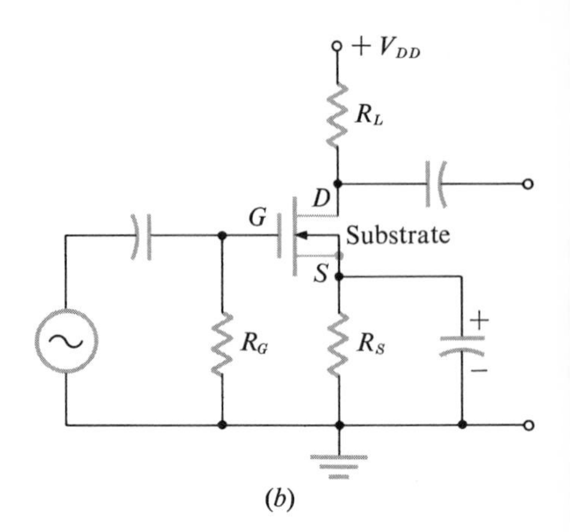

FIG. 13-5 (*a*) A junction FET with source bias used in the basic common-source FET amplifier and (*b*) the *N*-channel depletion-type MOS-FET used in the basic common-source amplifier with source bias.

some set amount of current into a specific impedance. As shown in Fig. 13-5, all that is needed to match the input of the FET to the output impedance of a signal source is a resistance from gate to ground. Since it is in parallel with the input impedance to the FET, the net result is a circuit input impedance essentially equal to the resistor value.

THE MOS-FET

The first crude experiments with FETs preceded the conventional junction transistor by many years. However, they were not commercially feasible until the development of the Planar diffusion process discussed in Chap. 7.

Figure 13-6 shows a cross-sectional comparison of the MOS-FET and the junction FET as they are actually made by the diffusion process. The manufacturing details differ, but the basic pattern is the same. A substrate of *P*-type silicon is masked, a pattern placed on the chip, and the silicon dioxide layer on the surface of the chip is etched through at the pattern. The chip is then exposed to an atmosphere of *N*-type dopants in the furnace, and *N*-type diffusion takes place, developing the *N* channel.

For the junction FET this process is repeated, and another diffusion of *P*-type material takes place to form the gate, as indicated in Fig. 13-6*a*. For the MOS-FET, after the first diffusion of the channel, a silicon dioxide or glass layer is evaporated onto the surface of the chip. Another masking takes place to lay out the source, drain, and gate contacts, and then etching takes place through the glass layer at the source and drain. The glass layer is very thin, but is an excellent

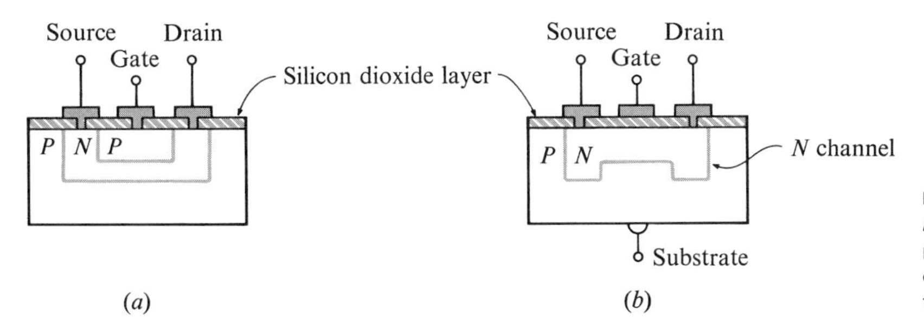

FIG. 13-6 (*a*) Cross sections of an *N*-channel FET made by the diffusion process and (*b*) an *N*-channel depletion-type MOS-FET made by the diffusion process.

insulation. The gate metallized layer covers the general area between the source and drain. The metallization process makes contacts at the source and drain that go through to the *N* channel.

When it is connected into a circuit such as that of Fig. 13-5 the MOS-FET will operate just like the junction FET. However, it is extremely delicate; the glass insulating layer is very thin, and sudden high-static voltages placed on the gate by handling will puncture the layer and destroy the device.

THE DEPLETION-TYPE MOS-FET

How does the MOS-FET work without a reverse-biased junction? The answer is simple. If a negative bias is placed between the gate and the source, the electrostatic field associated with the gate voltage extends into the channel even though no current flows. The negative field repels any electrons from the channel, thus depleting it of carriers. Hence this type of MOS-FET is known as a *depletion-type* MOS-FET.

The junction FET is limited in operation to gate voltages that are reverse biased, or at least forward biased to no more than 0.5 V. Otherwise gate current will flow, lowering the input impedance. This is not the case with the depletion-type MOS-FET. Since there is no junction, there will be no gate current even if the gate is forward biased. This means that larger signals may be applied to the gate of the MOS-FET—a fact that extends its usefulness considerably.

Of course, there are some difficulties in handling the MOS-FET. Walking across the floor or carpet or even rubbing a hand across a polyester shirt or smock builds up an electrostatic charge on the body, and as you may have observed when sliding across plastic seat covers, the sparks have considerable voltage. Thus the insulated gate of the MOS-FET is easily destroyed by handling. For this reason it is usually transported with all terminals shorted together until it is connected into a circuit. Projects dealing with MOS-FETs are not generally for the beginner unless he is prepared for substantial replacement costs. Moreover, these devices are frequently very expensive, although production techniques have reduced the price considerably. In fact

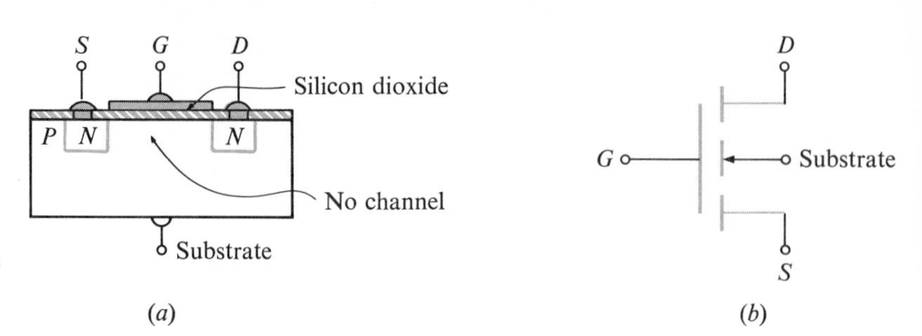

FIG. 13-7 (*a*) Cross section of an enhancement-type MOS-FET showing the absence of the *N*-channel which must be enhanced into the region between the source and drain and (*b*) the symbol for the *N*-channel enhancement-type MOS-FET.

there have been some successful attempts to produce low-cost MOS-FETs on plastic substrates.

THE ENHANCEMENT-TYPE MOS-FET

Instead of diffusing a complete channel, as with the depletion-type MOS-FET, the channel may be left out and only the source and drain areas diffused. In this case a valuable effect is achieved. Note in Fig. 13-7 that even though the channel is missing, a channel is electrostatically simulated for a positive gate bias. The positive voltage between the gate and the source attracts electrons into the area between the source and drain until the channel is enhanced sufficiently to carry current. This device is referred to as an *enhancement-type MOS-FET*.

Of course, if the bias drops below the threshold voltage needed to enhance the channel, all current ceases. And if the bias goes negative, of course, then no current flows. In many applications it is desirable to have no current flowing in a circuit until the voltage reaches a preset point. Hence this device, with its extremely high input impedance, has many uses.

THE FET AS AN AMPLIFIER

Like other transistors, the FET has three terminals and may be designed in three configurations: the common-source circuit, the common-drain circuit, and the common-gate circuit. These correspond to the common-emitter, the common-collector, and the common-base circuits, respectively.

Let us first consider the common-source circuit, which is designed almost exactly like the common-cathode-tube circuit. As before, approximately one-half the supply voltage should be across the device and one-half should be across the load resistance. Remember that the FET is a voltage-controlled device and cannot be handled exactly like the transistor. Note that the characteristic curves in Fig. 13-8 show the load line corresponding to the resistance in series with the FET, and that the curves correspond to changes in gate-source voltage V_{GS}.

Since the pinchoff voltage of the FET selected is about 2 to 4 V, a supply voltage greater than 4 V is needed. The breakdown voltage of the 2N5246 device is about 30 V, and so the supply voltage V_{DD} must be less than 30 V. The gate-source bias voltage must be between 0 and −2.0 V for operation in the linear range of the device.

As we shall see later, the voltage gain of this device is greatest when the gate voltage is near zero and the drain current is near I_{DSS}. Obviously for operation near these limits the signal must be very small. The design must of necessity place the Q point of the circuit lower on the set of curves than maximum.

The value of I_{DSS} for the 2N5246 is about 3.5 mA. This value will vary somewhat from one device to the next, but let us assume these parameters here. The variation in pinchoff voltage from one device to the next is also a factor, but if the design is for a Q point midway between the best and the worst V_P and I_{DSS} stated by the manufacturer, the amplifiers constructed will work for small signals.

In the drain characteristic curves of Fig. 13-8 note that the load line corresponds to a resistance of 5.6 kΩ. This resistance is divided into the 620-Ω source resistance and the 5-kΩ load resistance. The gate-voltage curve $V_{GS} = -1.0$ V intersects the load line at about 1.5

FIG. 13-8 (*a*) An experimental FET common-source amplifier and (*b*) drain characteristics for a 2N5246 FET with load line, input, and output curves plotted to show a voltage gain of about 8.

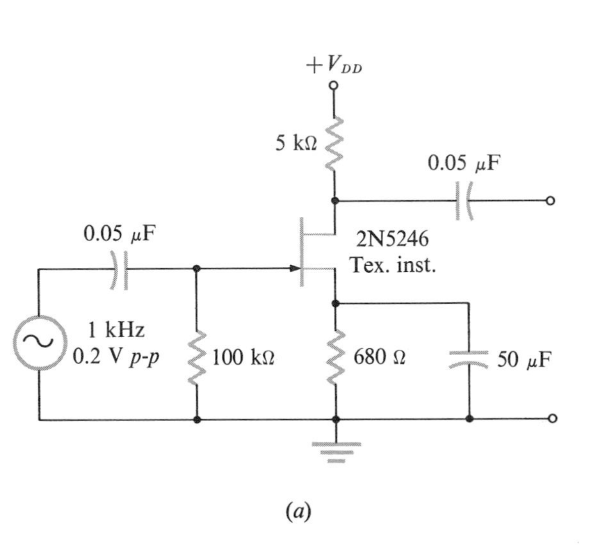

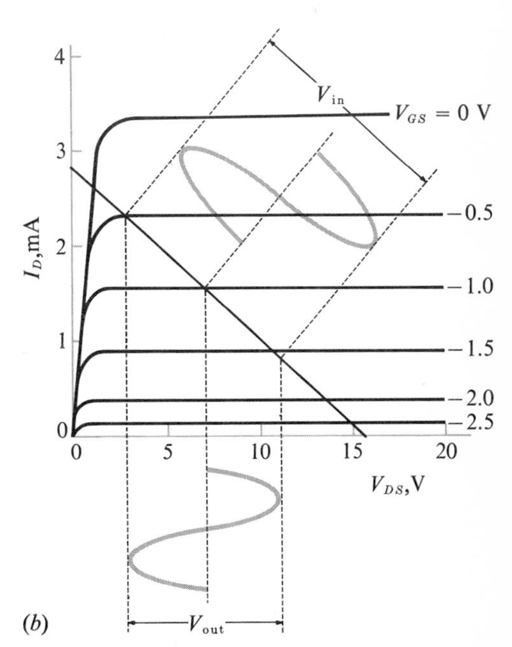

mA of drain current. Let us see how this fits into the picture, and how this circuit differs from the transistor design.

The Q point for $V_{GS} = 1$ V and $I_D = 1.5$ mA is needed to compute the size of the source resistance needed. This is, in fact, the place to start the design. The gate must be more negative than the source by 1.0 V. This can be achieved if the gate is held at 0 V and the source is held at 1.0 V. That is to say, if the source is more positive than the gate, the gate must be negative with respect to the source. As the drain current flows through the source resistance, there will be a voltage present equal to $I_D R_S$, about 1 V.

As a matter of fact, the bias voltage does not have to be exactly −1.0 V, and either a 620-Ω or a 680-Ω resistor will work in about the same way. If the bias voltage varies somewhat from the design Q point, the drain current will vary accordingly.

TRANSADMITTANCE

The load resistance indicated in Fig. 13-8 is 5 kΩ, but it may be any value from 4.7 to 5.6 kΩ. Generally the higher value will give a higher voltage gain. However, the higher the value of the load resistance, the lower the drain current, and hence the lower the voltage gain. A compromise must be worked out by means of a factor called the *transadmittance* Y_{fs}, sometimes referred to as the *transconductance* G_{fs}. In any case, admittance is to impedance as conductance is to resistance. Transadmittance must be computed for each drain current used, as will be seen presently.

The voltage distribution for the amplifier in Fig. 13-8 is, from Ohm's law,

$$V_S = I_D R_S = 1.5 \text{ mA} \times 680 \ \Omega = 1.02 \text{ V} \tag{13-2}$$

$$V_{R_L} = I_D R_L = 1.5 \text{ mA} \times 5 \text{ k}\Omega = 7.5 \text{ V} \tag{13-3}$$

$$\begin{aligned} V_{DS} &= V_{DD} - (V_S + V_{R_L}) \\ &= 16 \text{ V} - (1.02 + 7.5 \text{ V}) \\ &= 16 \text{ V} - 8.52 \text{ V} = 7.48 \text{ V} \end{aligned} \tag{13-4}$$

$$\begin{aligned} V_D &= V_{DS} + V_S \\ &= 7.48 \text{ V} + 1.02 \text{ V} = 8.5 \text{ V} \end{aligned} \tag{13-5}$$

With the dc bias conditions of the FET amplifier set by the drain current and the various resistances, let us consider the amplification of an ac small-signal voltage. If a signal voltage of 0.2 V *p-p* with a frequency between 1 and 10 kHz is applied to the gate resistance, there should be an output between the drain and ground. It should be an amplified and inverted form of the input voltage. As long as the input signal is kept small, the output signal should have very little if any distortion.

The size of the output signal voltage at the drain depends on the transadmittance Y_{fs} and the size of the load resistance R_L. It should again be emphasized that the same subscripts that applied to transistors also apply to FETs. The second letter in the subscript tells what amplifier configuration is being described, and lowercase denotes ac parameters. In other words, Y_{fs} is the forward transadmittance in a common-source amplifier.

But what is transadmittance? Just as admittance is the inverse of impedance, that is, $Y = i/v$ or $\Delta I/\Delta V$, so transadmittance is a comparison. But Y_{fs} is a comparison of the change in output current I_D compared to the change in the gate-source or input voltage that caused the output change,

$$Y_{fs} = \frac{\Delta I_D}{\Delta V_{GS}} \tag{13-6}$$

With a little trial and error Y_{fs} can be determined easily from Eq. (13-6). However, since it changes for every different value of drain current, another relationship is also of value,

$$Y_{fs} = \frac{2I_{DSS}}{V_P}\left(1 - \frac{|V_{GS}|}{V_P}\right) \tag{13-7}$$

where $|V_{GS}|$ = magnitude of V_{GS} without regard to polarity

This relationship allows us to predict an approximate transadmittance for the FET from known constants. The output signal voltage for the circuit of Fig. 13-8 may be determined in this way. Since $I_{DSS} \cong 3.5$ mA, $V_P \cong 3$ V, and $V_{GS} = 1$ V,

$$Y_{fs} = \frac{2 \times 3.5 \text{ mA}}{3 \text{ V}}\left(1 - \frac{1 \text{ V}}{3 \text{ V}}\right) = \frac{7 \text{ mA}}{3 \text{ V}}(1 - 0.33)$$

$$= 2.33 \times 10^{-3} \times 0.667 = 1.56 \text{ mS} \tag{13-8}$$

This indicates that the transadmittance is only about two-thirds as great at this bias as it is at zero gate bias.

With the actual transadmittance established, the approximate voltage gain can be predicted as

$$A_v \cong -Y_{fs}R_L$$
$$= 1.56 \times 10^{-3} \times 5 \times 10^3 \cong -7.8 \tag{13-9}$$

This approximation agrees rather well with the voltage gain of 8 found experimentally. This means that the 0.2 V *p-p* input signal should produce an output signal voltage of 7.8 × 0.2 V = 1.56 V *p-p*.

This approximation of the voltage gain is possible because of the very high dynamic impedance of the FET, as indicated by the flat slope of the characteristic curves. The dynamic resistance is in parallel with the load resistance, but it is so large in comparison that their parallel sum is equal to the load resistance R_L. This is the approximation used by most engineers in designing with FETs. There is so much variation between FETs of the same type that these results are quite acceptable. It should also be noted that the voltage gain for the common-source FET amplifier is negative; as indicated by Eq. (13-9), the output is 180° out of phase with the input.

The circuit of Fig. 13-9 provides a means of determining Y_{fs} for a wide variety of bias voltages and drain currents. The drain current depends on the gate-source voltage and is limited by the load resistance and the size of I_{DSS}. Notice that gate bias, rather than source bias, is used for the experimental setup. This avoids some of the confusion of the apparent interaction of the source and load resistances.

For any particular Q point and load-resistance value computed for a particular FET, the circuit may be quickly set up by adjusting the gate voltage and the load-resistance potentiometer. The dc drain

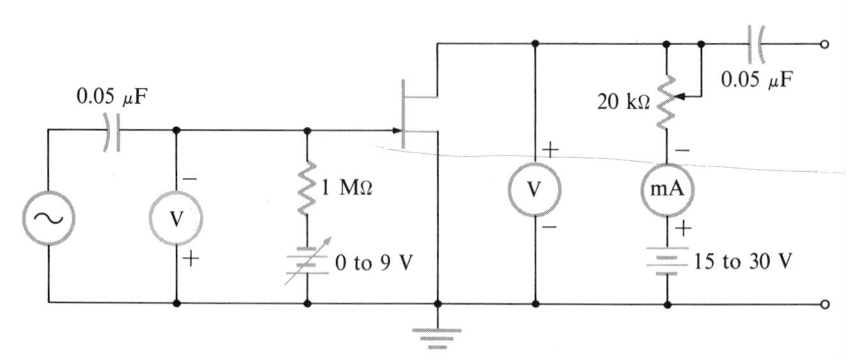

FIG. 13-9 A circuit for experimental determination of transadmittance Y_{fs}.

current I_D can be measured on the milliammeter. However, the ac drain current I_d must be determined either indirectly as the difference between dc measurements or directly by measuring the ac voltage across the load resistance and applying Ohm's law. In either case Eq. (13-6) should be used to compute Y_{fs}. The peak-to-peak values of both V_{in} and V_{out} converted to I_{out} should be used in place of the delta (Δ) values in Eq. (13-6).

VOLTAGE-GAIN DEPENDANCE ON SUPPLY VOLTAGE

Generally speaking, the voltage gain of an FET small-signal amplifier increases with an increase in supply voltage V_{DD}. The reasons are apparent from the load line plotted on the drain characteristic curves. The higher voltage allows us to use larger values of load resistance with the same gate bias and drain current to establish the Q point. The input voltage is thus limited in its swing by the gate voltage V_{GS}, but the output voltage is allowed to swing almost to the limit of the supply voltage minus the pinchoff voltage.

Figure 13-10 shows two load lines plotted on the same set of drain characteristic curves. Note that the larger supply voltage of the second load line allows a much larger output voltage for the same size of input signal voltage. This, of course, is in agreement with Eq. (13-9), since the bias and the drain current have not changed. Note also that the positive excursion of the input signal is limited, because too large a signal would drive the drain voltage into pinchoff and result in severe distortion. This factor makes it necessary to operate the FET

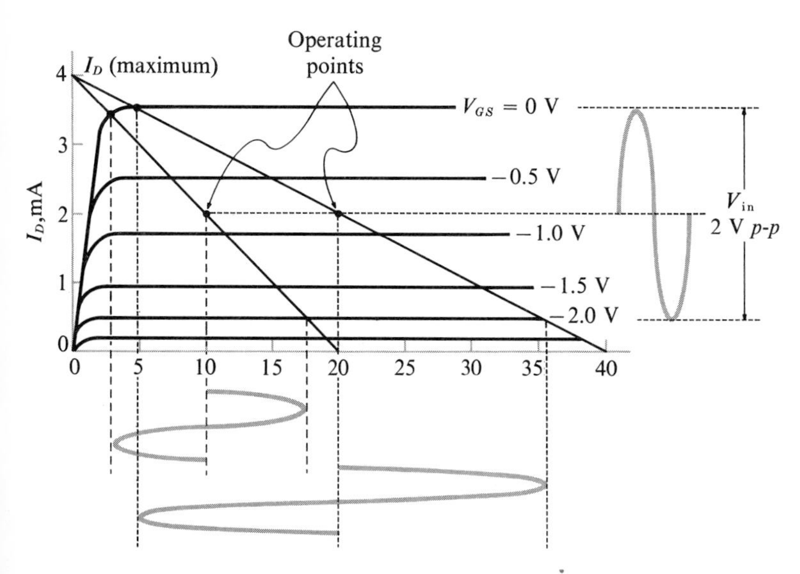

FIG. 13-10 Supply voltage and load resistance and their influence on voltage gain at one drain current.

at as high a drain voltage as possible without approaching the breakdown point or using excessive current. Excess drain current has a tendency to generate noise.

Higher voltage gain may be obtained by using high drain voltage and large load resistances, but there is always the possibility of lessening the stability of the operating point and diminishing the available bandwidth. One should always experiment with a circuit in order to determine its optimum design limits.

VOLTAGE GAIN AND LOAD RESISTANCE

Equation (13-9) points out that for a particular value of Y_{fs} the voltage gain is determined by the load resistance used in the amplifier circuit. This can be made clearer by a short graphical analysis. Since the value of Y_{fs} is primarily a function of the gate-source voltage V_{GS}, it should remain essentially constant in a circuit unless something such as temperature change causes a change in bias. Figure 13-11 indicates how the same drain-supply voltage could be used with three different load resistances.

If the gate bias voltage is maintained either by a separate bias supply or by a source bias resistance, the drain current should remain constant for this discussion. When the load resistances are changed and source biasing is employed, the value of the source resistance may have to be adjusted in order to maintain the bias point.

From Fig. 13-11 we see that limiting of the output signal occurs when the output current reaches its limit of I_{DSS}. This happens sooner

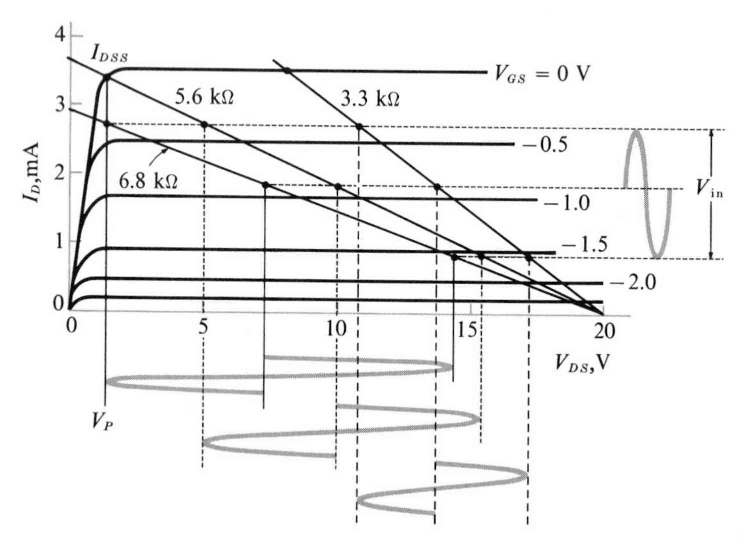

FIG. 13-11 Voltage-gain comparison for three different load resistances.

for the smaller load resistance and after a much larger excursion of the signal when the larger load resistance is used.

In experimenting with a circuit of this sort remember that the same small input signal voltage should be used with each size of load resistance. In fact, a whole range of load resistances could be used, with the gain plotted as a function of load-resistance value. The results should be as shown in Fig. 13-12.

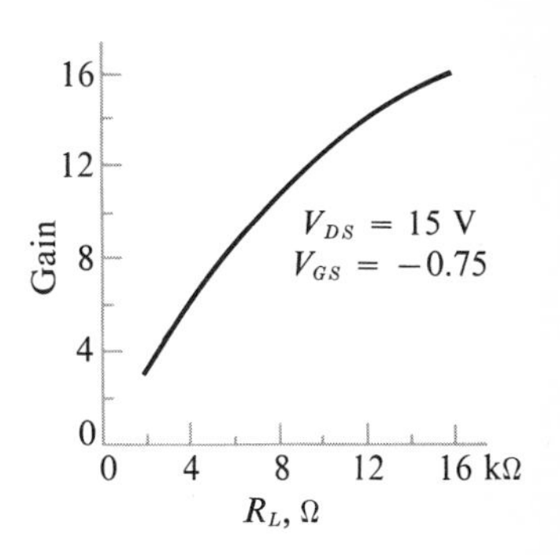

FIG. 13-12 A comparison of voltage gain to load resistance for the circuit of Fig. 13-8 and curves of Fig. 13-11.

One of the most baffling problems for most beginning students in electronics is determining the size of load resistance to use. This is actually not difficult if a few basic rules are followed and you know beforehand what you are trying to do.

The purpose of the load resistance is to divide the supply voltage so that the output signal may swing just as far in the positive as the negative direction with a minimum amount of distortion. For this reason, at the Q point of operation one-half the supply voltage should be across the load resistance. When source bias is used, this point changes slightly, but not much if the supply voltage is large. The best size of resistance to use depends on how much maximum current can flow and on the supply voltage.

If characteristic curves are available or can be plotted, they should be used. Otherwise Ohm's law must be used in conjunction with the published value of I_{DSS} and the pinchoff voltage. The latter method is not accurate, because there is a great deal of variation from one device to the next even in expensive devices.

First, choose the supply voltage that is to be used. This is limited generally to what is available. If possible, use voltages greater than 15 V, preferably closer to the breakdown voltage. Do not exceed the breakdown voltage. When the device is biased to cutoff, as it is when the gate swings to an amount equal to the pinchoff voltage, no current will pass through the FET. At this point the FET resistance is very high in relation to the load resistance, and the whole supply voltage is across the FET. This establishes one point of the load line, as indicated in Fig. 13-13.

The maximum current flow is the other point to be determined for the load line. The drain current is limited by I_{DSS} when the operation is in the constant-current region. It is limited by the size of the load resistance if R_L is so large that the maximum current is less than I_{DSS}. If the characteristic curve for $V_{GS} = 0$ V is observed, a point to the right of I_{DSS} should be chosen, such as point B in Fig. 13-13. The line drawn between points A and B and extended to intersect the Y axis is the *load line*.

The value of resistance represented by the load line is indicated by the slope of the line. That is, the line represents the ratio of voltage to current, which is resistance. Take the point of intersection of the X axis, which is the supply voltage V_{DD}, and divide this value by the

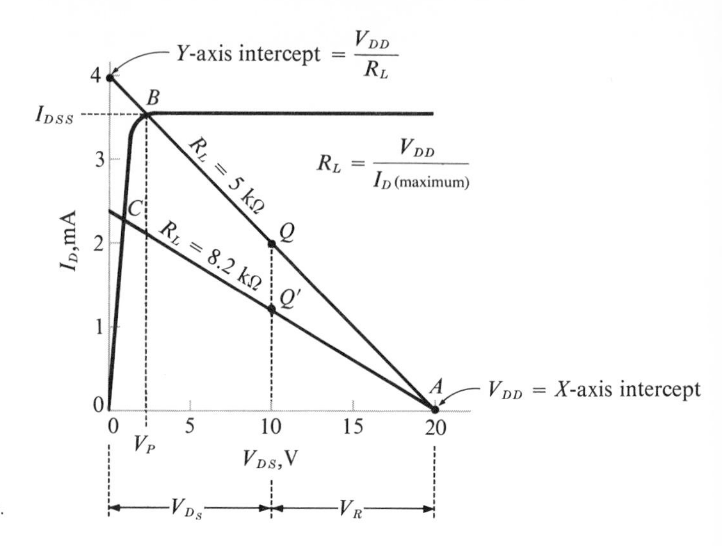

FIG. 13-13 Development of the load-line limits.

value of the drain current indicated at the intersection of the load line with the *Y* axis,

$$R_L = \frac{V_{DD}}{I_{D,\,\text{max}}} \tag{13-10}$$

where $I_{D,\,\text{max}}$ = the *Y*-axis intercept

This method will yield a good, usable value of load resistance; use the closest standard value.

Once the value of R_L has been determined from Eq. (13-10), draw a perpendicular line down from the intersection of the load line with the $V_{GS} = 0$ V curve. This is the *saturation voltage* of the FET; no more than this amount of current can flow, and so the voltage across the device cannot drop below the pinchoff voltage. In reality, if the gate is driven into a positive bias, it will draw current, and distortion will result. What we need to establish, then, is the limit beyond which distortion will result.

After the limits of the load line have been established, the output-voltage swing must be determined. Note that as the gate bias voltage is driven downward, the characteristic curves are closer and closer together. For this reason, the FET should not be driven too close to cutoff. Select a *Q* point on the load line such that an equal distance in either direction will not result in saturation or in cutoff. Draw another vertical line downward from this *Q* point to the intersection with the

X axis. This represents the voltage from the source to the drain terminal. The voltage to the right of this intersection represents the voltage across the resistance. The Q point tells how much drain current will be used and what the bias voltage V_{GS} will be.

Load line A to C represents a larger value of load resistance, where the current I_{DSS} can never be reached. At this larger resistance the voltage gain will be higher, but the input signal voltage must be smaller, or distortion will result. The gate-source voltage must be larger and the drain current smaller to reestablish the Q point. The transadmittance will also be smaller than before. These basic rules may be applied to establish any size of load resistance that is needed. Thus it is possible to start with the desired size of gain and work back to R_L.

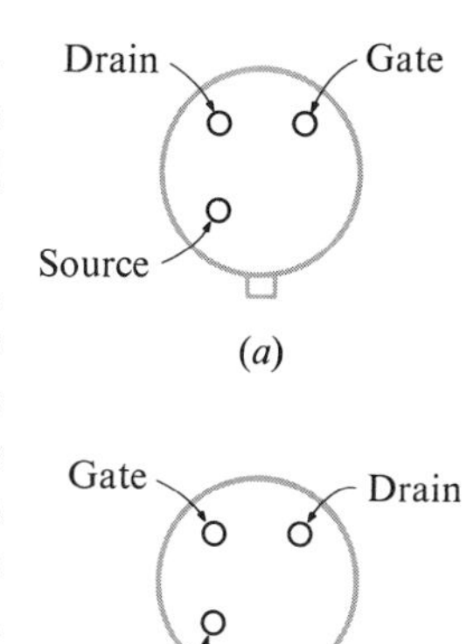

FIG. 13-14 Common metal-can FET basing arrangements: (*a*) bottom view *N* channel; (*b*) bottom view *P* channel.

THE P-CHANNEL FET

Circuit design for the *N*-channel FET and the *P*-channel FET is essentially the same. The only difference is that the supply-voltage terminals must be reversed so that the drain is negative with respect to the source.

Most FETs are manufactured symmetrically; that is, the pattern on the silicon chip is laid out so that the drain and the source are interchangeable. However, whether the circuit is symmetrical or not, the gate of the junction FET is normally reverse biased.

The basing arrangement for the *P*-channel FET is also generally different from the *N*-channel FET. Figure 13-14 shows the two most commonly used basing arrangements. There are other arrangements, however, and this should always be checked if the device number is known. In any case, the source terminal in metal-cased FETs is almost always the one to the left of the tab as seen from the bottom. In certain plastic-cased FETs the source is the terminal to the left as seen from the flat side with the pins pointing down. Other plastic-cased FETs, such as the 2N5246 made by Texas Instruments, are based as shown in Fig. 13-15*a*. In some cases a fourth lead is used as a connection to the case, which allows the case to act as a shield.

THE MOS-FET AMPLIFIER

For all practical purposes, the depletion-type MOS-FET functions much like the junction FET in a circuit. However, the MOS-FET, whether *N* channel or *P* channel, requires extreme care in handling to avoid destruction of the unit by static discharge. Always keep the leads shorted together until the MOS-FET is placed in the circuit.

The common-source MOS-FET amplifier in Fig. 13-15*b* is typical. The substrate terminal is actually a gate terminal and may be connected

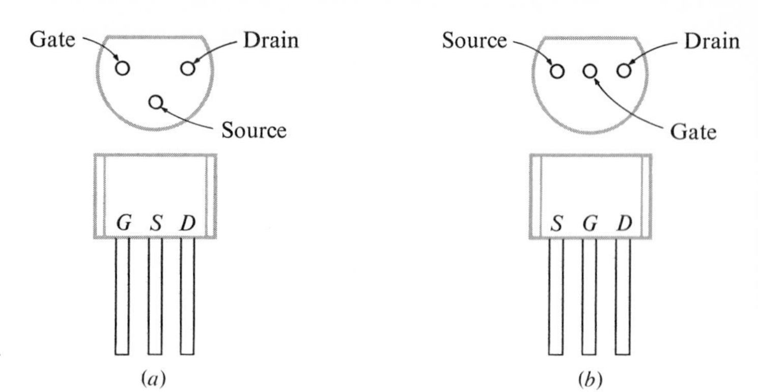

FIG. 13-15 Basing arrangements used by Texas Instruments Incorporated in their plastic-case line: (*a*) *N* channel used for the 2N5246; (*b*) *N* channel used for the 2N5248.

internally to the source lead, but in most cases there are four leads, or terminals. Consult the manufacturer's base layout for the MOS-FET, just as you would for the junction FET.

A detailed consideration of such factors as the square-law effect, distortion, and other circuit arrangements are given in the references at the end of the chapter. The discussion in this chapter is sufficient to provide an understanding of most modern solid-state amplifiers that use bipolar transistors in all but the input stage, where a high impedance is required.

TYPES 2N5245 THRU 2N5247
N-CHANNEL EPITAXIAL PLANAR SILICON FIELD-EFFECT TRANSISTORS

TYPES 2N5245 THRU 2N5247
BULLETIN NO. DL-S 6810917, SEPTEMBER 1968

N-CHANNEL SILECT† FIELD-EFFECT TRANSISTORS FOR VHF AMPLIFIER AND MIXER APPLICATIONS

- **High Power Gain . . . 10 dB Min at 400 MHz**
- **High Transconductance . . . 4000 μmho Min at 400 MHz (2N5245, 2N5247)**
- **Low C_{rss} . . . 1 pF Max**
- **High $|y_{fs}|/C_{iss}$ Ratio (High-Frequency Figure-of-Merit)**
- **Drain and Gate Leads Separated for High Maximum Stable Gain**
- **Cross-Modulation Minimized by Square-Law Transfer Characteristic**
- **For Use in VHF Amplifiers in FM, TV, and Mobile Communications Equipment**

mechanical data

These transistors are encapsulated in a plastic compound specifically designed for this purpose, using a highly mechanized process‡ developed by Texas Instruments. The case will withstand soldering temperatures without deformation. These devices exhibit stable characteristics under high-humidity conditions and are capable of meeting MIL-STD-202C method 106B. The transistors are insensitive to light.

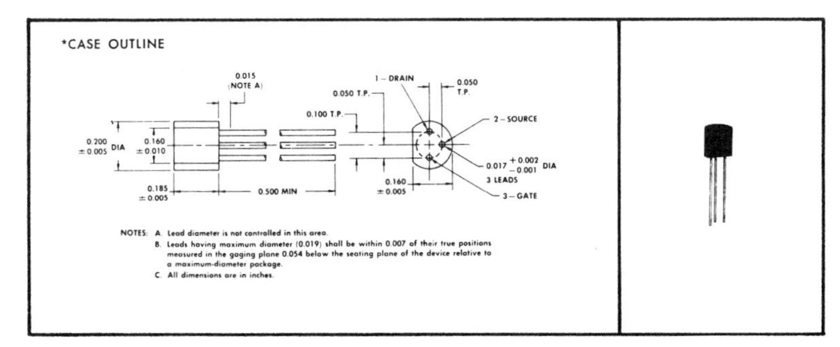

*absolute maximum ratings at 25°C free-air temperature (unless otherwise noted)

Drain-Gate Voltage	30 V
Reverse Gate-Source Voltage	−30 V
Continuous Forward Gate Current	50 mA
Continuous Device Dissipation at (or below) 25°C Free-Air Temperature (See Note 1)	360 mW
Continuous Device Dissipation at (or below) 25°C Lead Temperature (See Note 2)	500 mW
Storage Temperature Range	−65°C to 150°C
Lead Temperature 1/16 Inch from Case for 10 Seconds	260°C

NOTES: 1. Derate linearly to 150°C free-air temperature at the rate of 2.88 mW/°C.

2. Derate linearly to 150°C lead temperature at the rate of 4 mW/°C. Lead temperature is measured on the gate lead 1/16 inch from the case.

*Indicates JEDEC registered data

†Trademark of Texas Instruments

‡Patented by Texas Instruments and other patents pending.

TEXAS INSTRUMENTS
INCORPORATED
POST OFFICE BOX 5012 • DALLAS, TEXAS 75222

6703

Data sheets for 2N5245 through 2N5247 *(courtesy Texas Instruments Incorporated).*

TYPES 2N5245 THRU 2N5247
N-CHANNEL EPITAXIAL PLANAR SILICON FIELD-EFFECT TRANSISTORS

***electrical characteristics at 25°C free-air temperature (unless otherwise noted)**

PARAMETER		TEST CONDITIONS	2N5245		2N5246		2N5247		UNIT
			MIN	MAX	MIN	MAX	MIN	MAX	
$V_{(BR)GSS}$	Gate-Source Breakdown Voltage	$I_G = -1\ \mu A$, $V_{DS} = 0$	−30		−30		−30		V
I_{GSS}	Gate Reverse Current	$V_{GS} = -20$ V, $V_{DS} = 0$		−1		−1		−1	nA
		$V_{GS} = -20$ V, $V_{DS} = 0$, $T_A = 100°C$		−0.5		−0.5		−0.5	µA
$V_{GS(off)}$	Gate-Source Cutoff Voltage	$V_{DS} = 15$ V, $I_D = 10$ nA	−1	−6	−0.5	−4	−1.5	−8	V
I_{DSS}	Zero-Gate-Voltage Drain Current	$V_{DS} = 15$ V, $V_{GS} = 0$, See Note 3	5	15	1.5	7	8	24	mA
$\|y_{fs}\|$	Small-Signal Common-Source Forward Transfer Admittance	$V_{DS} = 15$ V, $V_{GS} = 0$, $f = 1$ kHz	4.5	7.5	3	6	4.5	8	mmho
$\|y_{os}\|$	Small-Signal Common-Source Output Admittance	$V_{DS} = 15$ V, $V_{GS} = 0$, $f = 1$ kHz		0.05		0.05		0.07	mmho
C_{iss}	Common-Source Short-Circuit Input Capacitance	$V_{DS} = 15$ V, $V_{GS} = 0$, $f = 1$ MHz		4.5		4.5		4.5	pF
C_{rss}	Common-Source Short-Circuit Reverse Transfer Capacitance			1		1		1	pF
$Re(y_{is})$	Small-Signal Common-Source Input Conductance	$V_{DS} = 15$ V, $V_{GS} = 0$, $f = 100$ MHz		0.1		0.1		0.1	mmho
$Im(y_{is})$	Small-Signal Common-Source Input Susceptance			3		3		3	mmho
$Re(y_{os})$	Small-Signal Common-Source Output Conductance			0.075		0.075		0.1	mmho
$Im(y_{os})$	Small-Signal Common-Source Output Susceptance			1		1		1	mmho
$Re(y_{is})$	Small-Signal Common-Source Input Conductance	$V_{DS} = 15$ V, $V_{GS} = 0$, $f = 400$ MHz		1		1		1	mmho
$Im(y_{is})$	Small-Signal Common-Source Input Susceptance			12		12		12	mmho
$Re(y_{fs})$	Small-Signal Common-Source Forward Transfer Conductance		4		2.5		4		mmho
$Re(y_{os})$	Small-Signal Common-Source Output Conductance			0.1		0.1		0.15	mmho
$Im(y_{os})$	Small-Signal Common-Source Output Susceptance			4		4		4	mmho

NOTE 3: This parameter must be measured using pulse techniques. $t_p = 100$ ms, duty cycle $\leq 10\%$.

***operating characteristics at 25°C free-air temperature**

PARAMETER		TEST CONDITIONS	2N5245		UNIT
			MIN	MAX	
G_{ps}	Small-Signal Common-Source Neutralized Insertion Power Gain	$V_{DS} = 15$ V, $I_D = 5$ mA, $f = 100$ MHz, $R_G' = 1\ k\Omega$, See Figure 1	18		dB
		$V_{DS} = 15$ V, $I_D = 5$ mA, $f = 400$ MHz, $R_G' = 1\ k\Omega$, See Figure 1	10		
NF	Spot Noise Figure	$V_{DS} = 15$ V, $I_D = 5$ mA, $f = 100$ MHz, $R_G' = 1\ k\Omega$, See Figure 1		2	dB
		$V_{DS} = 15$ V, $I_D = 5$ mA, $f = 400$ MHz, $R_G' = 1\ k\Omega$, See Figure 1		4	

*Indicates JEDEC registered data

6704

TEXAS INSTRUMENTS
INCORPORATED
POST OFFICE BOX 5012 • DALLAS, TEXAS 75222

TYPES 2N5245 THRU 2N5247
N-CHANNEL EPITAXIAL PLANAR SILICON FIELD-EFFECT TRANSISTORS

*PARAMETER MEASUREMENT INFORMATION

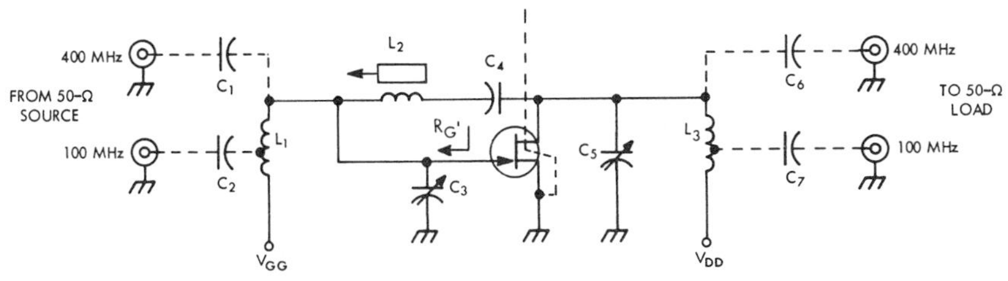

CIRCUIT COMPONENT INFORMATION						
CAPACITORS			COILS			
	100 MHz	400 MHz		100 MHz	400 MHz	
C_1	not used	1.8 pF	L_1	8.5 T, #16 copper, tapped 2.5 T from bottom, 3/8″ ID, 1 1/4″ long	1.25 T, #20 copper, 3/16″ ID, 3/8″ long	
C_2	7 pF	not used				
C_3	1 – 12 pF	0.8 – 8 pF	L_2	15 T, #20 enameled copper, close-wound, 1/4″ ID	4 T, #20 enameled copper, close-wound, 3/16″ ID	
C_4	1000 pF	27 pF				
C_5	1 – 12 pF	0.8 – 8 pF				
C_6	not used	1 pF	L_3	13.5 T, #16 copper, tapped 5 T from bottom, 3/8″ ID, 1 1/4″ long	0.5 T, #20 copper, 1/2″ ID, no length	
C_7	3 pF	not used				

FIGURE 1 — SCHEMATIC AND COMPONENT INFORMATION FOR 100-MHz AND 400-MHz NEUTRALIZED INSERTION POWER GAIN AND SPOT NOISE FIGURE TEST CIRCUITS

*Indicates JEDEC registered data

TYPICAL CHARACTERISTICS

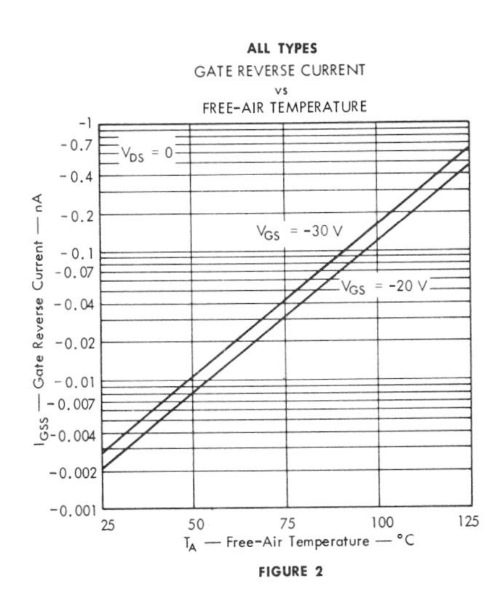

FIGURE 2

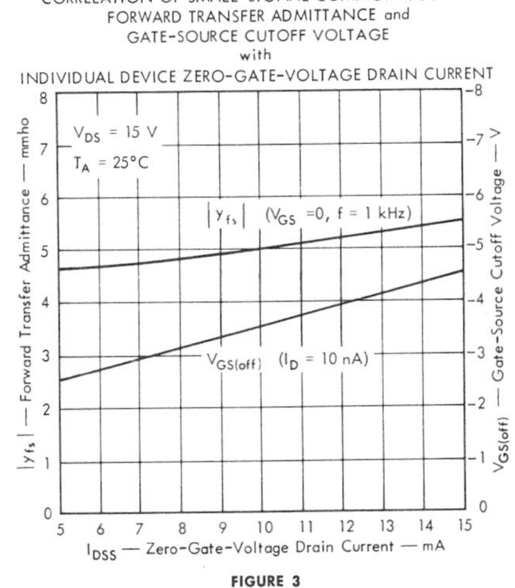

FIGURE 3

TYPES 2N5245 THRU 2N5247
N-CHANNEL EPITAXIAL PLANAR SILICON FIELD-EFFECT TRANSISTORS

2N5245 TYPICAL CHARACTERISTICS

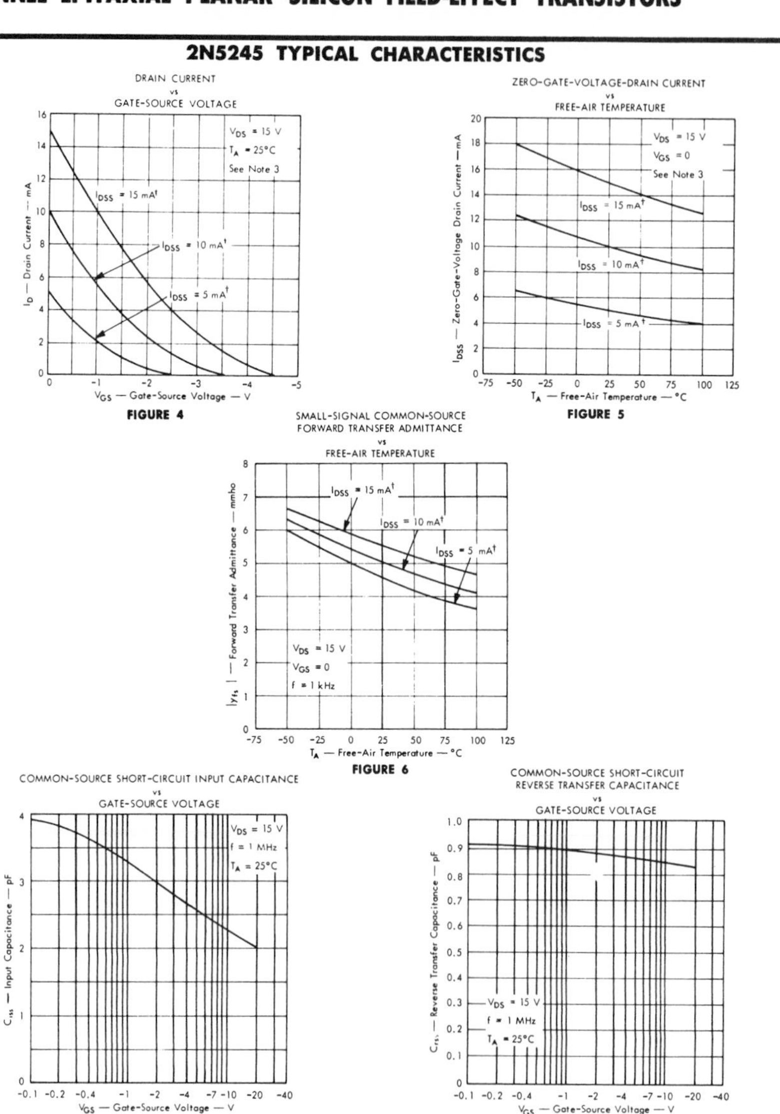

NOTE 3: This parameter must be measured using pulse techniques. $t_p = 300\ \mu s$, duty cycle $\leq 2\%$.

†Data is for devices having the indicated values of I_{DSS} at $V_{DS} = 15$ V, $V_{GS} = 0$, and $T_A = 25°C$.

TEXAS INSTRUMENTS
INCORPORATED
POST OFFICE BOX 5012 • DALLAS, TEXAS 75222

TYPES 2N5245 THRU 2N5247
N-CHANNEL EPITAXIAL PLANAR SILICON FIELD-EFFECT TRANSISTORS

2N5245 TYPICAL CHARACTERISTICS

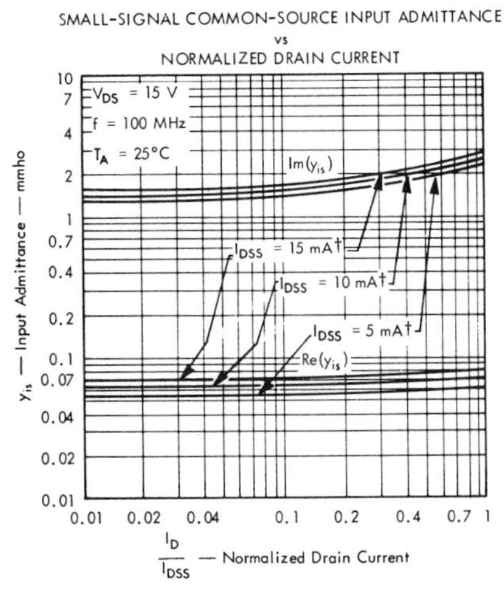

FIGURE 9

SMALL-SIGNAL COMMON-SOURCE
FORWARD TRANSFER ADMITTANCE
vs
NORMALIZED DRAIN CURRENT

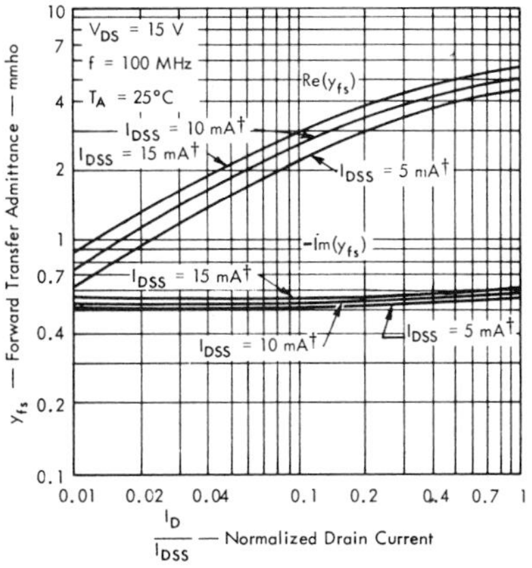

V_{DS} = 15 V
f = 100 MHz
T_A = 25°C
$Re(y_{fs})$
I_{DSS} = 10 mA†
I_{DSS} = 15 mA†
I_{DSS} = 5 mA†
$-Im(y_{fs})$
I_{DSS} = 15 mA†
I_{DSS} = 10 mA†
I_{DSS} = 5 mA†
y_{fs} — Forward Transfer Admittance — mmho
$\frac{I_D}{I_{DSS}}$ — Normalized Drain Current

FIGURE 10

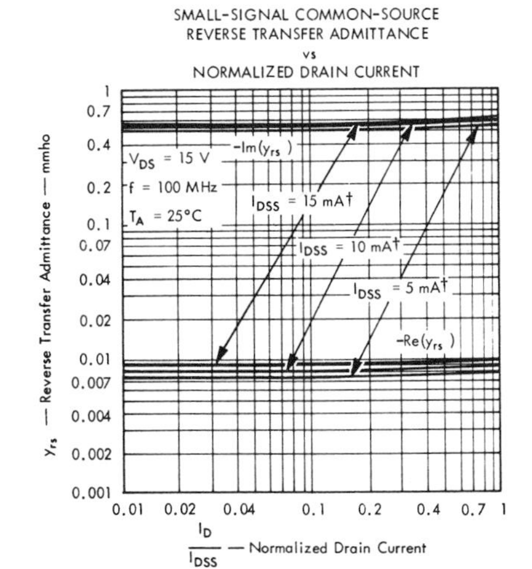

FIGURE 11

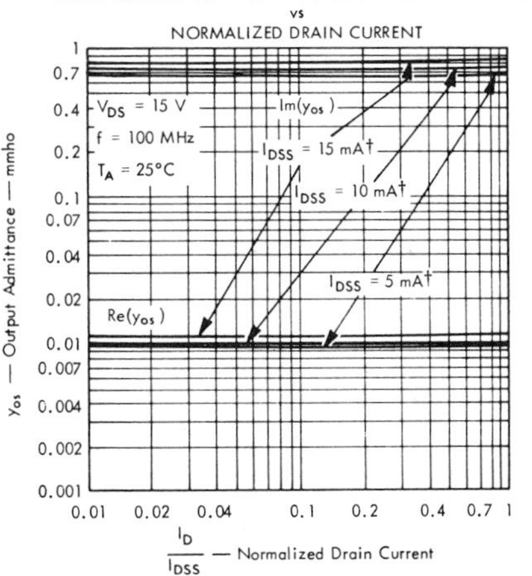

FIGURE 12

†Data is for devices having the indicated values of I_{DSS} at V_{DS} = 15 V, V_{GS} = 0, and T_A = 25°C.

TYPES 2N5245 THRU 2N5247
N-CHANNEL EPITAXIAL PLANAR SILICON FIELD-EFFECT TRANSISTORS

2N5245 TYPICAL CHARACTERISTICS

SMALL-SIGNAL COMMON-SOURCE INPUT ADMITTANCE vs FREQUENCY

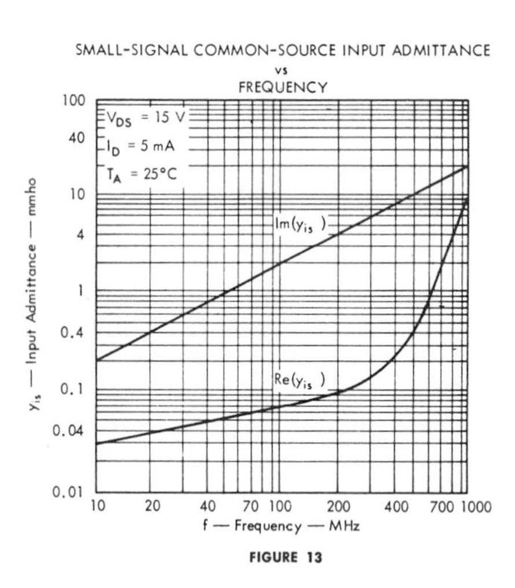

FIGURE 13

SMALL-SIGNAL COMMON-SOURCE FORWARD TRANSFER ADMITTANCE vs FREQUENCY

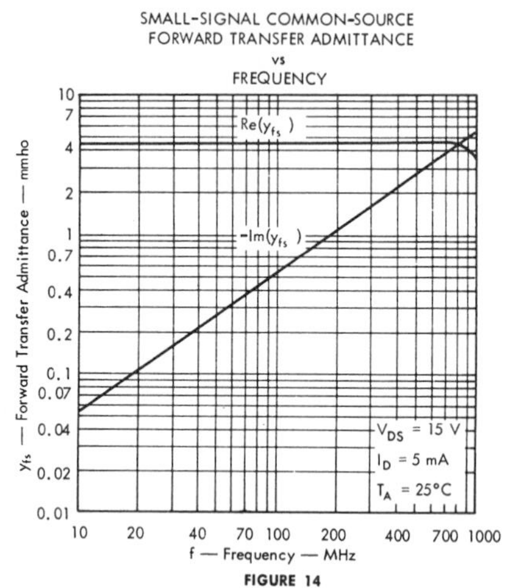

FIGURE 14

SMALL-SIGNAL COMMON-SOURCE REVERSE TRANSFER ADMITTANCE vs FREQUENCY

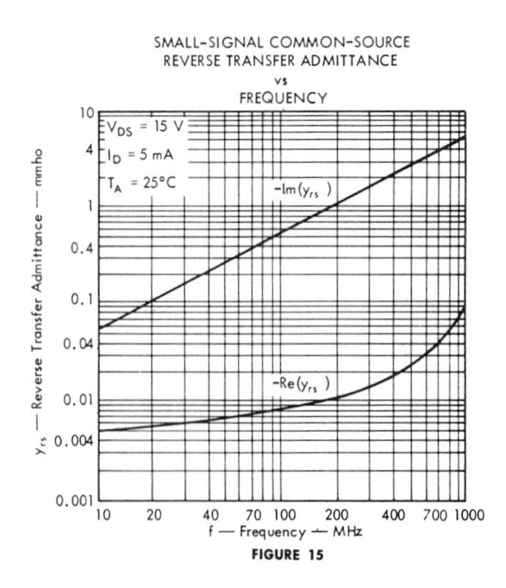

FIGURE 15

SMALL-SIGNAL COMMON-SOURCE OUTPUT ADMITTANCE vs FREQUENCY

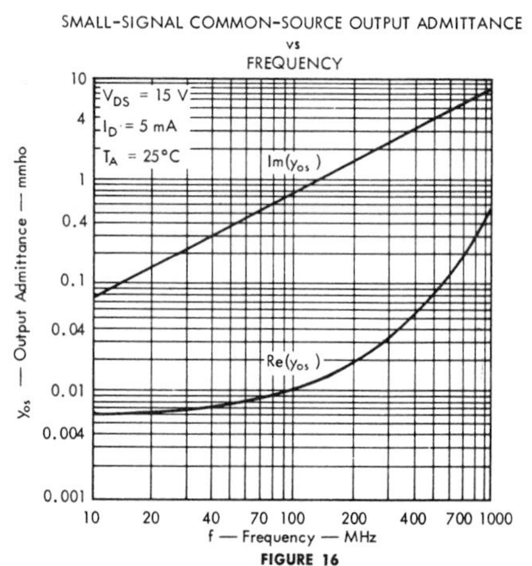

FIGURE 16

TEXAS INSTRUMENTS
INCORPORATED
POST OFFICE BOX 5012 • DALLAS, TEXAS 75222

TYPES 2N5245 THRU 2N5247
N-CHANNEL EPITAXIAL PLANAR SILICON FIELD-EFFECT TRANSISTORS

2N5245 TYPICAL CHARACTERISTICS

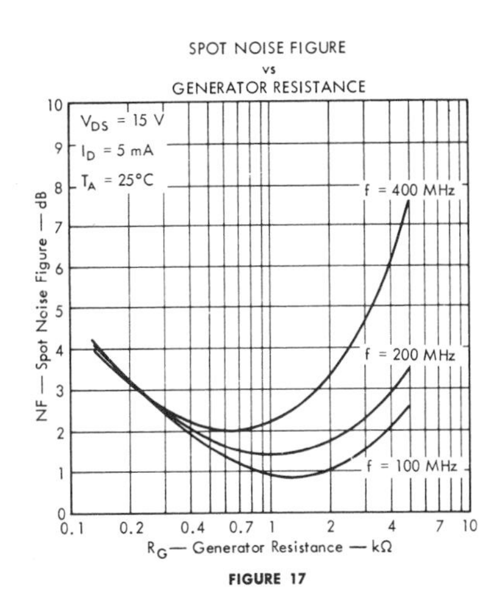

FIGURE 17

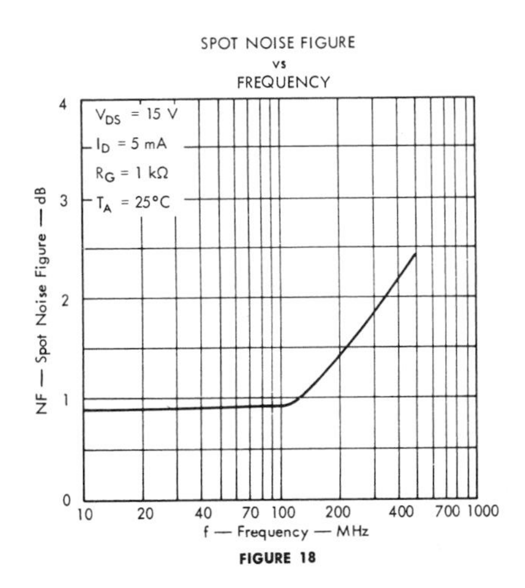

FIGURE 18

TYPES 2N5245 THRU 2N5247
N-CHANNEL EPITAXIAL PLANAR SILICON FIELD-EFFECT TRANSISTORS

TYPICAL APPLICATION DATA

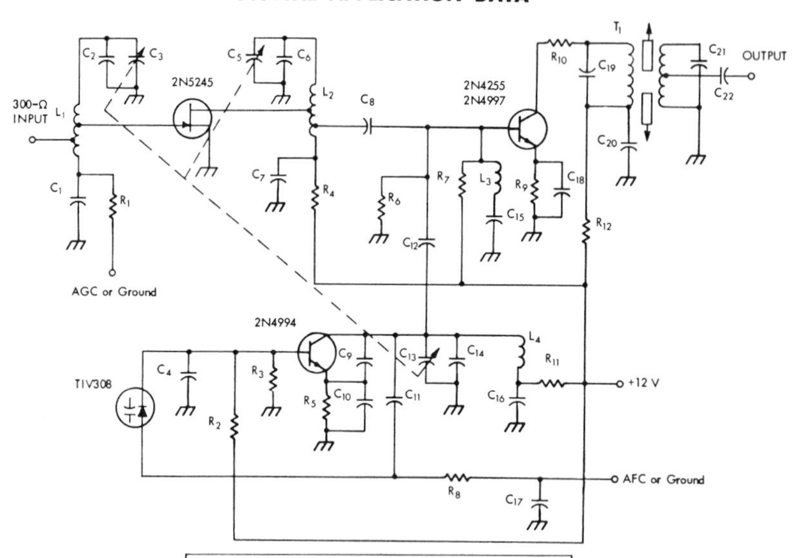

TYPICAL TUNER PERFORMANCE AT $f_0 = 98$ MHz	
Image Rejection (119.4 MHz)	47 dB
$f_0 + \frac{1}{2}$ IF Rejection (103.35 MHz)	73 dB
Sensitivity for 30 dB $\frac{S+N}{N}$ (±75-kHz deviation)	2.3 μV
Sensitivity for 30 dB $\frac{S+N}{N}$ (±22.5-kHz deviation)	3.4 μV
Voltage Gain from Input to Primary of IF Transformer	37 dB

CIRCUIT COMPONENT INFORMATION

CAPACITORS

C_1: 0.001 μF
C_2: 10 pF
C_3: †
C_4: 0.001 μF
C_5: †
C_6: 10 pF
C_7: 0.001 μF
C_8: 12 pF
C_9: 4.7 pF
C_{10}: 6.8 pF
C_{11}: 4.7 pF
C_{12}: 1.2 pF
C_{13}: †
C_{14}: 10 pF
C_{15}: 240 pF
C_{16}: 0.001 μF
C_{17}: 0.1 μF
C_{18}: 0.01 μF
C_{19}: 47 pF
C_{20}: 0.01 μF
C_{21}: 100 pF
C_{22}: 0.01 μF

†Three-gang, 6–21 pF each, with trimmers.

RESISTORS

R_1: 27 kΩ
R_2: 10 kΩ
R_3: 2.7 kΩ
R_4: 330 Ω
R_5: 1 kΩ
R_6: 2.7 kΩ
R_7: 10 kΩ
R_8: 330 kΩ
R_9: 820 Ω
R_{10}: 120 Ω
R_{11}: 330 Ω
R_{12}: 330 Ω
All resistors ½ W, ten percent tolerance

TRANSFORMER

T_1: 10.7 MHz IF transformer

COILS

L_1: 2.5 T, #16 bus, ¼″ ID, carbonyl "E" core, tapped at 1 T and 2 T from bottom
L_2: 4 T, #16 bus, ¼″ ID, air core, tapped at 1.3 T and 1 T from bottom
L_3: 1 μH
L_4: 3 T, #16 bus, ¼″ ID, carbonyl "E" core

FIGURE 19 — TYPICAL FM TUNER

SUMMARY OF CONCEPTS

1. The basis of the FET is a doped channel through which current must flow, which is bounded by two *NP* junctions and their depleted regions.
2. The size of the FET channel depends on the width of the depleted regions.
3. Reverse bias on both junctions can deplete the channel of current carriers and pinch off current flow.
4. The entrance to the FET channel is called the source and the exit is called the drain.
5. The depleted regions in the channel form a wedge shape due to the voltage gradient that results from channel current.
6. Pinchoff voltage is the drain-source voltage at which the rate of change in drain current approaches zero and the drain current becomes constant.
7. When the gate-source voltage is equal to $-V_P$, the drain current approaches zero; that is, the current is cut off.
8. The main operational difference between *N*-channel and *P*-channel FETs is reversal of the supply-voltage polarity.
9. The input impedance of a junction FET is extremely high and is limited only by the leakage current, which is normally less than 1 nA for high-quality devices.
10. The depletion-type MOS-FET operates in a circuit much like the junction FET, but with a much higher input impedance.
11. The MOS-FET has a thin glass insulating layer between the gate contact and the channel, which accounts for its extremely low gate leakage current and input impedances up to 10^{15} Ω.
12. The MOS-FET will work with a positive as well as negative gate-source bias, since there is no junction to be concerned about.
13. The enhancement-type MOS-FET depends on a forward bias from gate to source to enhance a channel in order for current to flow; hence a slight threshold bias voltage is needed for operation.
14. The design of the FET ac small-signal amplifier is similar to that of a vacuum-tube amplifier.
15. The output signal from an FET amplifier is limited by the pinchoff voltage of the FET.

16. The use of a higher supply voltage or a larger load resistance will increase the gain of the FET amplifier.

17. The voltage gain of the FET amplifier may be computed from the transadmittance Y_{fs} and the size of the load resistance R_L ($A_v = Y_{fs}R_L$).

18. The transadmittance varies with the drain current, which is a function of the bias voltage [$Y_{fs} = (2I_{DSS}/V_P)(1 - |V_{GS}|/V_P)$].

GLOSSARY

Depletion-type MOS-FET An insulated-gate FET that has a normal current channel, but whose gate is insulated from the channel by a layer of silicon dioxide. Control is by electrostatic depletion of the channel with changes in gate voltage.

Drain The current exit terminal from the channel of the FET.

Enhancement-type MOS-FET An insulated-gate FET that has no true channel, but depends on the enhancement of a channel by attraction of carriers into the channel area by a bias voltage on the gate. Operation is limited to bias voltages above a threshold voltage.

Gate The current control terminal on the FET. It may be a reverse-biased junction with the channel, or insulated from the channel and connected only by the electrostatic field.

Pinchoff The drain-source voltage that pinches off the current to a constant-current level, so that the rate of current increase approaches zero.

Source The terminal at which current enters the channel of the FET.

Transadmittance Transfer admittance, the ratio of output signal current to input signal voltage for an FET amplifier [$Y_{fs} = \Delta I_D / \Delta V_{GS} = 2I_{DSS}/V_P (1 - |V_{GS}|/V_P)$]. It is similar to the transconductance of a vacuum tube, but is the reciprocal of impedance, not of resistance.

REFERENCES

Altman, Lawrence: "New MOS Technique Points Way to Junctionless Devices," *Electronics*, May 11, 1970, pp. 112–118.

Applications of the Silicon Planar II MOS-FET, Application Bulletin APP-109, Fairchild Semiconductor, Mountain View, Calif., 1964.

Brazee, James G.: *Semiconductor and Tube Electronics*, Holt, Rinehart and Winston, Inc., New York, 1968, pp. 236–348.

Characteristics of Unipolar Field-effect Transistors, Application Tip, Siliconix, Inc., Sunnyvale, Calif., 1963.

Electronics: Circuit Designer's Casebook, vols. I and II, McGraw-Hill Book Company, New York, 1969, 1970.

FET Circuit Ideas, Application Tip, Siliconix, Inc., Sunnyvale, Calif., 1966.

FET Theory, Application Notes 1–3, Amelco Semiconductor, Mountain View, Calif., 1962–1963.

Garner, Louis E.: "Meet Mr. FET," *Popular Electronics*, Feb., 1967, pp. 47–55.

Noll, Edward M.: *FET Principles, Experiments, and Projects*, The Bobbs-Merrill Company, Inc., Indianapolis, 1968.

Preferred Semiconductors and Components from Texas Instruments, Texas Instruments Inc., Catalog CC202, Dallas, Tex., 1970, pp. 6703–6710.

The Relationship between V_P, I_{DSS}, and g_{fs}, Application Tip, Siliconix, Inc., Sunnyvale, Calif., 1964.

Sevin, Leonce J.: *Field-effect Transistors*, Texas Instruments, Inc., Electronics Series, McGraw-Hill Book Company, New York, 1965.

—— "A Simple Expression for the Transfer Characteristic of FETs," *Electronic Equipment Engineering*, vol. 11, pp. 58–61, August, 1963.

Sherwin, James S.: "Build Better Source Followers 10 Ways," *Electronic Design*, vol. 18:12, p. 80, June 7, 1970.

—— "Liberate Your FET Amplifier," *Electronic Design*, vol. 18:11, p. 78, May 24, 1970.

Shockley, W.: "A Unipolar Field-effect Transistor," *Proc. IRE*, vol. 40, pp. 1365–1376, November, 1952.

Silicon "Top" Epitaxial Field-effect Transistors, Application Notes, vol. 1, Dickson Electronics Corp., Scottsdale, Ariz., 1968, pp. 1–12.

REVIEW QUESTIONS

1. What are the primary differences between the bipolar transistor and the junction FET?

2. What factors control the amount of current in the FET channel?

3. Why does the depleted region in an operating junction FET take on a wedge shape?

4. What is meant by the drain pinchoff voltage?

5. How does the voltage gradient of the channel cause pinchoff?

6. What is the difference between the gate-source voltage needed to close off the FET channel and the drain pinchoff voltage?

7. How does the depletion-type MOS-FET differ from the junction FET?

8. What precautions must be taken in handling MOS-FETs?

9. Why does the FET have such a high input impedance in comparison to the bipolar transistor?

10. At what drain current is the voltage gain highest in an FET amplifier circuit?

11. What factors must be considered in determining the forward transadmittance of an FET?

12. In what ways may gate-source bias be provided in an FET amplifier circuit?

13. Why does a larger load resistance generally provide a larger voltage gain in an FET amplifier?

14. Why should the supply voltage always be larger than the pinchoff voltage of the FET?

15. What is the advantage of using as high a supply voltage as possible for an FET amplifier?

16. What common basing arrangements are used for FETs in metal cases?

17. Why do some FETs have four leads instead of three?

18. How would a *P*-channel circuit be converted to an *N*-channel circuit?

PROBLEMS

1. If the pinchoff voltage for a junction FET is 4.0 V, at what gate-source voltage will the drain current be essentially zero?

2. On the basis of the characteristic curves of Fig. 13-4, what is the drain current if the gate-source voltage is −0.5 V and the supply voltage is 30 V?

3. In a circuit like that of Fig. 13-5, if the gate-channel resistance is 1×10^9 and a gate-ground resistance of 470 kΩ is used, what is the circuit input impedance?

4. In a circuit like that of Fig. 13-8, what source resistance would be needed to produce a source bias of −1.5 V? (Use the characteristic curves of Fig. 13-8*b*.)

5. With the current determined in Prob. 4, what size of load resistance is needed to yield a drain-source voltage of about 8 V? Use a standard-value resistance and assume $V_{DD} = 17$ V.

6. In the amplifier of Prob. 4, if the input signal is 0.6 V *p-p* and the output signal is 4.8 V *p-p*, what is the voltage gain?

7. If an amplifier has an input-signal voltage of 0.4 V *p-p* and an output-signal drain current of 0.64 mA *p-p*, what is the circuit transadmittance?

8. When the supply voltage is 25 V, the source voltage is 2.0 V, and the load voltage is 12.0 V, what is the drain-source voltage?

9. In Prob. 8, if the drain current is 3.0 mA, what is the FET channel resistance?

10. If the transadmittance of a certain FET is 4.5 mS, what is the voltage gain if the load resistance is 5.6 kΩ?

11. If the pinchoff voltage of an FET is 3.5 V and $I_{DSS} = 4.0$ mA, and this FET is used in an amplifier biased at −1.0 V, what is the transadmittance?

12. When the signal voltage measured on a scope is 9.4 V *p-p* and the load resistance is 4.7 kΩ, what is the signal current?

13. In Fig. 13-10, what load resistance is represented by the load line that extends from 20 V to 4 mA?

14. In Fig. 13-10, what is the resistance represented by the load line that extends from 40 V to 4 mA?

15. For the conditions shown in Fig. 13-10, if $V_P = 3.0$ V, what is the transadmittance?

16. In Fig. 13-11, what are the approximate voltage gains for each load line?

17. If the leakage current in an *N*-channel FET is 1.0 nA, what is the voltage across a gate resistance of 470 kΩ?

CHAPTER 14

OBJECTIVES

1. To discuss the gains possible with more than one active device
2. To compare the various types of interstate coupling
3. To consider the impedance-matching and supply-voltage problems in coupling
4. To develop design criteria for cascade amplifiers
5. To consider the frequency limitations of multistage amplifiers
6. To explore the use of the FET and integrated circuitry in cascade amplifiers

CASCADE AMPLIFIERS

Now that we have covered the fundamentals of single-stage amplifiers, bias stabilization, gain, impedance matching, and the other aspects of transistor operation, we can turn to the fun of solid-state design. It is when we actually begin to put a system together that it begins to have meaning.

The single-stage amplifier is certainly limited in that it rarely provides the desired voltage gain or current gain at the proper conditions. Moreover, it is always necessary to match impedances to a signal transducer on the input and to some power transducer on the output. The cascade amplifier is a significant part of the solution to such problems.

Advances in solid-stage design have put the multistage amplifier into a tiny integrated-circuit chip. The whole package needed to produce several Watts of high-fidelity sound is available in integrated-circuit form. However, to understand the operation of such circuits we must first analyze the system in its discrete form.

INTERSTAGE COUPLING

Voltage gains of several hundred are possible with a single-stage transistor amplifier. However, when the amplifier is stabilized, the voltage gain frequently drops to as low as 10 or 20, depending on the design. For this reason two or more stages are often *cascaded* to increase the overall gain. For example, if one stage of amplification is 15 and the second is 20, the two stages acting together as a cascade amplifier have a total gain of $15 \times 20 = 300$. Two transistors each having a current gain h_{fe} of 100 would in cascade have a current gain of 10,000!

Frequently there is an impedance-matching problem if a single-stage amplifier is used. An emitter-follower stage is often used in front of a common-emitter amplifier to provide the necessary high input impedance. Sometimes, when a very low input impedance is needed, a common-base circuit is used.

In phonograph preamplifiers, where a high input impedance must be coupled to a high voltage gain and a low output impedance, there is also a need for frequency compensation. This may require three or four stages of amplification, some direct and some *RC* coupled. Where RF amplifiers are used, matching between stages must be considered along with selected frequency bandpass. This can be accomplished by means of a tuned *LC* or transformer coupling. Within any electronic system no single stage of amplification operates by itself. It must frequently operate with and be coupled to dozens of other active subsystems.

Three methods of coupling are in common use, although others are available: *RC* coupling, direct coupling, and transformer coupling. Each type has specific uses. For example, in a common-emitter or common-source amplifier the output from the collector or drain is at a much higher voltage than the input at the base or gate. If these two points are connected directly, the voltage differences will immediately stabilize by voltage and current division. This, of course, upsets the circuit and can damage it. In this case *RC* or capacitor coupling will supply the needed isolation.

RC COUPLING

Two independent stages of amplification may be connected together as shown in Fig. 14-1. The only change in the circuitry is the addition of a capacitor between the collector of the first stage and the base of the second stage. There are a multitude of considerations in selecting specific values for the required gain and frequency bandpass, but the concept itself is simple.

At the collector of the first transistor in Fig. 14-1 there is a voltage of about 8.5 V. At the base of the second transistor the voltage is about

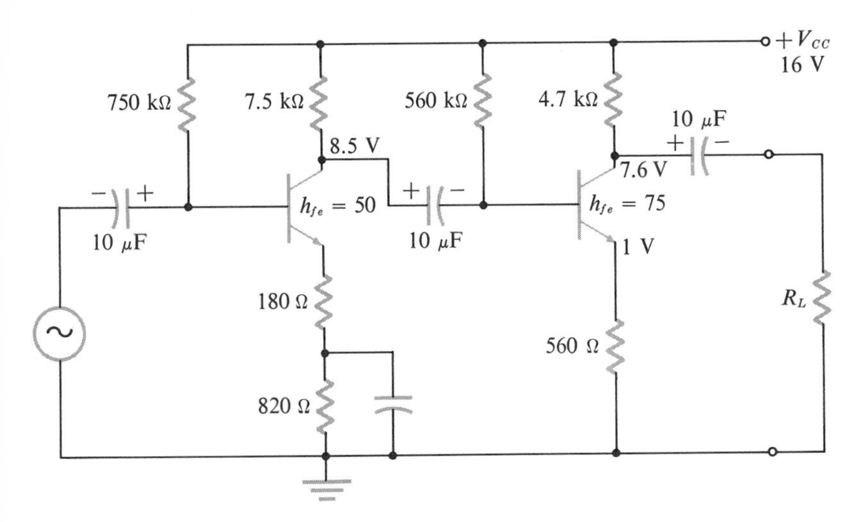

FIG. 14-1 A two-stage capacitively coupled amplifier.

1.7 V. This means that a potential difference exists across the capacitor in the amount of 8.5 V − 1.7 V = 6.8 V. The capacitor that isolates these two voltages must be able to withstand this difference constantly, as well as occasional excursions up to 16 V.

The choice of the coupling capacitor is by logic. It must withstand the potential difference without puncturing and with minimum attenuation of the signal. Since the reactance of the capacitor rises as the frequency of the signal drops, the capacity must be computed at the lowest signal frequency to be used. For an audio amplifier this is usually 20 to 50 Hz.

Without going into the reasons at this point, the reactance of the coupling capacitor should be equal to or less than the summation of the output impedance of the first stage and the input impedance of the second stage at the low-frequency half-power point. Since the capacitive reactance X_C is given by

$$X_C = Z_{\text{out}} + Z_{\text{in}} = \frac{1}{2\pi f C} \tag{14-1}$$

the *coupling capacitance* C_C is given by the variation

$$C_C = \frac{1}{2\pi f(Z_{\text{out}} + Z_{\text{in}})} \tag{14-2}$$

The output impedance from stage 1 is approximately the load resistance of 7.5 kΩ. The input impedance to stage 2 is approximately 39 kΩ, which is the parallel summation of 560 kΩ $\|[h_{ie} + (75 \times 560)]$. Substitution of these values into Eq. (14-2) gives us the actual capacitance needed. The next larger common size is generally used. Thus, since $1/2\pi = 0.159$,

$$C_C = \frac{0.159}{20 \text{ Hz } (7.5 \text{ k}\Omega + 39 \text{ k}\Omega)} = \frac{0.159}{930 \times 10^3}$$

$$= 0.17 \times 10^{-6} \text{ F} = 0.17\ \mu\text{F} \cong 0.2\ \mu\text{F} \qquad (14\text{-}3)$$

When the signal is in the midrange of frequencies, 1 to 10 kHz, the coupling capacitors have such small reactances that they act as short circuits to the signal.

DIRECT COUPLING

Direct coupling is an outgrowth of the concept of voltage-divider bias. The transistor acts as one of the resistances in the divider string. If the voltage at the collector of the first transistor is determined, the voltage at the base of the second transistor is designed to be this same value. If the two points to be connected are of the same potential, they may be connected together directly without any need for isolation.

As shown in Fig. 14-2, there is no need for a base-current-limiting resistance for the second stage. The tiny base current passes through the load resistance of the first stage. Since the collector current is so much larger than the base current, the collector voltage remains essentially constant when the two circuits are connected.

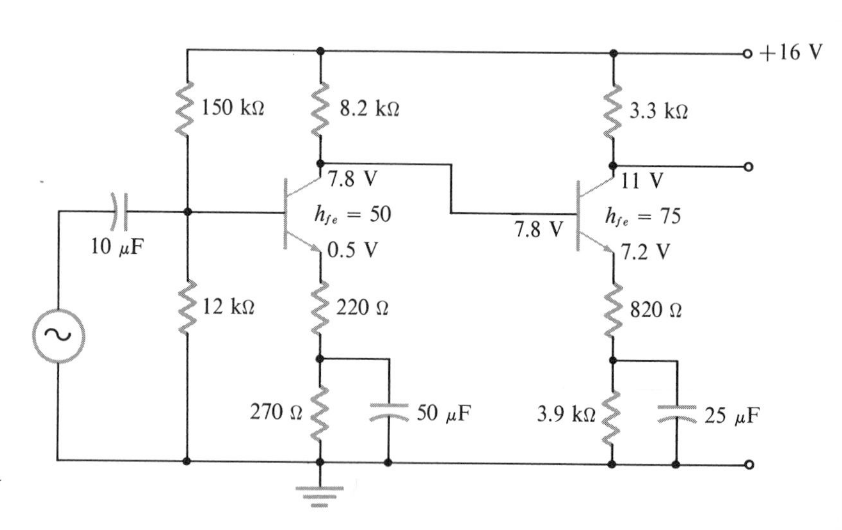

FIG. 14-2 A two-stage direct-coupled transistor amplifier.

The major concern in the direct-coupled amplifier is that the second stage must not load down the ac signal generated in the first stage or change operating points with changes in temperature. This sometimes requires that the collector supply voltage of the first stage be lower than that of the second stage. The voltage difference is usually provided by a simple bypassed series resistance, but it is rarely needed except where more than two stages are directly coupled.

When a two-stage amplifier is directly coupled, the frequency response is down to direct current because any voltage change in the input is immediately followed by a change at the output. The high-frequency response is limited by the junction and wiring capacitances distributed throughout the circuit, since they act to shunt the high-frequency signals to ground by their low impedances.

Perhaps the simplest form of direct coupling is an *NPN* transistor for the first stage coupled to a *PNP* transistor for the second. Since they act in a complementary fashion, they tend to be self-controlling. Figure 14-3 shows how complementary transistors may be used in a circuit. In this situation a three-stage amplifier may be used without the usual power-supply level problems. The overall gain is the product of the three individual gains.

The Darlington circuit Frequently a circuit design calls for very high current gain and high input impedance. With the *Darlington configuration* of direct or compound coupling, shown in Fig. 14-4, both these needs can be fulfilled. Note that the base current of the second transistor is the emitter current of the first. The two collectors are usually connected together to prevent circuit current loss. In some

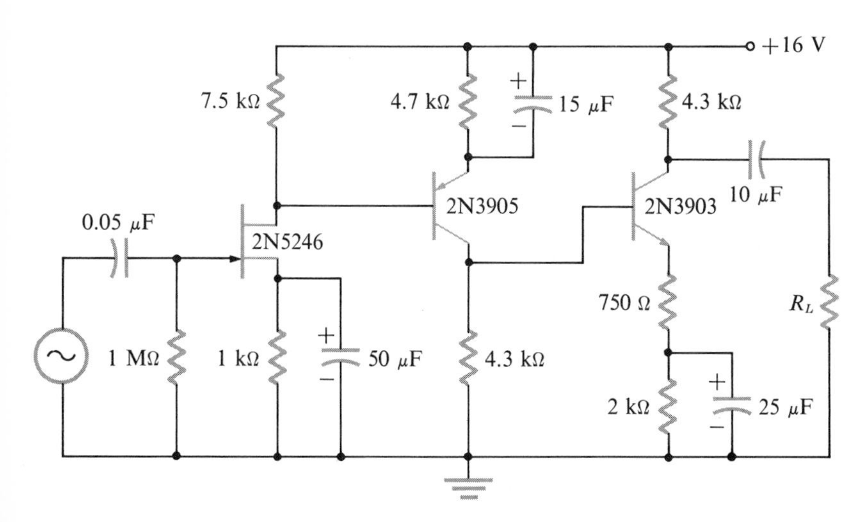

FIG. 14-3 A three-stage direct-coupled amplifier using complementary transistors and an FET first stage for high-impedance input.

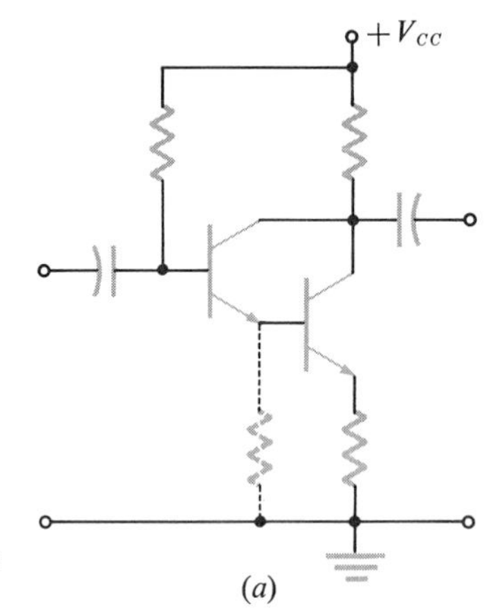

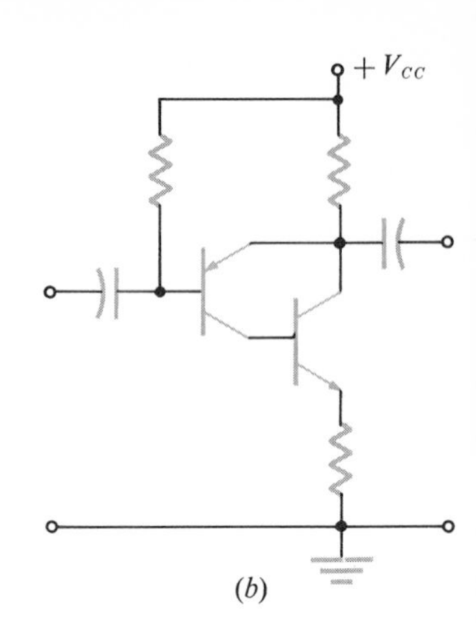

FIG. 14-4 The Darlington amplifier configuration. (*a*) The normal *NPN* pair with emitter bias on the first transistor; (*b*) the *PNP-NPN* amplifier showing how the *PNP* collector is the effective emitter.

cases for added bias stabilization, the emitter of the first transistor is returned to ground through a separate resistance.

The Darlington configuration is often employed in power amplifiers because it permits the use of complementary transistors for automatic phase inversion. Figure 14-4*b* indicates how the collector of the *PNP* transistor is the effective emitter in terms of electron current. A negative-going signal at the base of the input transistor will produce a positive input to the *NPN* transistor in stage 2.

TRANSFORMER COUPLING

Transformer coupling is generally used in RF amplifiers because it provides good impedance matching and some gain and allows for a tuned bandpass. Figure 14-5 shows a typical two-stage intermediate-frequency amplifier for use in a superhet radio. Circuits such as this will operate properly only when they are laid out on an etched circuit board or in a compact chassis. The usual clip-lead experimental layout involves too much wiring capacitance and inductance for proper operation.

The dc biasing for transformer coupling is much the same as for the capacitor coupling. There must still be some current limiting in the base circuits, or collector currents may become excessive. The ac load impedance is developed by the RF collector currents, which generate very high inductive reactances.

Note in Fig. 14-5 that the collector of each transistor connects into the coupling transformer at a primary tap. This allows for proper impedance matching as well as for proper loading. Note that the load impedance Z_L reflected back to the primary, and thus to the collector,

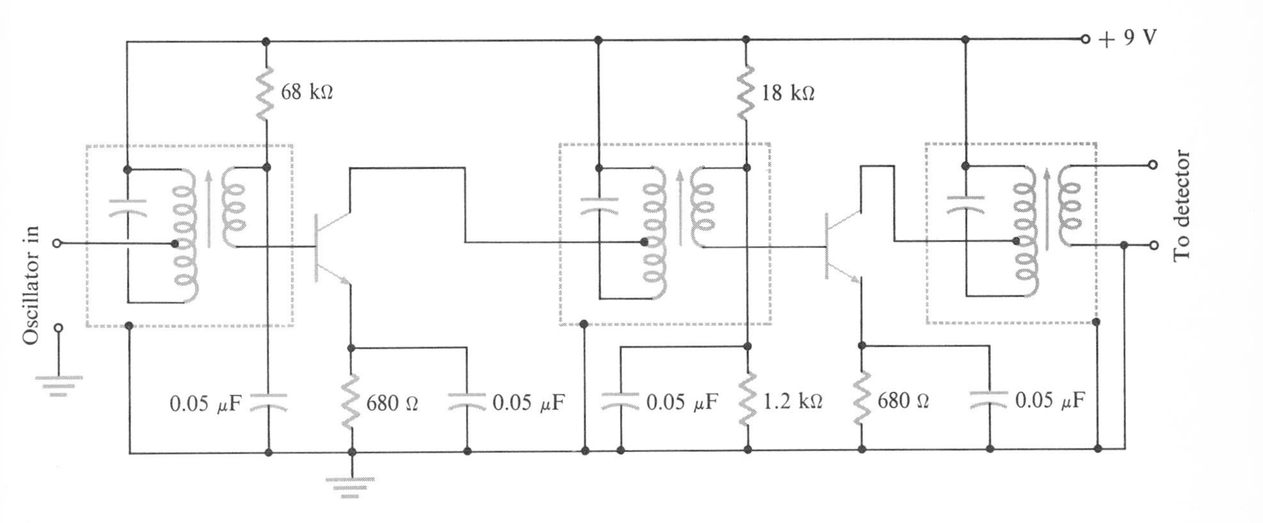

FIG. 14-5 A typical transformer-coupled intermediate-frequency amplifier.

depends on the turns ratio squared and the load R_L connected to the secondary. That is,

$$Z_L = \left(\frac{N_P}{N_S}\right)^2 R_L \tag{14-4}$$

where Z_L = reflected impedance
R_L = secondary load
N_P = number of turns on the primary
N_S = number of turns on the secondary

The transformers are tuned to the right frequency range by means of a fixed capacitor in parallel with the primary and an adjustable core which will vary the inductance.

THE AC LOAD LINE

So far in our discussions of voltage amplification we have computed the gain on the basis of the dc load line. However, when any amplifier is connected to its proper external load, unless the load impedance is very large, it will reduce the impedance seen by the transistor. This in turn reduces the voltage gain of the circuit.

Consider the circuit of Fig. 14-6. The input impedance to the second stage, which is the external load, is simulated by the resistance R_{in}. This is, of course, seen by the first transistor as in parallel with its

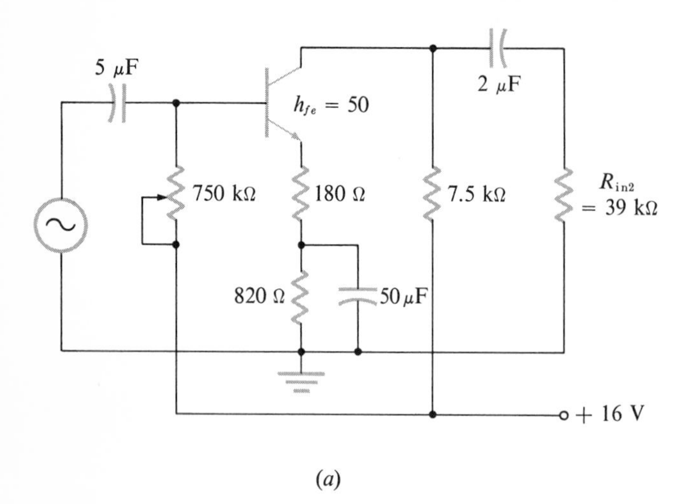

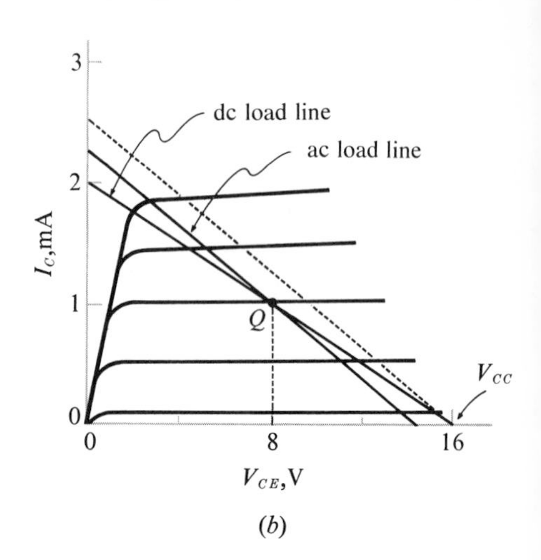

FIG. 14-6 (*a*) A circuit for simulating a two-stage *RC* coupling for computation of the ac load line and (*b*) construction of the ac load line at V_{CC} with a parallel transfer to the *Q* point.

load resistance. Ideally the input impedance to the second stage should be at least 10 times greater than the output impedance of the first stage. In reality this is rarely possible; in fact it is often less than 5 times greater. The consequence is a lower voltage gain than would be predicted for a first stage not loaded by the second.

With a first-stage output impedance of about 7.5 kΩ, which is acting in parallel with an input impedance of 39 kΩ for the second stage, the load impedance seen by the first transistor is about 6.3 kΩ. That is,

$$R_{L,\text{ ac}} = R_L \parallel R_{\text{in}} = \frac{R_L R_{\text{in}}}{R_L + R_{\text{in}}}$$

$$= \frac{7.5\text{ k}\Omega \times 39\text{ k}\Omega}{7.5\text{ k}\Omega + 39\text{ k}\Omega} = 6.3 \qquad (14\text{-}5)$$

Since the ac load line must pass through the *Q* point of the amplifier on the characteristic curves, it is usually easier to plot the line's slope at the supply voltage and then move it to the *Q* point, as shown in Fig. 14-6*b*. A load line representing 6.3 kΩ when the supply voltage is 16 V would intersect the *Y* axis at 2.55 V. Once this line is constructed, construct a line parallel to it but passing through the *Q* point.

Whenever the ac load line is not the same as the dc load line the voltage gain must be computed on the basis of the ac load impedance.

Remember that the voltage gain is about the same as the resistance gain with R_E unbypassed. The resistance gain is approximately $R_L/h_{ib} + R_E$, and so

$$A_{v1} \cong \frac{-R_L}{h_{ib} + R_E} \cong \frac{-7.5\ \text{k}\Omega}{30 + 180} = -36 \tag{14-6}$$

$$A_{v2} \cong \frac{-R_{L,\,\text{ac}}}{h_{ib} + R_E} \cong \frac{-6.3\ \text{k}\Omega}{30 + 180} = -30 \tag{14-7}$$

RC COUPLING VERSUS DIRECT COUPLING

To clarify the relationship between *RC* and direct coupling, two single-stage amplifiers should be designed, checked, and then coupled together with a capacitor. When the *RC* coupling has been verified, the capacitor should be removed, along with the base resistance for the second stage. The two stages should then be coupled directly and checked out again.

Let us begin with the following design parameters:

Supply voltage = 16 V
Two silicon transistors with $h_{FE} = 100$ at $I_C = 1$ mA
Input impedance to stage 1 at least 5 kΩ
Voltage gain of stage 1 at least 60
Voltage gain of stage 2 at least 4
Collector voltage of stage 1 = base voltage of stage 2

With these parameters understood, establish the transistor terminal voltages and the desired currents. Voltages should be chosen to reflect standard resistance values and impedance matching.

AMPLIFIER 1

Since direct-coupled amplifiers are subject to drift, the first stage should be quite stable, perhaps with voltage-divider biasing. If the transistor is to use a collector current of 1.0 mA, the base current will be 10 μA. In order to maintain the required 5 kΩ of input impedance an emitter resistance will have to be left at least partially unbypassed. In addition, the base-ground input resistance will have to be large enough not to reduce the input impedance developed by the emitter resistance. This requires at least 1.0 V at the base.

Since the junction voltage of a silicon transistor is a little bit over 0.6 V at 1 mA, the emitter resistance should be about 470 Ω. This will give an emitter voltage of about 0.5 V and a base voltage of about 1.1 V.

The supply voltage of 16 V must be divided at the base to give about 1.1 V at a bleeder current 10 times the base current. The gain h_{FE} of the transistor is 100, and so the base current at a collector current of 1.0 mA will be 10 μA, or 0.01 mA. A bleeder current of 0.1 mA would be needed. From Ohm's law, the base-ground resistance should be 11 kΩ; the nearest standard value is 12 kΩ.

The voltage across the other resistance in the base circuit of Fig. 14-7*a* is 16 V − 1.1 V = 14.9 V. Hence at 0.1 mA this resistance would be about 150 kΩ.

The input impedance must be at least 5 kΩ, and so the parallel summation of 12 and 150 kΩ must be determined. They form an input resistance of about 10 kΩ, which will be in parallel with the transistor input impedance. Thus if 5 kΩ of input impedance is to be maintained, the input impedance to the transistor must be at least 10 kΩ: 10 kΩ‖10 kΩ = 5 kΩ.

The h_{fe} value of the transistor is about 100, and so approximately 100 Ω of the emitter resistance should be left unbypassed to yield the 10-kΩ input impedance [$Z_{\text{in}} = h_{ie} + (1 + h_{fe})R_E$]. Since the emitter resistance is set at about 470 Ω, if a 360- and a 120-Ω resistance are substituted and the 360-Ω resistance is properly bypassed, the input impedance and the dc bias will be preserved.

One-half the supply voltage should be across the load resistance, and at 1.0 mA the nearest standard value is 8.2 kΩ. This completes the design of the first-stage amplifier, as summarized in Fig. 14-7.

FIG. 14-7 Two-stage amplifier design; (*a*) the input-stage amplifier, (*b*) the second-stage amplifier, and (*c*) approximate design voltages for the two-stage amplifier.

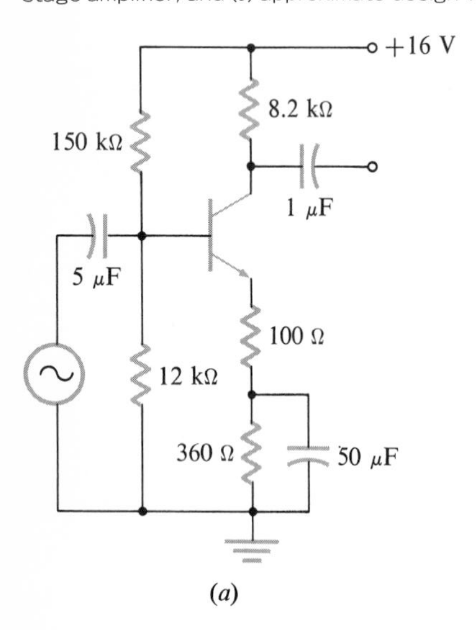

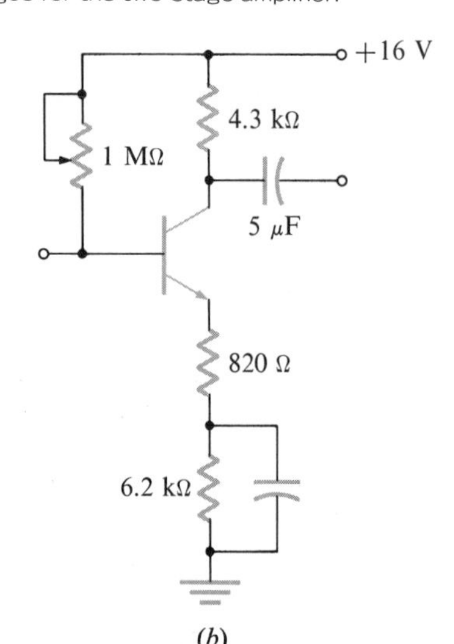

	T_1	T_2
V_E	0.5 V	7.2 V
V_B	1.1 V	7.8 V
V_C	7.8 V	11.7 V
V_{R_L}	8.2 V	4.3 V

(*c*)

AMPLIFIER 2

According to the design parameters, the base voltage of stage 2 must be the same as the collector voltage of stage 1: 7.8 V. The emitter voltage will be about 0.6 V lower, or about 7.2 V. At 1.0 mA of current the emitter resistance should be about 7.2 kΩ. However, so that the input impedance of stage 2 does not appreciably load stage 1, it should be about 10 times larger than the stage 1 output impedance. This would require about 82 kΩ of input impedance.

The fact that $h_{fe} = 100$ suggests the same technique used in amplifier 1—leaving part of the emitter resistance unbypassed. At least 820 Ω is needed, which leaves about 6,380 Ω to be bypassed. The nearest standard value of 6.2 kΩ should be used. It will allow a small amount of extra emitter current, but this will not matter except for a slight raise in the voltage gain.

The base resistance, which will be removed in the second part of the experiment, should be either an 820-kΩ resistance or a potentiometer adjusted to give the proper voltages.

The load resistance should have one-half the remaining 8.8 V across it. This would be 4.4 V, which at 1.0 mA would require a resistance of 4.4 kΩ. The nearest value is 4.3 kΩ.

Connect each of the amplifiers as shown in Fig. 14-7. If transistors with $h_{fe} = 100$ are not available, use some other value and adjust the R_B values to give the voltages shown in Fig. 14-7.

Apply power to each circuit. Measure each voltage and record the values. If the base voltage of amplifier 2 is not the same as the collector voltage of amplifier 1, adjust the base potentiometer until they match. Measure and record the voltage gain and input impedances to each stage and compare the measured values with those predicted.

Disconnect the power from both circuits. Connect the coupling capacitor from the collector of amplifier 1 to the base of amplifier 2. Reconnect the power, and repeat the previous measurements. Did the input impedance change significantly? Why? Did the total circuit gain compare favorably with the product of the two individual gains? Was there appreciable loading of the first stage when the second stage was attached?

Disconnect the power from the circuit. Remove the coupling capacitor and the base-current-adjusting potentiometer from amplifier 2. Connect the collector of amplifier 1 directly to the base of amplifier 2. Repeat the same experiments as before.

Were the results with direct coupling significantly different from those with capacitor coupling? If so, how do you account for these changes in terms of the basic concepts? (There should be very little difference.)

THE DIFFERENTIAL AMPLIFIER CIRCUIT

The workhorse of most linear integrated circuits is the differential amplifier circuit. It is used at the input of practically every operational amplifier and other circuits because of its flexibility and its ability to screen out a common-mode signal from two separate signals.

Because no two transistors are ever identical or operate identically, the differential amplifier had to await the development of integrated circuitry. It requires two matched transistors, which can be achieved to very close tolerance with integrated circuitry because all the transistors are on the same silicon chip. Previously, transistors were matched by hand, with matched pairs priced as high as $45. Now whole operational amplifier integrated circuits using the differential input are available for only a dollar or so.

Figure 14-8 shows two common forms of the differential amplifier, which differ only in the method by which constant current is developed.

As its name implies, the differential amplifier amplifies the difference between two signals. To do this the circuit requires a constant current. If the constant current is 4 mA, and the amplifiers are matched exactly, a current of 2.0 mA will flow through each transistor. If one amplifier has a higher input voltage than the other, more current will pass through that transistor, leaving the remainder of the 4 mA for the second transistor. The voltages at the collectors will reflect the

FIG. 14-8 The differential amplifier: (*a*) two inputs and a resistive balance and constant-current source; (*b*) a constant-current (FET) diode as the constant-current source.

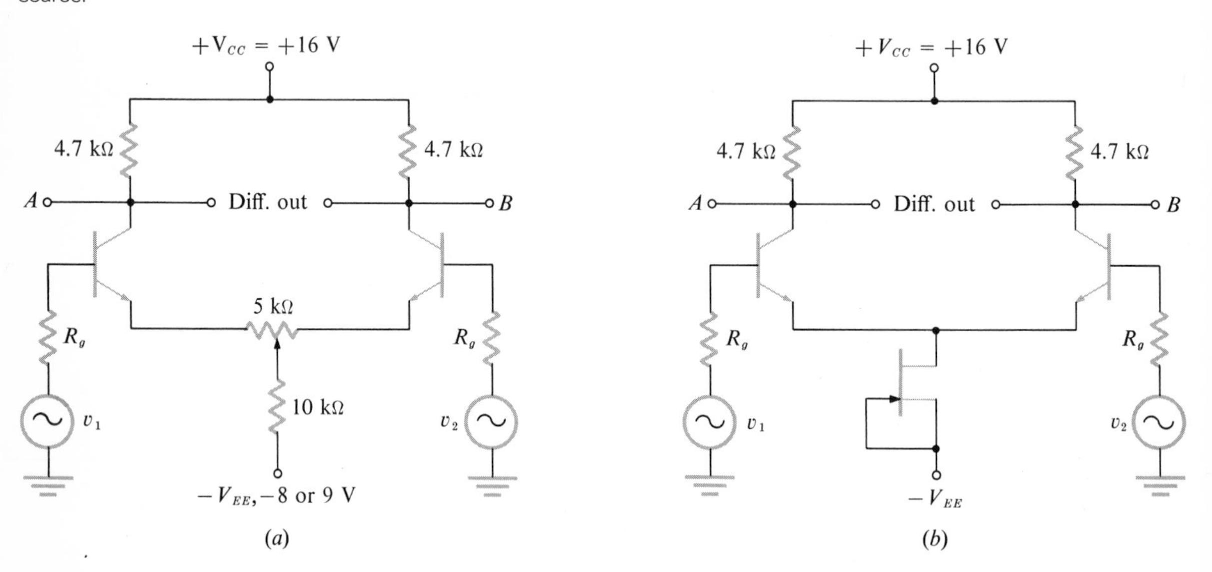

amount of difference between the two inputs. The difference in the outputs of the two collectors will be an amplified version of the difference between the two inputs. In fact, the output from either of the collectors to ground will be an amplified version of the difference between the two signals. However, the collector-collector output is just twice as large as that between the one collector and ground.

In Fig. 14-8, for example, an input at v_1 will be amplified and appear inverted at A and noninverted at B. An input at v_2 will appear noninverted at A and inverted at B. The opposite input is considered as zero, and so the circuit still amplifies the difference between the two generators. In both cases, however, the generator or a resistance must be connected, since the bias for the amplifier is provided by the generator. If only one generator is used, as in Fig. 14-9 or Fig. 14-10*b*, the base of the second transistor is simply returned to ground. Ground, remember, is at some point between $-V_{EE}$ and $+V_{CC}$.

CONSTANT-CURRENT GENERATION

The differential amplifier relies heavily on a constant current for error-free operation. When the current increases in one side, it is diminished in the other side by the same amount. Constant current may be provided by a very large resistance and a greatly negative power supply, as in Fig. 14-8*a*. However, this method is wasteful in many respects and is not too accurate. A far simpler and more efficient method is to use the dynamic resistance of a transistor operating in its constant-current region. This is what is usually done in most integrated-circuit operational amplifiers.

Since the advent of the FET, a form known as the constant-current diode has been developed. It consists of an FET whose gate is shorted directly to the source and operating beyond the pinchoff voltage. Figure 14-8*b* shows how it should be connected. It allows a very high dynamic resistance, but keeps the dc resistance relatively low. This tends to overcome the error amplification and increases the common-mode rejection.

COMMON-MODE REJECTION

In addition to its vast applications as a dc amplifier, the differential amplifier has a quality that allows it to reject noise. Hum and noise are frequent problems, especially in computer operation. If they are present on both inputs to a differential amplifier, the noise that is common to both inputs will be cancelled, so that only the difference will be amplified. For this reason differential amplifiers, or systems with differential amplifiers at the input stage, are frequently rated according to their common-mode rejection.

EMITTER-FOLLOWER, COMMON-BASE OPERATION

The differential amplifier is actually a variation of the emitter-follower and common-base amplifiers. In Fig. 14-9 the input v_i is amplified and appears at A in the normal common-emitter inverted form. It also appears at B noninverted and amplified. This is because of the emitter coupling between the two stages of the amplifier. The first stage acts as an emitter follower, coupling into the second emitter with a voltage gain of 1. The second stage is a grounded-base amplifier with a very high voltage gain. In neither of these amplifiers is the signal inverted.

The constant-current-diode FET is included in Fig. 14-9 for simplicity and to help the emitter-follower gain approach 1. The nearer to constant the current, the better the operation.

If the output is to be taken from only one side of the differential amplifier, since the current is constant there is no need for two load resistances to limit the current. The first stage is going to act as an emitter follower anyway. The output is taken from the second collector and ground as indicated in Fig. 14-10.

The differential amplifier may be used for noninverting or inverting operation or for single input, as shown in Fig. 14-10*b*.

DIFFERENTIAL VOLTAGE GAIN

The voltage gain of a differential amplifier depends on where the output is taken. If the output is single ended—that is, if it is taken from only one of the collectors—the voltage gain is approximately equal to one-half the resistance gain,

$$A_v = \frac{R_L}{2h_{ib}} \qquad (14\text{-}8)$$

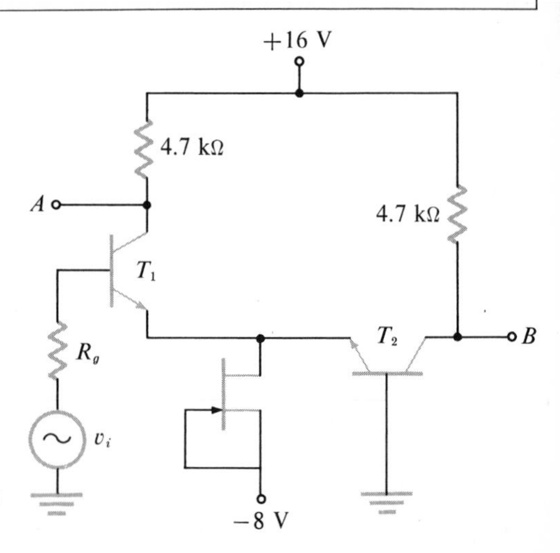

FIG. 14-9 The differential amplifier as a combination emitter-follower and common-base circuit.

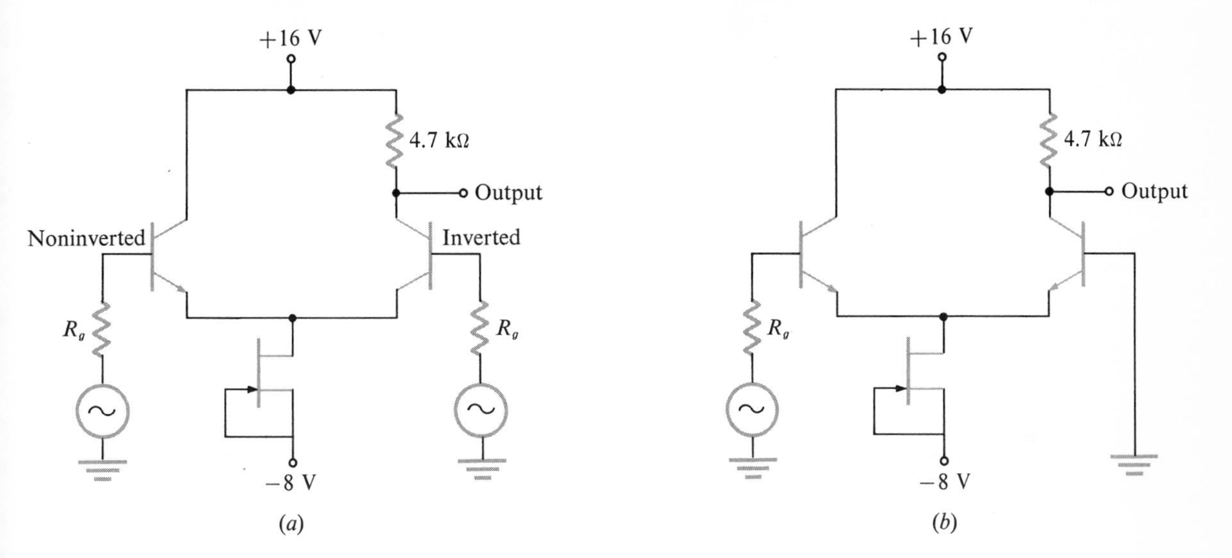

FIG. 14-10 The differential amplifier with only one load resistance: (*a*) two inputs in use; (*b*) one input in use, with the other grounded.

If the differential amplifier has a balanced output—that is, if it is taken from the collector collector—the signal voltage increases in one collector as it decreases in the other. Therefore, as expected, the voltage gain is twice as great as for single-ended operation

$$A_v = \frac{R_L}{h_{ib}} \tag{14-9}$$

If a small resistance is used in the emitter leads for stabilization, it must also affect the voltage gain, since it is difficult to bypass. Thus

$$A_v = \frac{R_L}{h_{ib} + R_E} \tag{14-10}$$

FREQUENCY RESPONSE

With two or more stages of amplification cascaded, an amplifier has a very high gain in the middle frequency range but is usually lacking at

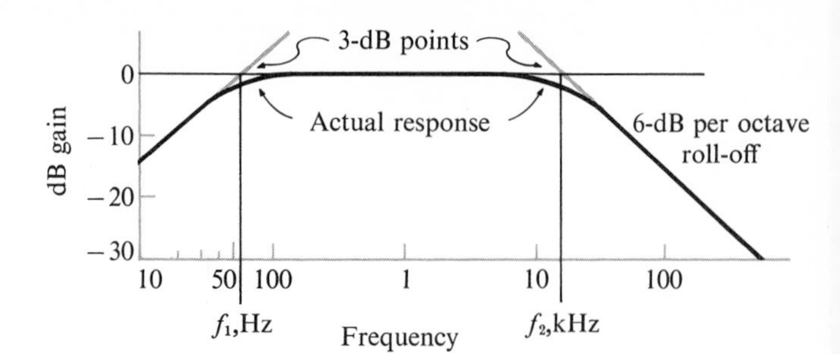

FIG. 14-11 The frequency-response curve for a typical two-stage amplifier, indicating the low-frequency cutoff f_1 and the high-frequency cutoff f_2.

the lower and upper frequencies. The *frequency response* of the amplifier is a measure of its adequacy to amplify signals in a wide band of frequencies. Frequency response generally refers to that band of frequencies that falls between the two frequencies where the voltage gain has dropped to 0.707 times its midfrequency gain. Recall that 0.707 is the sine of 45°. That is, because of the capacitance in the circuit, the signal current and voltage are out of phase by 45° at the lower and upper frequency limits of the frequency-response range. At frequencies closer to midfrequency, the voltage and current are less out of phase and do not tend to cancel each other, and so the gain is higher.

With each stage of amplification, then, the middle range gets the full amplification and the lower and upper ends get less. The response curve is said to *roll off* at a rate of −6 decibels (dB) per octave. Stated another way, this means that each time the frequency doubles, the gain drops to a point 0.5 times its previous value, or −6 dB. Figure 14-11 illustrates a typical two-stage-amplifier response curve. By the same reasoning, each time the frequency drops to one-half its former frequency the gain is also reduced by 6 dB.

THE DECIBEL

The human ear and other devices respond to sound in a logarithmic fashion rather than linearly. For this reason, when sound power level doubles, the ear is just able to detect a difference in sound intensity. Sound power is therefore rated on a logarithmic scale, with the unit of the ratio of two powers being the Bel. Since the Bel is too large a unit for practical use, the decibel, or 1/10 Bel, is used instead.

The formula for determining the number of decibels in the ratio of two power levels is

$$\text{No. of dB} = 10 \log_{10} \frac{P_{\text{out}}}{P_{\text{in}}} = 20 \log_{10} \frac{V_{\text{out}}}{V_{\text{in}}} \qquad (14\text{-}11)$$

For example, if the power output is twice the input, this is computed as

$$\text{No. of dB} = 10 \log_{10} \frac{2}{1} = 10 \log_{10} 2 = 10 \times 0.3$$

$$= 3 \text{ dB} \qquad (14\text{-}12)$$

Similarly, if the output power is one-half the input power, the power gain is −3 dB; that is, the output power is 3 dB less than the input power.

Recall that the logarithm of a number is actually the exponent of 10 that equals the number; for example, $10^{0.3} = 2$. Recall also that

$$\frac{1}{2} = 1/10^{0.3} = 10^{-0.3} \qquad \text{and} \qquad \log_{10} 10^{-0.3} = -0.3$$

When the power level has dropped to one-half its former value, the decibel gain must then be $10 \times -0.3 = -3$ dB.

In terms of voltage gains, the decibel gain must be equal to 20 log (V_{out}/V_{in}), because according to the power formula, the power is equal to the voltage squared divided by the resistance,

$$P = VI = V\frac{V}{R} = \frac{V^2}{R} \qquad (14\text{-}13)$$

When two power levels are to be compared and the voltages and resistances are known, the decibel development usually takes the form

$$\text{No. of dB} = 10 \log \frac{V_{out}^2/R_{out}}{V_{in}^2/R_{in}} = 10 \log \frac{V_{out}^2}{R_{out}} \frac{R_{in}}{V_{in}^2}$$

$$= 10 \log \frac{V_{out}^2}{V_{in}^2} \frac{R_{in}}{R_{out}}$$

$$= 10 \log \frac{V_{out}^2}{V_{in}^2} + 10 \log \frac{R_{in}}{R_{out}} \qquad (14\text{-}14)$$

Therefore, if only the voltage gain is to be considered,

$$\text{No. of dB} = 10 \log \frac{V_{out}^2}{V_{in}^2} \qquad (14\text{-}15)$$

Logarithm table

N	0	1	2	3	4	5	6	7	8	9
10	0000	0043	0086	0128	0170	0212	0253	0294	0334	0374
11	0414	0453	0492	0531	0569	0607	0645	0682	0719	0755
12	0792	0828	0864	0899	0934	0969	1004	1038	1072	1106
13	1139	1173	1206	1239	1271	1303	1335	1367	1399	1430
14	1461	1492	1523	1553	1584	1614	1644	1673	1703	1732
15	1761	1790	1818	1847	1875	1903	1931	1959	1987	2014
16	2041	2068	2095	2122	2148	2175	2201	2227	2253	2279
17	2304	2330	2355	2380	2405	2430	2455	2480	2504	2529
18	2553	2577	2601	2625	2648	2672	2695	2718	2742	2765
19	2788	2810	2833	2856	2878	2900	2923	2945	2967	2989
20	3010	3032	3054	3075	3096	3118	3139	3160	3181	3201
21	3222	3243	3263	3284	3304	3324	3345	3365	3385	3404
22	3424	3444	3464	3483	3502	3522	3541	3560	3579	3598
23	3617	3636	3655	3674	3692	3711	3729	3747	3766	3784
24	3802	3820	3838	3856	3874	3892	3909	3927	3945	3962
25	3979	3997	4014	4031	4048	4065	4082	4099	4116	4133
26	4150	4166	4183	4200	4216	4232	4249	4265	4281	4298
27	4314	4330	4346	4362	4378	4393	4409	4425	4440	4456
28	4472	4487	4502	4518	4533	4548	4564	4579	4594	4609
29	4624	4639	4654	4669	4683	4698	4713	4728	4742	4757
30	4771	4786	4800	4814	4829	4843	4857	4871	4886	4900
31	4914	4928	4942	4955	4969	4983	4997	5011	5024	5038
32	5051	5065	5079	5092	5105	5119	5132	5145	5159	5172
33	5185	5198	5211	5224	5237	5250	5263	5276	5289	5302
34	5315	5328	5340	5353	5366	5378	5391	5403	5416	5428
35	5441	5453	5465	5478	5490	5502	5514	5527	5539	5551
36	5563	5575	5587	5599	5611	5623	5635	5647	5658	5670
37	5682	5694	5705	5717	5729	5740	5752	5763	5775	5786
38	5798	5809	5821	5832	5843	5855	5866	5877	5888	5899
39	5911	5922	5933	5944	5955	5966	5977	5988	5999	6010
40	6021	6031	6042	6053	6064	6075	6085	6096	6107	6117
41	6128	6138	6149	6160	6170	6180	6191	6201	6212	6222
42	6232	6243	6253	6263	6274	6284	6294	6304	6314	6325
43	6335	6345	6355	6365	6375	6385	6395	6405	6415	6425
44	6435	6444	6454	6464	6474	6484	6493	6503	6513	6522
45	6532	6542	6551	6561	6571	6580	6590	6599	6609	6618
46	6628	6637	6646	6656	6665	6675	6684	6693	6702	6712
47	6721	6730	6739	6749	6758	6767	6776	6785	6794	6803
48	6812	6821	6830	6839	6848	6857	6866	6875	6884	6893
49	6902	6911	6920	6928	6937	6946	6955	6964	6972	6981
50	6990	6998	7007	7016	7024	7033	7042	7050	7059	7067
51	7076	7084	7093	7101	7110	7118	7126	7135	7143	7152
52	7160	7168	7177	7185	7193	7202	7210	7218	7226	7235
53	7243	7251	7259	7267	7275	7284	7292	7300	7308	7316
54	7324	7332	7340	7348	7356	7364	7372	7380	7388	7396

N	0	1	2	3	4	5	6	7	8	9
55	7404	7412	7419	7427	7435	7443	7451	7459	7466	7474
56	7482	7490	7497	7505	7513	7520	7528	7536	7543	7551
57	7559	7566	7574	7582	7589	7597	7604	7612	7619	7627
58	7634	7642	7649	7657	7664	7672	7679	7686	7694	7701
59	7709	7716	7723	7731	7738	7745	7752	7760	7767	7774
60	7782	7789	7796	7803	7810	7818	7825	7832	7839	7846
61	7853	7860	7868	7875	7882	7889	7896	7903	7910	7917
62	7924	7931	7938	7945	7952	7959	7966	7973	7980	7987
63	7993	8000	8007	8014	8021	8028	8035	8041	8048	8055
64	8062	8069	8075	8082	8089	8096	8102	8109	8116	8122
65	8129	8136	8142	8149	8156	8162	8169	8176	8182	8189
66	8195	8202	8209	8215	8222	8228	8235	8241	8248	8254
67	8261	8267	8274	8280	8287	8293	8299	8306	8312	8319
68	8325	8331	8338	8344	8351	8357	8363	8370	8376	8382
69	8388	8395	8401	8407	8414	8420	8426	8432	8439	8445
70	8451	8457	8463	8470	8476	8482	8488	8494	8500	8506
71	8513	8519	8525	8531	8537	8543	8549	8555	8561	8567
72	8573	8579	8585	8591	8597	8603	8609	8615	8621	8627
73	8633	8639	8645	8651	8657	8663	8669	8675	8681	8686
74	8692	8698	8704	8710	8716	8722	8727	8733	8739	8745
75	8751	8756	8762	8768	8774	8779	8785	8791	8797	8802
76	8808	8814	8820	8825	8831	8837	8842	8848	8854	8859
77	8865	8871	8876	8882	8887	8893	8899	8904	8910	8915
78	8921	8927	8932	8938	8943	8949	8954	8960	8965	8971
79	8976	8982	8987	8993	8998	9004	9009	9015	9020	9025
80	9031	9036	9042	9047	9053	9058	9063	9069	9074	9079
81	9085	9090	9096	9101	9106	9112	9117	9122	9128	9133
82	9138	9143	9149	9154	9159	9165	9170	9175	9180	9186
83	9191	9196	9201	9206	9212	9217	9222	9227	9232	9238
84	9243	9248	9253	9258	9263	9269	9274	9279	9284	9289
85	9294	9299	9304	9309	9315	9320	9325	9330	9335	9340
86	9345	9350	9355	9360	9365	9370	9375	9380	9385	9390
87	9395	9400	9405	9410	9415	9420	9425	9430	9435	9440
88	9445	9450	9455	9460	9465	9469	9474	9479	9484	9489
89	9494	9499	9504	9509	9513	9518	9523	9528	9533	9538
90	9542	9547	9552	9557	9562	9566	9571	9576	9581	9586
91	9590	9595	9600	9605	9609	9614	9619	9624	9628	9633
92	9638	9643	9647	9652	9657	9661	9666	9671	9675	9680
93	9685	9689	9694	9699	9703	9708	9713	9717	9722	9727
94	9731	9736	9741	9745	9750	9754	9759	9763	9768	9773
95	9777	9782	9786	9791	9795	9800	9805	9809	9814	9818
96	9823	9827	9832	9836	9841	9845	9850	9854	9859	9863
97	9868	9872	9877	9881	9886	9890	9894	9899	9903	9908
98	9912	9917	9921	9926	9930	9934	9939	9943	9948	9952
99	9956	9961	9965	9969	9974	9978	9983	9987	9991	9996

However, since the log of a number squared is 2,

$$\text{No. of dB} = 10 \times 2 \log \frac{V_{\text{out}}}{V_{\text{in}}} = 20 \log A_v \qquad (14\text{-}16)$$

This means that when the voltage gain has dropped to 0.707 times its midfrequency gain, the decibel gain will be −3 dB:

$$\begin{aligned} \text{No. of dB} &= 20 \log \frac{0.707}{1.000} = 20 \log 7.07 \times 10^{-1} \\ &= 20 \log 7.07 + \log 10^{-1} = 20(0.85 - 1) \\ &= 20(-0.15) = -3 \text{ dB} \end{aligned} \qquad (14\text{-}17)$$

This −3-dB gain is both the low- and the high-frequency cutoff point previously mentioned.

The point at which the voltage gain is *one-half* its midfrequency gain was stated as −6 dB with a roll-off of −6 dB per octave,

$$\begin{aligned} \text{No. of dB} &= 20 \log 0.5 = 20 \log 5 \times 10^{-1} \\ &= 20 \log 5 + \log 10^{-1} = 20(0.7 - 1) \\ &= -6 \text{ dB} \end{aligned} \qquad (14\text{-}18)$$

In Fig. 14-11 the corner frequency points are at −3 dB and the roll off is at a *rate* of −6 dB per octave.

FREQUENCY LIMITATIONS

The low-frequency response is governed primarily by how low the reactance of the coupling and bypass capacitors can be made at the lowest frequency needed. This requires rather large capacitors when the impedances to be coupled or bypassed are low. Large capacitors are usually of the electrolytic variety, and they vary by as much as 50 percent in value and also tend to be rather leaky. When a leaky capacitor is used the isolation tends to be poor in coupling and the bias will drift.

The high-frequency response is limited by three factors that are difficult to work with. The first is the transistor high-frequency cutoff point due to junction capacitances and carrier transit time through the base. The second, called the *Miller effect*, is an amplification of the collector-base capacitance, especially a problem in the common-

TABLE 14-1 Decibel equivalents

VOLTAGE GAIN	POWER GAIN	DECIBEL GAIN
1.00	1.00	0
1.12	1.26	1
1.26	1.59	2
1.41	2.00	3
1.58	2.51	4
1.78	3.16	5
2.00	4.00	6
2.51	6.31	8
3.16	10.00	10
4.00	15.90	12
5.00	25.00	14
6.31	40.00	16
7.95	63.10	18
10.00	100.00	20
17.80	316.00	25
31.60	1,000.00	30
1×10^{2}	1×10^{4}	40
1×10^{3}	1×10^{6}	60
1×10^{4}	1×10^{8}	80
1×10^{5}	1×10^{10}	100
0.891	0.795	−1
0.795	0.631	−2
0.708	0.502	−3
0.631	0.398	−4
0.562	0.316	−5
0.502	0.251	−6
0.398	0.158	−8
0.316	0.100	−10
0.100	0.010	−20
0.010	0.0001	−40
1×10^{-3}	1×10^{-6}	−60
1×10^{-4}	1×10^{-8}	−80
1×10^{-5}	1×10^{-10}	−100

emitter circuit. The third problem is that of distributed wiring and circuit-layout capacitance.

All the capacitances add together and appear to the signal as an easy shunt to ground. The higher the frequency, the lower the reactance and the better the shunt path. Good layout and good choice of operating voltages both help to extend the upper frequency limits of the amplifier. Generally the common-base amplifier is used when RF amplifiers must operate at very high frequencies, since the Miller effect is less in this circuit. Good choice of high-frequency transistors can also extend the upper frequency limit.

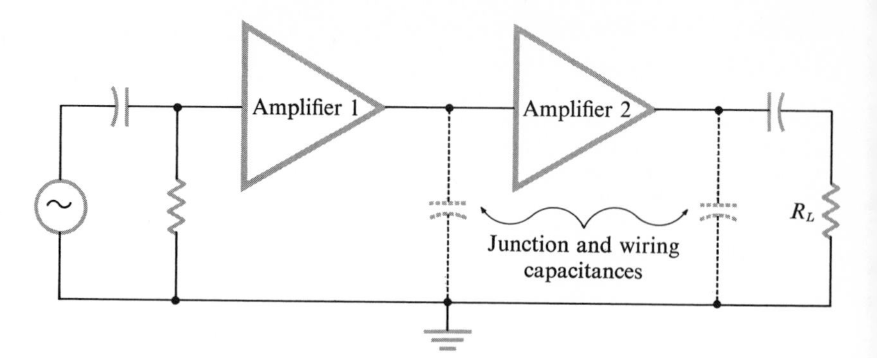

FIG. 14-12 The two-stage amplifier showing the coupling capacitors, which contribute to low-frequency gain loss, and the shunt capacitances, which are the major causes of high-frequency gain loss.

CURRENT-GAIN–BANDWIDTH PRODUCT

The current gain h_{fe} tends to roll off at about −6 dB per octave after some frequency at which the gain is down −3 dB. This frequency is known as the *β cutoff frequency*. If this cutoff frequency is multiplied by the midfrequency h_{fe} value, their product will be the frequency at which the current gain approaches 1. This frequency, usually denoted by f_T is the *gain-bandwidth product*. That is, this frequency, at which the gain is 1.0, represents a band of frequencies all the way from direct current multiplied by the gain of 1. If a certain bandwidth is desired for some design, the current gain for that bandwidth may be computed by dividing the gain-bandwidth product by the bandwidth.

THE PREAMPLIFIER

Most power amplifiers are driven by a *preamplifier*, a multistaged amplifier in which the frequency response has been changed from a flat response to conform to the output of a transducer. A phonograph preamplifier, for example, must respond with very high fidelity to low frequencies but attenuate the high frequencies. This is because the magnetic phonograph pickup cartridge has a low low-frequency output but a high high-frequency output from modern records. The records are "equalized" for recording purposes to a standard equalization curve established by the Record Industry of America Association (RIAA). The magnetic cartridge actually has a relatively flat frequency response, and so the preamplifier must correct the RIAA equalization to return the frequency response back to "flat."

The regulation or equalization of the response curve is done by selective *RC* feedback of part of the output signal to the input. The feedback is negative, which means that if a large signal is fed back at some frequency, the output at that frequency will be degenerated, or reduced. A simple capacitor feedback will produce a high-frequency rolloff because the capacitive reactance is reduced as the frequency is increased. The feedback therefore increases as the frequency rises, and hence the output falls off.

THE INTEGRATED-CIRCUIT PREAMPLIFIER

The Motorola Dual Stereo Preamp integrated circuit MC1303L is a perfect illustration of everything we have discussed so far. It is a multistage direct-coupled amplifier using a differential input, and it has a complementary Darlington pair output stage. It is a dual amplifier in a tiny integrated-circuit dual-inline package priced at less than \$5. Note that it is constructed in such a way as to allow for external feedback which will shape the frequency response to whatever curve is desired. The application notes included with the specifications show how this device may be connected as either a straight broadband amplifier, a phonograph preamp, or a tape-playback preamp.

Most operational amplifiers of this type require a dual power supply. This may be easily constructed if a center-tapped transformer is available. Figure 14-13 shows one very simple method of achieving a voltage of +13 and −13 V. Other voltages may be obtained by substituting the proper breakdown diodes for the ones given.

A number of companies market both the preamplifier and the power amplifier in integrated-circuit units. This is only one small part of the integrated-circuit usage in the commercial field. Color television utilizes integrated circuits by the millions in order to miniaturize the circuitry and reduce maintenance, as well as to simplify design.

FIG. 14-13 A dual power supply suitable for powering the Motorola MC1303L Dual Preamplifier integrated circuit.

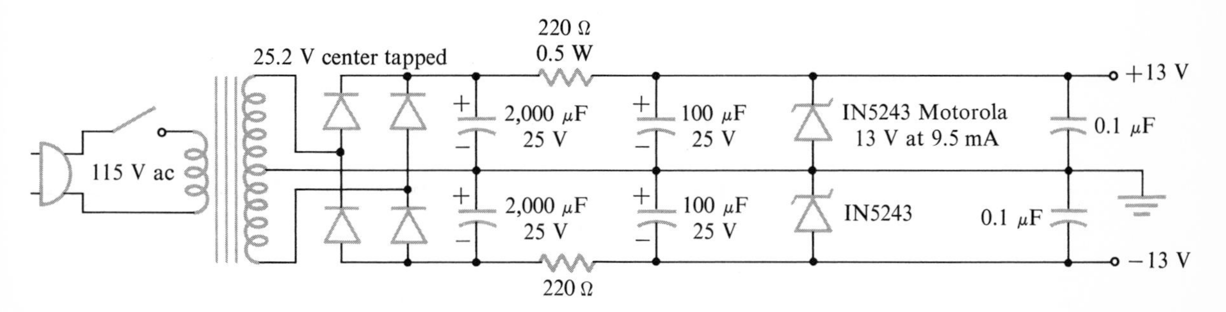

MC1303L

MONOLITHIC DUAL STEREO PREAMPLIFIER

. . . designed for amplifying low-level stereo audio signals with two preamplifiers built into a single monolithic semiconductor.

Each Preamplifier Features:

- Large Output Voltage Swing – 4.0 V(rms) min
- High Open-Loop Voltage Gain = 6000 min
- Channel Separation = 60 dB min at 10 kHz
- Short-Circuit-Proof Design

DUAL STEREO PREAMPLIFIER INTEGRATED CIRCUIT

MONOLITHIC SILICON EPITAXIAL PASSIVATED

MARCH 1969 – DS 9093 R1

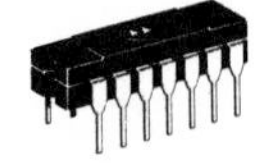

CERAMIC PACKAGE
CASE 632
TO-116

MAXIMUM RATINGS (T_A = +25°C unless otherwise noted

Rating	Symbol	Value	Unit
Power Supply Voltage	V^+ V^-	+15 -15	Vdc Vdc
Power Dissipation (Package Limitation) Derate above 25° C	P_D	625 5.0	mW mW/°C
Operating Temperature Range	T_A	0 to +75	°C

Maximum Ratings as defined in MIL-S-19500, Appendix A.

CIRCUIT SCHEMATIC

EQUIVALENT CIRCUIT

Data sheets: Motorola MC1303L Stereo Preamplifier Integrated Circuit (*courtesy Motorola Semiconductor Products Inc.*)

MC1303L

ELECTRICAL CHARACTERISTICS (Each Preamplifier) (V^+ = +13 Vdc, V^- = -13 Vdc, T_A = +25°C unless otherwise noted)

Characteristic Definitions (linear operations)	Characteristic	Symbol	Min	Typ	Max	Unit
$A_{VOL} = \frac{e_{out}}{e_{in}}$; e_{in}, e_{out}	Open Loop Voltage Gain	A_{VOL}	6,000	10,000	-	V/V
	Output Voltage Swing (R_L = 10 kΩ)	V_{out}	4.0	5.5	-	V(rms)
I_2, I_1	Input Bias Current $I_b = \frac{I_1 + I_2}{2}$	I_b	-	1.0	10	μA
I_2, I_1	Input Offset Current ($I_{io} = I_1 - I_2$)	I_{io}	-	0.2	0.4	μA
V_{io}, $V_{out} = 0$	Input Offset Voltage	V_{io}	-	1.5	10	mV
	DC Power Dissipation (Power Supply = ±13 V, V_{out} = 0)	P_D	-	-	400	mW
e_{in}, $e_{out\,1}$, $e_{out\,2}$	Channel Separation (f = 10 kHz)	$\frac{e_{out\,1}}{e_{out\,2}}$	60	70	-	dB

OUTLINE DIMENSIONS

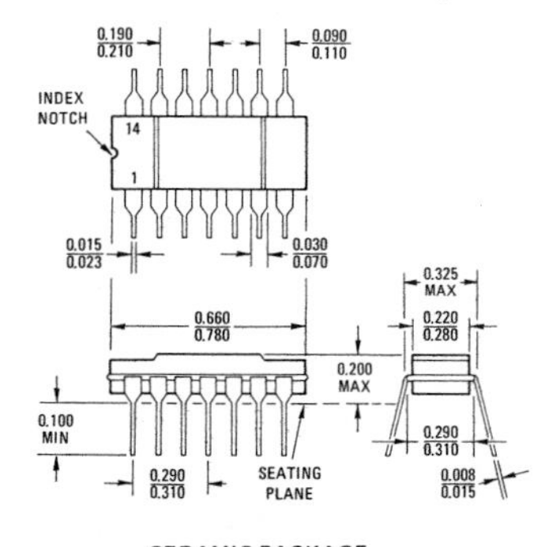

CERAMIC PACKAGE
CASE 632
TO-116

To convert inches to millimeters multiply by 25.4.
All JEDEC TO-116 dimensions and notes apply.

TYPICAL PREAMPLIFIER APPLICATIONS

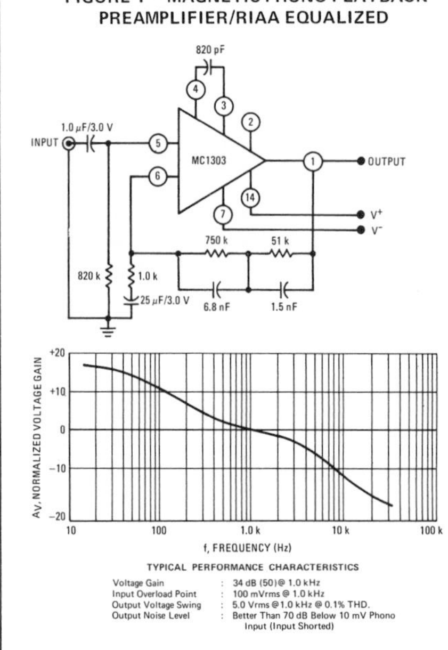

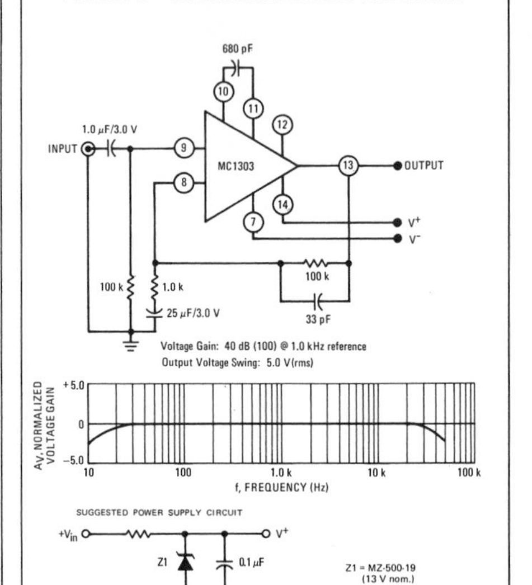

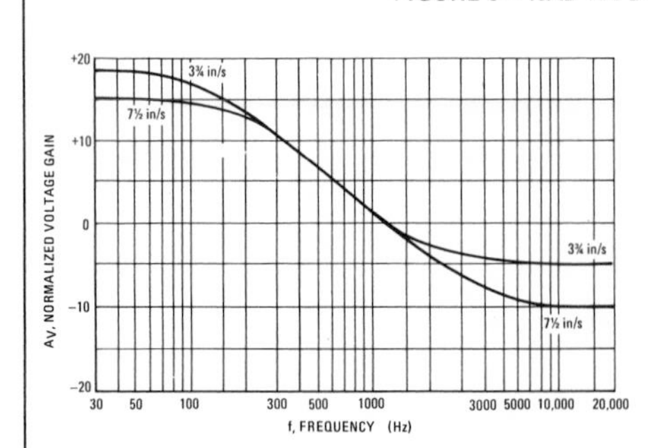

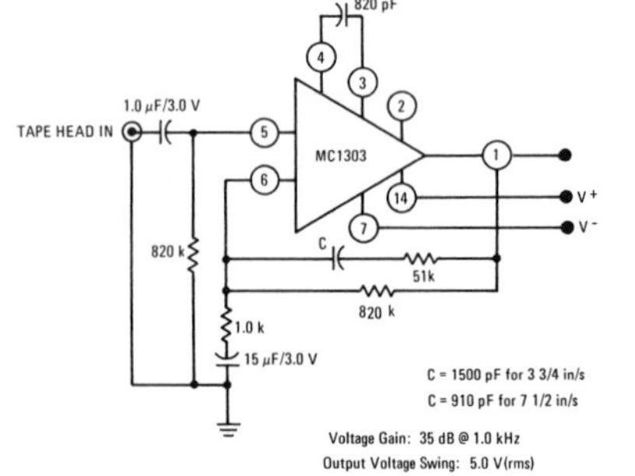

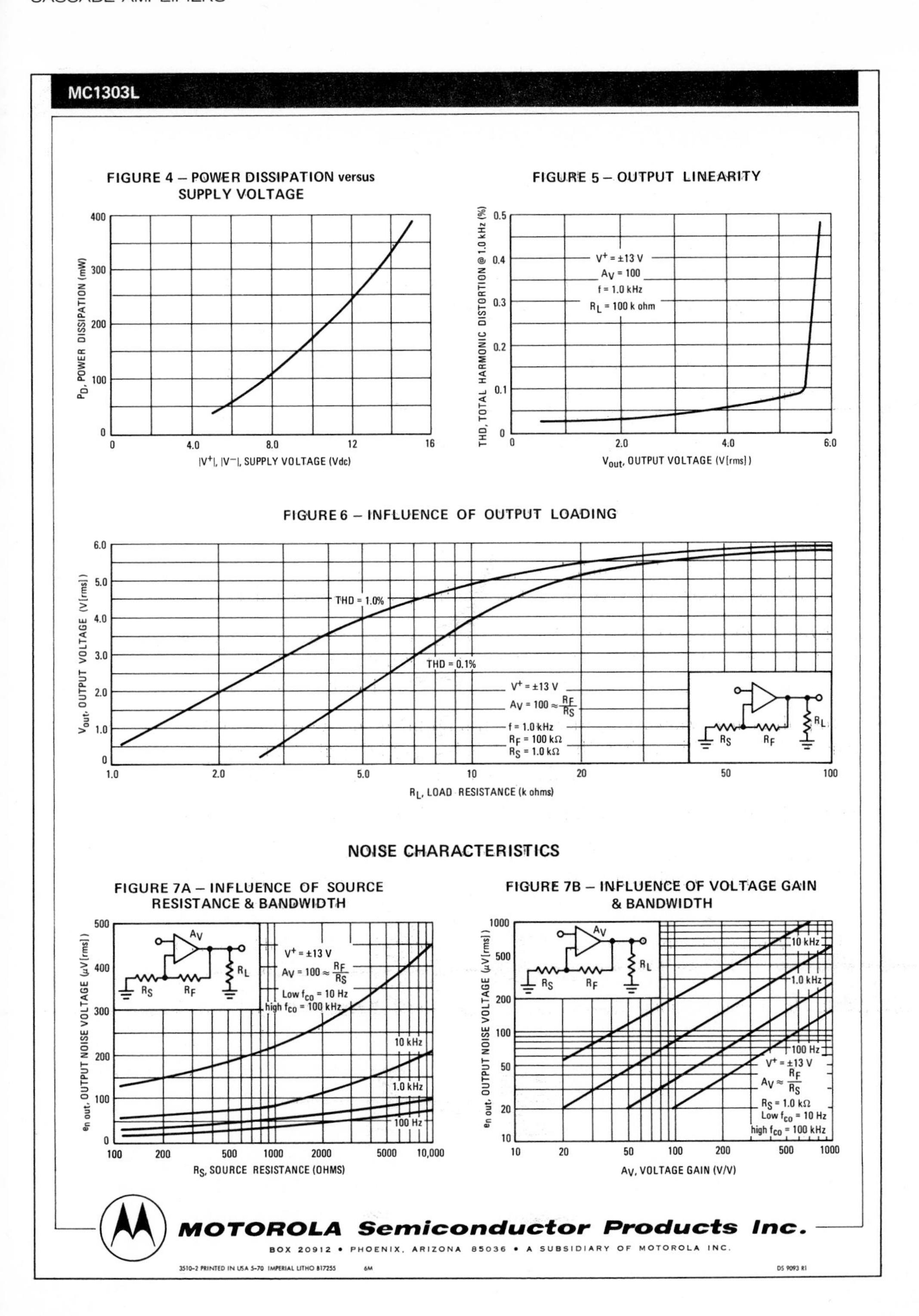
MC1303L
FIGURE 4 – POWER DISSIPATION versus SUPPLY VOLTAGE
P_D, POWER DISSIPATION (mW)
400
300
200
100
0
0
4.0
8.0
12
16
$|V^+|$, $|V^-|$, SUPPLY VOLTAGE (Vdc)
FIGURE 5 – OUTPUT LINEARITY
THD, TOTAL HARMONIC DISTORTION @ 1.0 kHz (%)
0.5
0.4
0.3
0.2
0.1
0
$V^+ = \pm 13$ V
$A_V = 100$
f = 1.0 kHz
$R_L = 100$ k ohm
0
2.0
4.0
6.0
V_{out}, OUTPUT VOLTAGE (V[rms])
FIGURE 6 – INFLUENCE OF OUTPUT LOADING
V_{out}, OUTPUT VOLTAGE (V[rms])
6.0
5.0
4.0
3.0
2.0
1.0
0
THD = 1.0%
THD = 0.1%
$V^+ = \pm 13$ V
$A_V = 100 \approx \frac{R_F}{R_S}$
f = 1.0 kHz
$R_F = 100$ kΩ
$R_S = 1.0$ kΩ
R_S
R_F
R_L
1.0
2.0
5.0
10
20
50
100
R_L, LOAD RESISTANCE (k ohms)
NOISE CHARACTERISTICS
FIGURE 7A – INFLUENCE OF SOURCE RESISTANCE & BANDWIDTH
$e_{n\ out}$, OUTPUT NOISE VOLTAGE (μV[rms])
500
400
300
200
100
0
A_V
R_S
R_F
R_L
$V^+ = \pm 13$ V
$A_V = 100 \approx \frac{R_F}{R_S}$
Low f_{co} = 10 Hz
high f_{co} = 100 kHz
10 kHz
1.0 kHz
100 Hz
100
200
500
1000
2000
5000
10,000
R_S, SOURCE RESISTANCE (OHMS)
FIGURE 7B – INFLUENCE OF VOLTAGE GAIN & BANDWIDTH
$e_{n\ out}$, OUTPUT NOISE VOLTAGE (μV[rms])
1000
500
200
100
50
20
10
A_V
R_S
R_F
R_L
10 kHz
1.0 kHz
100 Hz
$V^+ = \pm 13$ V
$A_V \approx \frac{R_F}{R_S}$
$R_S = 1.0$ kΩ
Low f_{co} = 10 Hz
high f_{co} = 100 kHz
10
20
50
100
200
500
1000
A_V, VOLTAGE GAIN (V/V)
MOTOROLA Semiconductor Products Inc.
BOX 20912 • PHOENIX, ARIZONA 85036 • A SUBSIDIARY OF MOTOROLA INC.
3510-2 PRINTED IN USA 5-70 IMPERIAL LITHO 817255 6M
DS 9093 R1

SUMMARY OF CONCEPTS

1. Capacitor coupling is used to isolate the collector voltage of one stage from the base voltage of the next stage.
2. Transformer coupling provides dc isolation as well as impedance matching.
3. Direct coupling may be used when two stages do not require dc isolation.
4. Direct coupling extends the frequency response down to dc, allowing amplification of changes in dc voltages.
5. The choice of coupling capacitor must be such as to have a very low reactance at the lowest frequency to be used. The reactance is set equal to the sum of the output impedance of the first stage and the input impedance of the second stage, at the lowest frequency.
6. Direct coupling uses the concept of voltage-divider bias in that the collector current of the first stage is significantly larger than the base current of the second stage.
7. The input impedance of the second stage should be sufficiently large so as not to load the first stage and reduce its voltage gain ($Z_{in} > 10 Z_{out}$).
8. Overall voltage gain is the product of the stage gains.
9. The Darlington configuration is a direct-coupled amplifier with a current gain equal to the product of the two current gains and a very high input impedance.
10. Darlington amplifiers may be made with *NPN* or *PNP* pairs or with complementary pairs of transistors.
11. Complementary transistors may be used in direct-coupled amplifier design to keep supply voltages to a reasonable level.
12. The FET may be used at the input of a cascade amplifier to achieve a high impedance and stable operation and be coupled directly to a bipolar transistor stage.
13. The reflected load impedance in a transformer-coupled amplifier is the product of the turns ratio squared and the external load resistance connected to the secondary winding [$Z_L = (N_P/N_S)^2 R_L$].
14. The gain of an amplifier is a function of the ac load rather than the dc load. The ac load is the parallel summation of the output impedance of the transistor, the load resistance, and the input impedance to the next stage or the external load impedance.

15. The ac load line must be plotted through the Q point on the characteristic curves.
16. The differential amplifier is a two-stage emitter-coupled dc amplifier. It can provide difference output at either collector or between collectors, an inverted output, or a noninverted output and is frequently used as the input stage of operational amplifiers.
17. The differential amplifier must operate with a constant-current source, which may be a large resistance in the emitter circuit or an FET or a transistor biased in the constant-current region.
18. The differential amplifier is rated for its common-mode rejection of noise present at both inputs.
19. Frequency response is the range of frequencies that fall between the −3-dB points on a gain-vs.-frequency plot.
20. The −3-dB point is the point at which the power has been reduced to one-half its former value.
21. The −6-dB point is the frequency at which the voltage gain has been reduced to one-half its midfrequency value.
22. The roll-off rate for a two-stage amplifier is about −6 dB per octave, that is, −6 dB each time the frequency doubles or is reduced to one-half.
23. Low-frequency gain is limited primarily by the coupling and bypass capacitances used in the circuitry.
24. High-frequency gain is limited primarily by the junction capacitances, the Miller effect, and wiring capacitance.
25. The gain-bandwidth product is the frequency at which a transistor's h_{fe} approaches 1.
26. Negative feedback through capacitances can shape the frequency response of an amplifier.

GLOSSARY

β cutoff frequency The high-frequency point at which the h_{fe} value of the transistor is down by 3 dB.

Cascade Devices connected one after another, as in a cascade amplifier, where the output of one circuit is the input to the next.

Common-mode signal A signal that is common to two inputs.

Constant-current diode An FET whose gate is connected to the source, effectively allowing a constant current to flow when it is operated above its pinchoff voltage.

Darlington amplifier A compound two-stage direct-coupled transistor amplifier in which the emitter of the first transistor is connected directly to the base of the second and the collector of the first is connected to the collector of the second. The input to the first transistor is the input to the Darlington pair.

Decibel A unit of comparison of two power levels, based on the logarithm of the power gain; one-tenth of a Bel.

Differential amplifier An emitter-coupled balanced dc amplifier capable of giving an output proportional to the difference between two signals.

Gain-bandwidth product The product of the current gain and the bandwidth. Usually the high-frequency limit of a transistor where the current gain has dropped to 1.

Half-power point The point on a frequency-response curve where the power level has dropped to one-half. The −3-dB point, with the midfrequency gain taken as the 0-dB reference point.

Interstage Generally the circuitry that couples two stages of amplification.

Miller effect The high-frequency-gain loss factor or effect caused by the effective amplification of the collector-base junction capacitance, which shunts part of the signal to ground.

Octave The interval between two frequencies where one frequency is twice the other.

Roll-off The rate of change of circuit gain with frequency, usually about −6 dB per octave.

Transducer A device which is capable of changing one form of energy into a more usable form; for example, a speaker or a microphone.

REFERENCES

Amos, S. W.: *Principles of Transistor Circuits,* 3d ed., Hayden Book Company, Inc., New York, 1965, chaps. 7–8.

Brazee, James G.: *Semiconductor and Tube Electronics*, Holt, Rinehart and Winston, Inc., New York, 1968, chaps. 10, 12.

Cutler, Phillip: *Electronic Circuit Analysis,* vol. 2, McGraw-Hill Book Company, New York, 1967, pp. 257–269.

——: *Semiconductor Circuit Analysis,* McGraw-Hill Book Company, New York, 1964, chap. 6.

Herrick, Clyde N.: *Electronic Circuits,* Charles E. Merrill Books, Inc., Columbus, Ohio, 1968, chaps. 7–9.

Horowitz, Mannie: *Practical Design with Transistors,* Howard W. Sams & Co., Inc., Indianapolis, 1968, chaps. 7–10.

The Integrated Circuit Data Book, Motorola Semiconductor Products Inc., Phoenix, Ariz., 1968, pp. 10-73–10-104.

Lenert, Louis H.: *Semiconductor Physics, Devices, and Circuits,* Charles E. Merrill Books, Inc., Columbus, Ohio, 1968, pp. 302–408.

Malvino, Albert Paul: *Transistor Circuit Approximations*, McGraw-Hill Book Company, New York, 1968, chap. 11.

Teeling, John: *An Integrated Circuit Stereo Preamplifier*, Application Note AN-420, Motorola Semiconductor Products Inc., Phoenix, Ariz., 1968.

Transistor Manual, 7th ed., General Electric Company, Syracuse, N.Y., 1964, chaps. 4, 11.

REVIEW QUESTIONS

1. What are some of the advantages of *RC* coupling over direct coupling? What are some disadvantages?

2. What are some of the advantages of direct coupling? What are some disadvantages?

3. Why is cascading of amplifier stages desirable?

4. What is an advantage of using transformer coupling of stages at radio frequencies?

5. What conditions must be met before two amplifiers may be direct coupled?

6. Explain why complementary transistors are sometimes used in multistage direct-coupled amplifiers.

7. What is a Darlington amplifier and what is its use?

8. What advantage is there in using an FET for the first stage of a multistage amplifier?

9. What is the difference between a dc load line and an ac load line?

10. In what way is the voltage gain of a circuit affected by the ac load line?

11. What factors must be known in order to design a two-stage direct-coupled amplifier?

12. How does a differential amplifier differ from an ordinary direct-coupled amplifier?

13. What are some of the ways in which a differential amplifier can operate?

14. What is meant by a constant-current source and why is it needed for a differential amplifier?

15. Give three methods of obtaining a constant current.

16. What is meant by common-mode rejection in a differential amplifier?

17. What is meant by frequency response?

18. What is a decibel and how is it related to an amplifier?

19. What factors control the low-frequency response of an amplifier?

20. What factors control the high-frequency response of an amplifier?

21. Describe the function of a preamplifier in terms of feedback and frequency response.

PROBLEMS

1. If the gain of the first stage is 30 and the gain of the second stage is 25, what is the overall gain of the amplifier?

2. If the output impedance of one stage is 6 kΩ and the input to the second stage is 40 kΩ, what is the ac load impedance seen by the first transistor?

3. In the amplifier of Prob. 2, what size of coupling capacitor would be needed if the lowest frequency required is 50 Hz and the voltage difference between the collector and the base is 8.5 V?

4. If the first stage of an amplifier is to have a dc load line the same as the ac load line, what must the approximate input impedance to the second stage be if the load resistance in the first stage is 4.7 kΩ?

5. How much of an emitter resistance should be left unbypassed if a transistor having an h_{fe} of 75 is to have an input impedance of 50 kΩ at 1.0 mA?

6. Draw a diagram of a two-stage complementary transistor amplifier and compute the part values required if h_{fe} is 75 for both transistors and the supply voltage is 14 V.

7. What is the reflected impedance to the primary of a coupling transformer that is loaded on the secondary by 2,000 Ω and has a turns ratio of 4:1?

8. If the ac load impedance is 5 kΩ, the input impedance is 5 kΩ, and h_{fe} is 100, what is the approximate voltage gain of the first-stage amplifier?

9. If the ac load impedance is 3.3 kΩ and the emitter resistance is 680 Ω, what is the approximate voltage gain?

10. If a differential amplifier has a source current of 1.8 mA and the current through one side is 1.1 mA, what is the current through the other side?

11. In Prob. 10, if the load resistors are both 6.8 kΩ, what will be the voltage across each one?

12. If the voltage gain of a differential amplifier is 50 and the voltage is 1.15 V at input 1 and 1.30 V at input 2, what is the output voltage between output 2 and ground?

13. If a differential amplifier has a sine wave of 2.7 V *p-p* at input 1 and an in-phase sine wave of 2.1 V *p-p* at input 2, what is the common-mode signal voltage?

14. In Fig. 14-9, what is the approximate voltage gain of transistor T_2? If the gain of T_2 is 110, what is the total voltage gain for an output at B? At A?

15. If the interstage equivalent impedance is 20 kΩ and the coupling capacitor is 0.5 μF, what is the low-frequency cutoff, the −3-dB point? [Use Eq. (14-2) rearranged as $f = 1/2\pi C\ Z_{eq}$]

16. If the voltage gain of a two-stage amplifier is −6 dB at 30 kHz, at what frequency will it be −12 dB?

17. If the power input to an amplifier is 15 mW and the output power is 30 mW, what is the decibel gain? What would be the gain if the output increased to 60 mW?

CHAPTER 15

OBJECTIVES

1. To apply known concepts of voltage and current amplification to power amplification
2. To analyze transformer-coupled and direct-coupled power amplifiers
3. To examine the design of class A, class B, and class AB amplifiers
4. To describe graphical methods of predicting the operation and power dissipation for a transistor amplifier
5. To investigate integrated-circuit design and applications of power-amplifier units

POWER AMPLIFIERS

The next step in our development of an electronic system is from the cascade amplifier to the power amplifier. It is the power amplifier that converts the large signal available from the voltage amplifiers into the power that is required to finish the job. This requires that the output device be able to handle fairly large currents, several amps at times. Creating a functional power amplifier is often a challenge to the imagination and ability of the designer. However, with careful adherence to basic concepts of electronics design, a simple amplifier can be developed that can evolve into a high-power unit with little distortion and few problems.

THE SINGLE-ENDED AMPLIFIER

An amplifier that has only one device driving the transducer that delivers the power is called a *single-ended amplifier*. This is in contrast to the *push-pull amplifier,* which has two output devices that operate alternately.

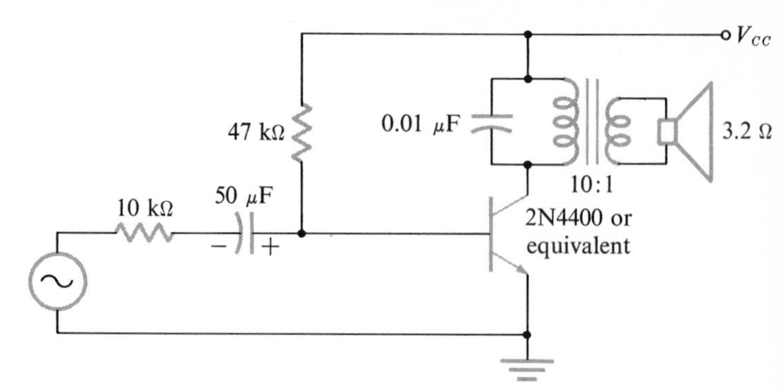

FIG. 15-1 A simple single-ended power amplifier.

Figure 15-1 shows the simplest form of the single-ended power amplifier. Note that its basic structure is that of the normal common-emitter amplifier, but the load resistance and output capacitor have been replaced with an output transformer. The transformer primary impedance acts as the collector load for the transistor.

CLASS A OPERATION

When an amplifier is designed to operate in the middle of its characteristic curve such that current flows during the entire cycle, the amplifier is called *class A*. This is the mode of operation that should give very little distortion to the signal. It does, however, have several important deficiencies. The efficiency is low, about 25 percent, and the transformer has dc flow at all times, which tends to saturate the core and limit the signal output. Because of its low efficiency, this design is fairly wasteful and is now avoided in most applications.

Figure 15-2 is a set of characteristic curves for a typical power transistor. Also shown are a curve representing the maximum power dissipation by the transistor at 25°C and the ac load line for the amplifier of Fig. 15-1. Note that this load line does not intersect the power-dissipation curve. If it did, the transistor would overheat and might be destroyed.

Investigation of Fig. 15-2 shows the Q point of operation to be such that the maximum amount of power is dissipated even while there is no signal being amplified. In choosing the output transformer for a single-ended power amplifier be sure that the transformer can handle at least twice the amount of current that will be passing through it at the Q point. It must not at any time during the cycle of operation become saturated. In other words, the maximum number of lines of force that the transformer core can develop is a function of some dc current. If this number develops at some current before the peak of the waveform, energy cannot be transferred to the secondary, with the result that the wave will be flattened or distorted.

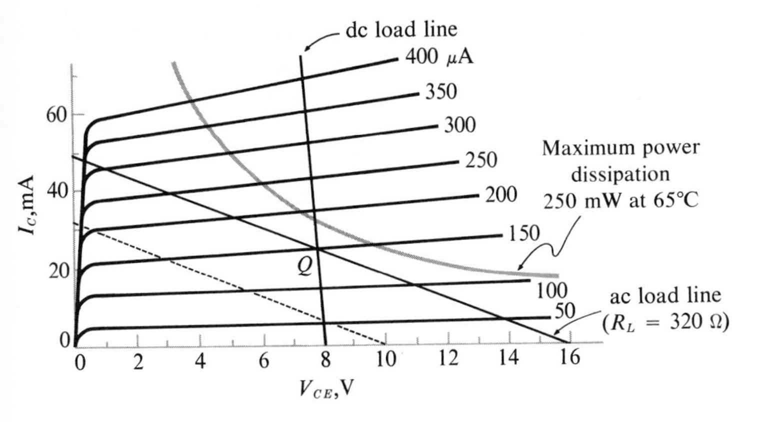

FIG. 15-2 Characteristics of a medium-power transistor showing the power-dissipation curve and the dc and ac load lines.

The dc load line cannot conveniently be plotted on the characteristic curves, because the dc resistance of the primary winding is usually less than 100 Ω, and this would put the *Y* intercept of the load line well above the end of the graph. However, the ac load impedance can be plotted easily and should pass through the desired dc *Q* point.

The *Q* point plotted in Fig. 15-2 uses about 25 mA of collector current at a collector-emitter voltage of about 7.5 V. The circuit is using 8 V × 0.025 A = 200 mW of power, well below the maximum allowed. The ac load line represents an ac, or dynamic, load impedance of about 320 Ω. This is just about what we would expect from an output transformer with a 10:1 turns ratio and loaded with a small 3.2-Ω speaker.

In plotting the ac load line there are a number of ways to start as we saw in Chap. 14. Perhaps the easiest procedure is to examine the available components. Note the speaker impedance and the turns ratio or input-output ratio of impedances for the transformer. With the impedance ratio known, the turns ratio may be determined by means of Eq. (14-4),

$$Z_L = \left(\frac{N_P}{N_S}\right)^2 R_L$$

Rearrangement yields

$$\frac{N_P}{N_S} = \sqrt{\frac{Z_L}{R_L}} = \sqrt{\frac{320\ \Omega}{3.2\ \Omega}} = \sqrt{100} = 10{:}1 \qquad (15\text{-}1)$$

Most of the small inexpensive speakers are either 3.2 or 4 Ω nominal impedance. Most better speakers are 8 Ω. Using Eq. (15-1), compute the turns ratio and the reflected impedance for that particular speaker. The transformer may be labeled 800/8 but also produce 320/3.2 Ω.

Once the impedance reflected to the transformer primary from the secondary is known, this value is to be used as the collector ac load impedance for the amplifier design. To obtain maximum power transfer and efficiency from the power amplifier it is good design practice to have the reflected impedance match the output impedance of the transistor being used. This can be approximated if a set of characteristic curves is available by computing the slope of the current curve at the Q point. However, for our purposes this will not be necessary. We are concerned about a functional power amplifier that will work at low power and with class A operation. For high-power design the direct-coupled class AB amplifier is preferred.

Pick some current as a reference point and compute what supply voltage would be needed to cause this much collector current to flow if the amount of reflected impedance were inserted in the line (see Fig. 15-2). Draw a temporary load line between these two points. Next, draw a line parallel with this ac load line, but through the Q point. Note that it extends beyond the actual supply voltage. Do not be concerned; this is as it should be. Remember that we are working with an inductance whose magnetic field will rise and fall. The energy put in on the positive half-cycle will come out as a collapsing field on the negative half-cycle. The current in the secondary winding will be ac.

With the ac load line established, the remaining part of our simple design is to compute the base-current-limiting resistance. From the Q point on the characteristic curves, the base current must be 150 μA. From inspection of Fig. 15-1 it is clear that nothing is being done about bias stabilization, and so the voltage at the emitter is 0 V. Therefore the voltage at the base will be about 0.7 V at 25 mA. The voltage across the base resistor will be

$$V_{CC} - V_B = 8\text{ V} - 0.7\text{ V} = 7.3\text{ V}$$

From Ohm's law, the base resistance is computed to be 48.6 kΩ, with the nearest standard value 47 kΩ.

The coupling or input capacitor must be quite large to prevent loss of gain, but a capacitance of 25 to 50 μF will be sufficient for this purpose.

This simple single-ended amplifier should now be constructed and tested. The base-limiting resistance, however, should be replaced with a potentiometer in order to control the current. Remember that the collector current will be limited only by the dc resistance of the transformer primary and the base current. Thus it is important that the base current be kept within reason.

POWER AND EFFICIENCY

In our simple single-ended amplifier above, the total power dissipated by the circuit was 200 mW. This was, however, the total dc power, and not signal power delivered to the load. The *efficiency* of the circuit is the measure of how well the dc power is converted into ac power. The theoretical maximum efficiency of a class A transformer-coupled power amplifier is 50 percent. This means that the power in the load, $P_L = 0.5\ P_{CC}$, is one-half the power delivered by the power supply. Generally, of course, the efficiency will not reach even this value.

From Fig. 15-2 it appears that the quiescent current is about one-half the maximum current available with maximum excitation. The quiescent collector voltage is about one-half the maximum possible voltage swing—that is, $V_{CE,\,\max} = 2V_{CC}$. If we assume the amplifier to be driven by some signal to its maximum limit, the maximum peak-to-peak signal voltage will be $2V_{CC}$ and the maximum peak-to-peak signal current will be $2I_{C,\,Q}$

Power is figured in rms units in comparing with dc power and so the maximum peak-to-peak units should be converted to rms units. Therefore

$$I_{\mathrm{rms}} = \frac{I_{p\text{-}p}}{2.82} = \frac{I_{p\text{-}p}}{2\sqrt{2}} \tag{15-2}$$

and

$$P_L = \left(\frac{I_{C,\,\max}}{2\sqrt{2}}\right)^2 R_L' = \frac{I^2_{C,\,\max}}{8} R_L' \tag{15-3}$$

However, since $P = VI$,

$$\begin{aligned} P_L &= \frac{I^2_{C,\,\max}}{8} R_L' = \frac{V_{CE,\,\max} I_{C,\,\max}}{8} \\ &= \frac{2V_{CC}\ 2V_{CC}/R_L'}{8} = \frac{V_{CC}^2}{2R_L'} \end{aligned} \tag{15-4}$$

Substituting the circuit values into Eq. (15-3), we have

$$P_L = \frac{16 \text{ V } p\text{-}p \times 50 \text{ mA } p\text{-}p}{8} = \frac{800 \text{ mW}}{8} = 100 \text{ mW} \tag{15-5}$$

The circuit power consumed was

$$P_{CC} = 8 \text{ V} \times 25 \text{ mA} = 200 \text{ mW}$$

$$= \frac{V_{CC}^2}{R_L'} = \frac{64}{320} = 200 \text{ mW}$$

The power consumed or dissipated by the transistor is

$$P_{CE} = V_{CE,Q} I_{C,Q}$$

$$= 8 \text{ V} \times 25 \text{ mA} = 200 \text{ mW} \tag{15-6}$$

The efficiency η of conversion of dc power to ac power is

$$\eta = \frac{P_L}{P_{CC}} = \frac{100 \text{ mW}}{200 \text{ mW}} = 0.5 = 50 \text{ percent} \tag{15-7}$$

Comparison of Eqs. (15-4) and (15-6) shows that the transistor operating in a class A transformer-coupled power amplifier dissipates twice as much power as it delivers to the load. In other words, if a power transistor is capable of dissipating 50 W of power, it will deliver only 25 W of class A power to the transformer primary.

Losses and derating factors We have said nothing so far about the efficiency of the transformer or the heating of the transistor and its leakage-current problems. Remember that silicon transistors have a small leakage current that doubles for every 10 to 12°C. The higher the collector current, the higher the temperature of the junction and the less efficient the transistor.

Most manufacturers of power transistors suggest a power *derating factor* for temperature changes. The amount of derating depends on the thermal efficiency of the case in allowing heat to flow out of the case into the air or into a heat sink. For example, Texas Instruments derates their *NPN* Silicon Power transistor TIP33, which has a plastic case, as follows:

Continuous device dissipation at or below 25°C *case* temperature = 80 W, derated at 0.64 W/deg
Continuous device dissipation at or below 25°C *free-air* temperature = 3.5 W, derated at 28 mW/deg up to 150°C

This example points out the need for using a heat sink with power transistors in order to use them at high power. The heat sink must be capable of dissipating the heat rapidly, with maximum heat flow from the transistor case to the sink. Any thermal resistance reduces the efficiency of the system. For this reason a good quality of silicone grease should be used between the transistor and the heat sink.

TRANSFORMER INPUT COUPLING

Since the maximum power transfer from one system to another occurs when the output impedance of the signal generator equals the input impedance to the amplifier, impedance matching is essential for power amplifiers. The input impedance to a power amplifier is extremely small, often less than 100 Ω. Hence transformer input coupling is frequently used for good matching. If the two impedances are known, the turns ratio for the input transformer can easily be computed.

Figure 15-3 shows a circuit similar to Fig. 15-1, but with an input transformer and bias stabilization. The voltage-divider bias is used to provide thermal stabilization and make the circuit less susceptible to variations in the current gains of various transistors. The resistance R_{B2} may be changed to a potentiometer so that the base voltage can be varied to observe the effect on gain.

The procedure for determining the proper primary impedance of

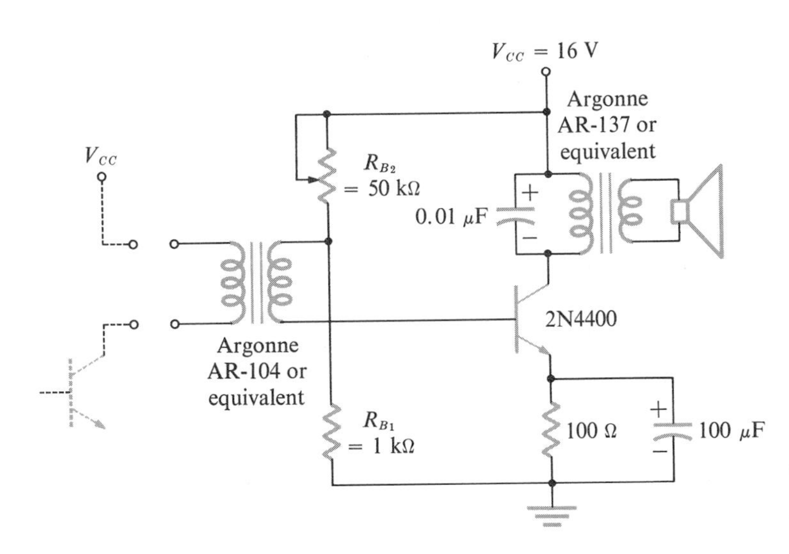

FIG. 15-3 A transformer-coupled single-ended power amplifier with stabilization.

the input, or driver, transformer is that for the output transformer, given by Eq. (14-4). Note, however, that the output dynamic impedance of a transistor voltage amplifier may be as high as 50 kΩ. This must be made to match the input impedance of the power amplifier, which may be 1 kΩ or less.

THE PUSH-PULL POWER AMPLIFIER

The push-pull power amplifier is far more efficient than the class A single-ended amplifier. Since each of the two transistors handles only one-half the total cycle, current flows through each transistor for only one half-cycle. The amplified version of the two half-cycles is recombined in the secondary of the output transformer to produce the whole cycle.

Actually, there are three classes of push-pull amplifiers, differing in the amount of the cycle each transistor is required to handle. The class B type requires each transistor to handle one half-cycle. The class C type requires each transistor to handle less than half the cycle. Both these classes have a great deal of distortion. To remove the distortion the class AB amplifier has each transistor amplify slightly more than one-half of the whole cycle. This means that the efficiency is reduced for the sake of distortion removal.

CLASS B OPERATION

The main advantage of the class B amplifier is that when there is no signal applied to the input there is no quiescent collector current, and hence no device dissipation. The main disadvantage is the severe distortion at low signal levels.

Figure 15-4 shows a transformer-coupled push-pull power amplifier. It is similar in structure to two of the circuits in Fig. 15-3 placed back to back. Of course, the input and output transformers would have

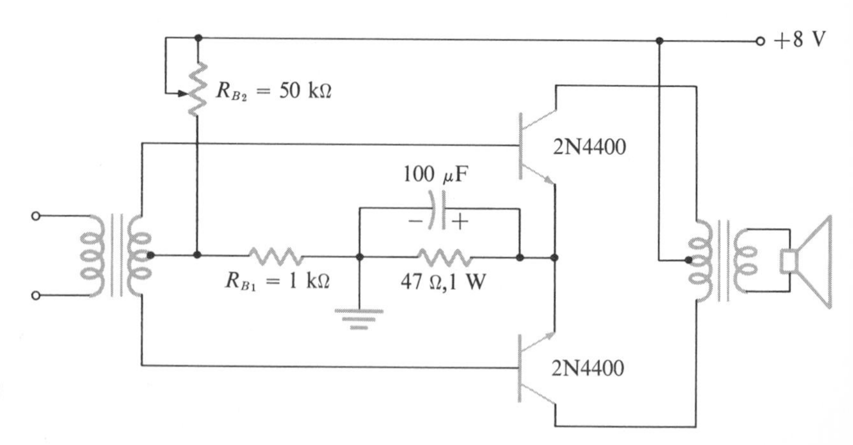

FIG. 15-4 A transformer-coupled push-pull amplifier.

a common primary and a common secondary, respectively. For operation, the biasing resistances must be such that the voltage at the base of each transistor is slightly less than the turn-on voltage, about 0.6 V above the emitters. Very little emitter current will pass until the junction potential is greater than 0.6 V. However, if the bias is left at zero, the input signal will produce no output unless it is greater than that needed to exceed the junction voltage.

The resistance of the transformer windings is negligible, and so a resistive biasing is required in most cases. The use of a single center-tapped transformer at the input means that each transistor can be fed a signal alternately from the same signal source. The center-tapped output transformer allows the same dc supply to be used for each transistor and produces a combined output cycle for the two half-cycles present in the alternate halves of the primary winding.

Figure 15-5 presents the class B amplifier graphically; Fig. 15-5*a* shows the collector characteristic analysis of one transistor, and Fig. 15-5*b* presents the combined transfer curves for both transistors.

The positive transition of the signal waveform puts the upper transistor of Fig. 15-4 into conduction, and the negative transition cuts off the upper transistor and turns on the lower one. The transition from off to on is nonlinear, as indicated by the transfer curve and the output

FIG. 15-5 (*a*) Collector characteristics; (*b*) push-pull transfer characteristics of a 2N4400 transistor with a 250-mW power-dissipation curve (input and output waveforms shown).

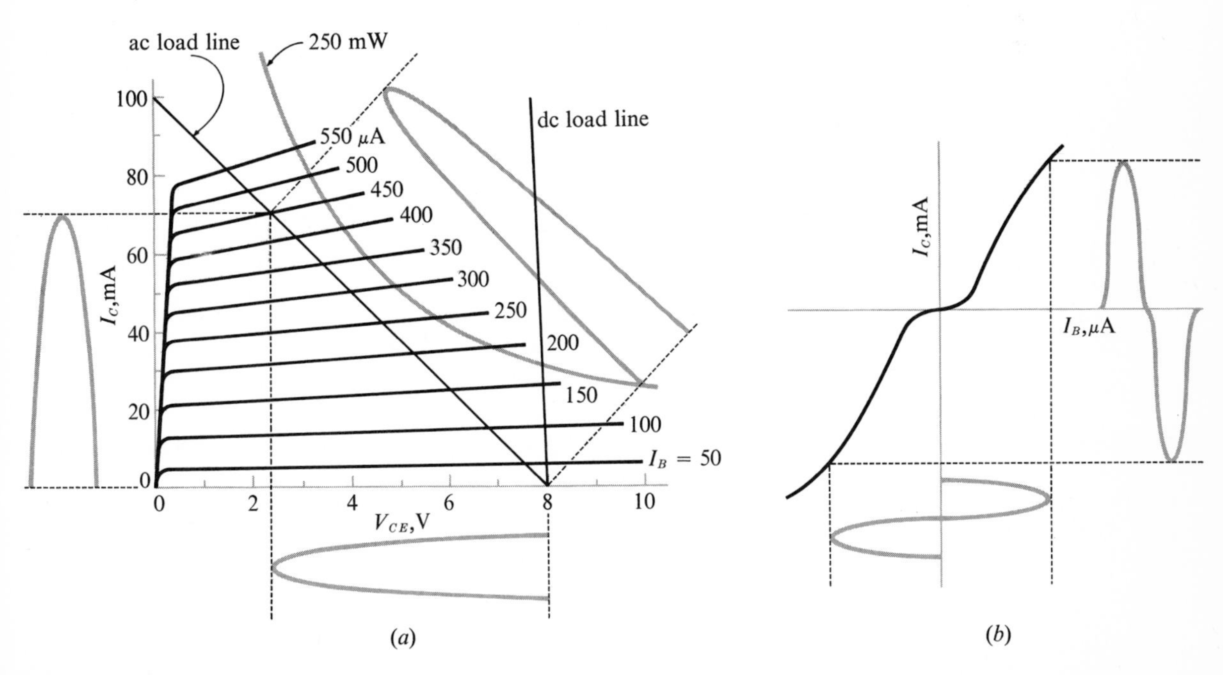

waveform. The crossover from one transistor to the other is not smooth, but introduces *crossover distortion*. As can be seen, this distortion is most severe at the low signal levels; large signals are rather presentable. The closer the combined transfer curves approach a straight line, the smaller the amount of distortion that is introduced.

Selection of the output transformer involves consideration of the center-tapped operation, where only half the winding must match the output impedance of the device. That is,

$$R'_{L,\,\mathrm{ac}} = \left(\frac{N_P}{2N_S}\right)^2 R_{L,S}$$

$$= \left(\frac{10}{2}\right)^2 \times 3.2 = 80\ \Omega \qquad (15\text{-}8)$$

If the same transistor used in Fig. 15-2 is used again, with a primary reflected impedance of 80 Ω from the output transformer, the ac load line should appear as in Fig. 15-5*a*. Since the load line does not intersect the maximum-power-dissipation curve, it is permissible and will not cause thermal runaway.

TRANSISTOR AND LOAD POWER

For the class B operation, since the quiescent current is nearly zero, the maximum signal drive should cause the half-sine-wave collector current to peak at the maximum load current,

$$I_{C,\,\max} = \frac{V_{CC}}{R'_{L,\,\mathrm{ac}}} \qquad (15\text{-}9)$$

Similarly, the collector voltage should drop from V_{CC} to near zero and then return to V_{CC}. The power delivered to the ac load is rms power, which for a half sine wave is $0.707 \times P_{\mathrm{rms}}$ [review Eq. (6-3)]. This is another way of saying that the rms power of a half-wave is one-half the peak power. For this reason the power of the load is, for each transistor,

$$P_{L1} = \left(\frac{I_{C,\,\max}}{2}\right)^2 R'_{L,\,\mathrm{ac}} = \left(\frac{V_{CC}}{2R'_{L,\,\mathrm{ac}}}\right)^2 R_L = \frac{V_{CC}^2}{4R'_{L,\,\mathrm{ac}}} \qquad (15\text{-}10)$$

Since there are two transistors delivering power to the load, the total ac power output is

$$P_{L,T} = 2\frac{V_{CC}^2}{4R_{L,\text{ ac}}} = \frac{V_{CC}^2}{2R_{L,\text{ ac}}} \tag{15-11}$$

The efficiency maximum for the class B power amplifier is 78.5 percent. The derivation is somewhat involved, but basically, power is delivered by the supply over the entire cycle, so that the average current value is

$$I_{Av} = \frac{2I_{C,\text{ max}}}{\pi} \tag{15-12}$$

Therefore

$$P_{CC} = V_{CC}I_{Av} = V_{CC}\frac{2I_{C,\text{ max}}}{\pi} \tag{15-13}$$

Substituting Eq. (15-9) for $I_{C,\text{ max}}$, we have

$$P_{CC} = V_{CC}\frac{2V_{CC}}{\pi R'_{L,\text{ ac}}} = \frac{2V_{CC}^2}{\pi R'_{L,\text{ ac}}} \tag{15-14}$$

Therefore the efficiency is

$$\eta = \frac{P_{L,T}}{P_{CC}} = \frac{V_{CC}^2/2R'_{L,\text{ ac}}}{2V_{CC}^2/\pi R'_{L,\text{ ac}}}$$

$$= \frac{V_{CC}^2}{2R'_{L,\text{ ac}}}\frac{\pi R'_{L,\text{ ac}}}{2V_{CC}^2} = \frac{\pi}{4} = 0.785 = 78.5 \text{ percent} \tag{15-15}$$

Of course, this is the theoretical maximum efficiency and is never actually reached. It requires maximum signal drive, which would very rarely occur.

The power dissipated by the transistor may be computed similarly for the peak signal. Inspection of Fig. 15-5*a* discloses that the ac load line approaches the maximum-power-dissipation curve at a point equal to $V_{CC}/2$ and $I_{C,\,\max}/2$. This, then, is the point of peak power dissipation for the transistor. The peak power amounts to

$$P_{CE,\,\max} = \frac{V_{CC}}{2}\frac{I_{C,\,\max}}{2} = \frac{V_{CC}}{2}\frac{V_{CC}}{2R'_{L,\,\mathrm{ac}}} = \frac{V_{CC}^2}{4R'_{L,\,\mathrm{ac}}} \qquad (15\text{-}16)$$

The peak power that must be dissipated by the transistor is just twice that peak power that can be delivered to the ac load impedance [see Eq. (15-11)].

DISTORTION

Although the transformer coupled class B amplifier does have a bad crossover distortion, second harmonics are cancelled in the output transformer. Even though one transistor handles one-half of each cycle, harmonics are always present and generate problems. The second harmonic is twice the frequency of the fundamental and is the one that causes most of the problems since it is the largest. It activates the off transistor, causing current to flow in both directions in the primary of the output transformer, which effectively cancels part of the signal. The secondary responds to the difference current.

In power transistors, once the optimum collector current is passed, the characteristic curves begin to crowd together. This crowding of h_{FB}, usually called *α crowding*, causes a drop in the forward current gain h_{FE}, as indicated in Fig. 15-6. For large-signal excursions

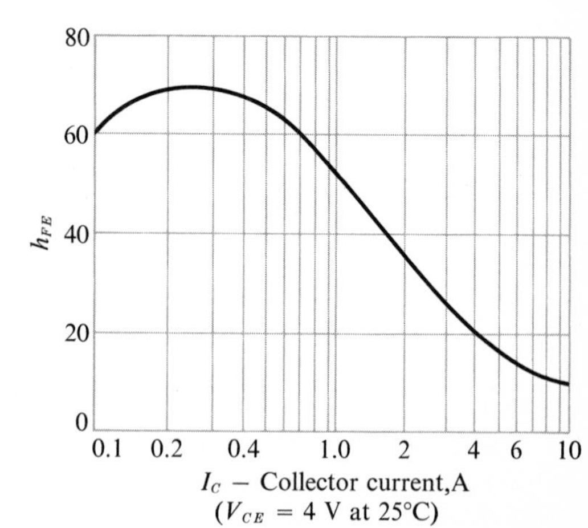

FIG. 15-6 The change in static forward-current-transfer ratio (h_{FE}) vs. collector current.

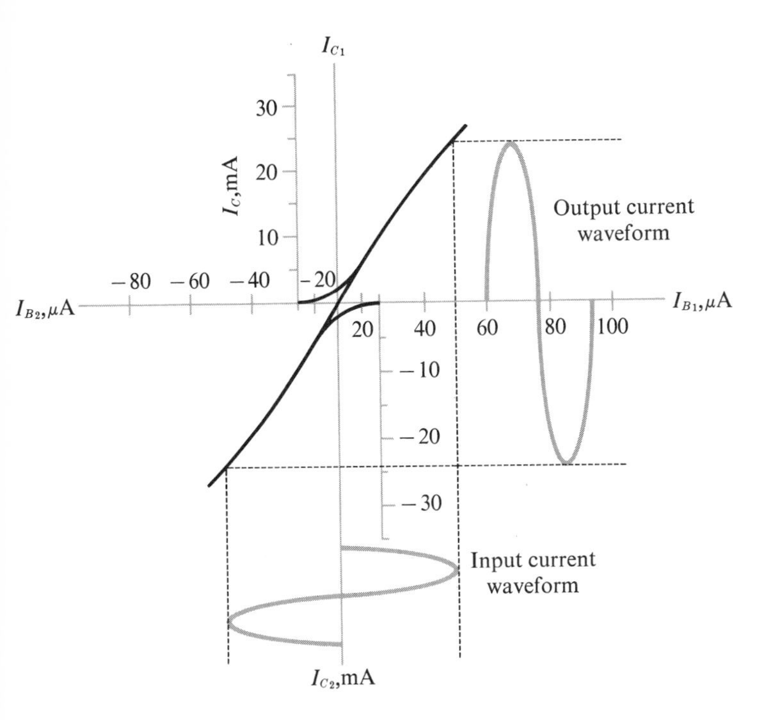

FIG. 15-7 The composite of two forward-transfer curves plotted for a class AB amplifier and showing reduced crossover distortion.

the signal becomes nonlinear, but most of the distortion is second harmonic and is cancelled in the class B amplifier. It is present, however, in the class A amplifier.

OPERATION CLASS AB

A compromise that retains the best features of the class A and class B amplifiers is the class AB. This is an amplifier exactly like Fig. 15-4, but with R_{B2} adjusted to give more collector current with no signal applied. The net result is to offset the two transfer-characteristic curves as shown in Fig. 15-7. The resulting combined transfer curve is essentially a straight line over most of the needed range of input and output.

The class AB amplifier is not as efficient as the class B, but it is relatively free of distortion and is much more efficient than the class A. It forms the basis for the high-fidelity amplifier. Some form of negative feedback is usually used to reduce any nonlinearity of amplification that may be present.

DIRECT-COUPLED POWER AMPLIFIERS

Whenever transformers or capacitors are used in coupling stages there are frequency-response problems. To prevent some of these and extend the frequency-response range, direct coupling is used. Figure 15-8

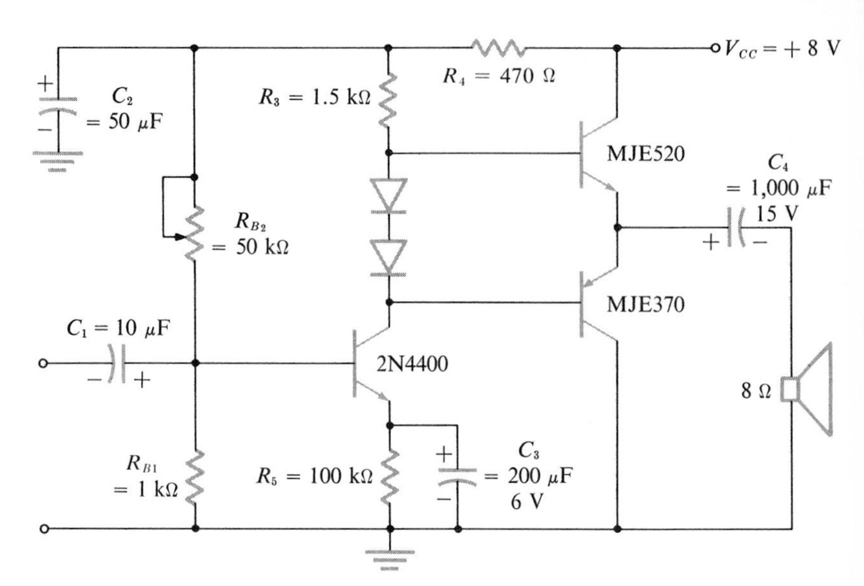

FIG. 15-8 A direct-coupled complementary-symmetry power amplifier. Resistance R_{B2} must be adjusted to give one-half the supply voltage across each transistor in the output.

shows one form of direct coupling used to achieve a class AB amplifier. Note that the two output transistors are operating in series, with the output to the speaker load taken from between them. Note also that the two output transistors are complementary. This is a common practice in inexpensive amplifiers now that low-cost matched complementary transistors are available.

In the class AB amplifier it is usually specified that the two output transistors be delivered signals 180° out of phase with each other, so that one transistor operates while the other is off. This is a holdover from vacuum-tube days. With transistors that operate in complementary symmetry there is no need for phase inversion, since the same signal will provide drive in one direction for the *NPN* and in the other direction for the *PNP* transistor. Thus the signal provided by the input voltage amplifier to the two output power amplifiers provides push-pull operation without another stage of phase inversion.

EMITTER-FOLLOWER ACTION

In order to deliver power to a low-impedance load such as a speaker, the output impedance of a circuit must also be very low. This is accomplished by the circuit of Fig. 15-8. The circuit is actually an emitter follower, with the *PNP* transistor acting as the emitter load resistance. This provides good current gain and also a higher input impedance. The large value of the output coupling capacitor is required to deliver the signal output to the low-impedance speaker without excessive attenuation for low-frequency signals.

The two diodes between the two bases of the output transistors provide enough forward bias to keep the transistors slightly on. A resistance could be substituted for the diodes, but the diodes track the emitter-base junction voltages of the output transistors and provide some thermal stabilization. The two emitter junctions place the two bases about 1.3 V apart, which is just about the same as the forward voltage across the two diodes. Since current can pass in only one direction through the diodes, they also act to isolate the two bases while providing the same signal to each.

The bypass capacitor C_2 is used to decouple the power supply from the input stage by bypassing R_4. Power amplifiers are subject to instability problems because of the low impedance of the power supply. The input stages should be decoupled from the power supply by a π network, or oscillation will probably result. Lead dress and part placement are also stability problems.

FEEDBACK AND OSCILLATION

Certain difficulties are always associated with high-gain circuitry, especially in terms of constant gain and frequency response. The open-loop gain without feedback is subject to the parameters of the devices used. In order to reduce the dependence of the circuit on the devices and increase the frequency response, a negative feedback is normally employed. That is, a small amount of the output signal is fed back, 180° out of phase, to a previous stage. This causes some cancellation of midfrequency gain.

Caution must be used in employing feedback. It should rarely be used over more than two stages. What is negative feedback at the midfrequencies may end up as positive feedback at low or high frequencies. This is because of the progressive phase shift of current and voltage due to the reactive elements in the circuit. The phase angle may be leading at some frequencies, whereas at other frequencies it may lag. If the collective-feedback phase shift reaches 0° as the closed-feedback-loop gain reaches 1, oscillation will occur.

The closed-loop gain may be observed intuitively to be less than the open-loop gain, that is, $A_v' < A_v$ under normal circuit conditions. The actual equation is

$$A_v' = \frac{A_v}{1 - BA_v} \qquad \text{where } B = \frac{\text{feedback voltage}}{\text{output voltage}} \qquad (15\text{-}17)$$

This is shown in diagram form in Fig. 15-9. The denominator of Eq. (15-17) is always greater than 1 for amplifier operation. Notice that the voltage gain is normally a negative number, since for a common-

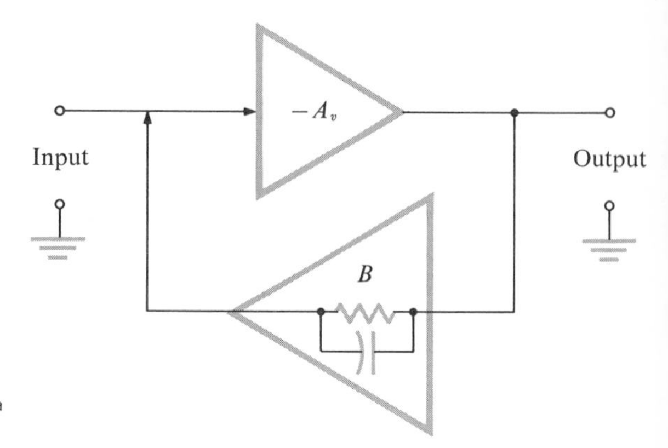

FIG. 15-9 Diagram of a common-emitter amplifier with selective frequency feedback from output to input of a one- or two-stage amplifier.

emitter amplifier the output is 180° out of phase with the input. This changes the denominator to

$$1 - B(-A_v) = 1 + BA_v > 1 \qquad (15\text{-}18)$$

If the voltage gain at some frequency is not negative and the expression BA_v should equal 1 at 0 or 360°, the denominator in Eq. (15-17) becomes zero, and the closed-loop gain A'_v would suddenly go toward infinity. This would immediately be followed by oscillation and instability.

To prevent oscillation in power amplifiers the negative feedback is generally kept at one or two stages. The high frequencies are allowed to roll off at some rate that reduces them as a problem source. Feedback is sometimes controlled by means of a resistance shunted by a small capacitor.

Since the speaker is an inductance, the load impedance seen by the amplifier tends to rise beyond the audio range. To prevent this occurrence and its resulting phase shift, the speaker is sometimes shunted by a 0.22-μF capacitor in series with a 22-Ω resistor.

THE DARLINGTON DRIVER

In most cases the circuit of Fig. 15-8 is insufficient to develop large amounts of power without significant problems. Over the years a number of alternatives have been developed to deal with most of the thermal and oscillation problems and to compensate for nonlinearities. One such circuit, shown in Fig. 15-10, uses a Darlington connection for each output transistor.

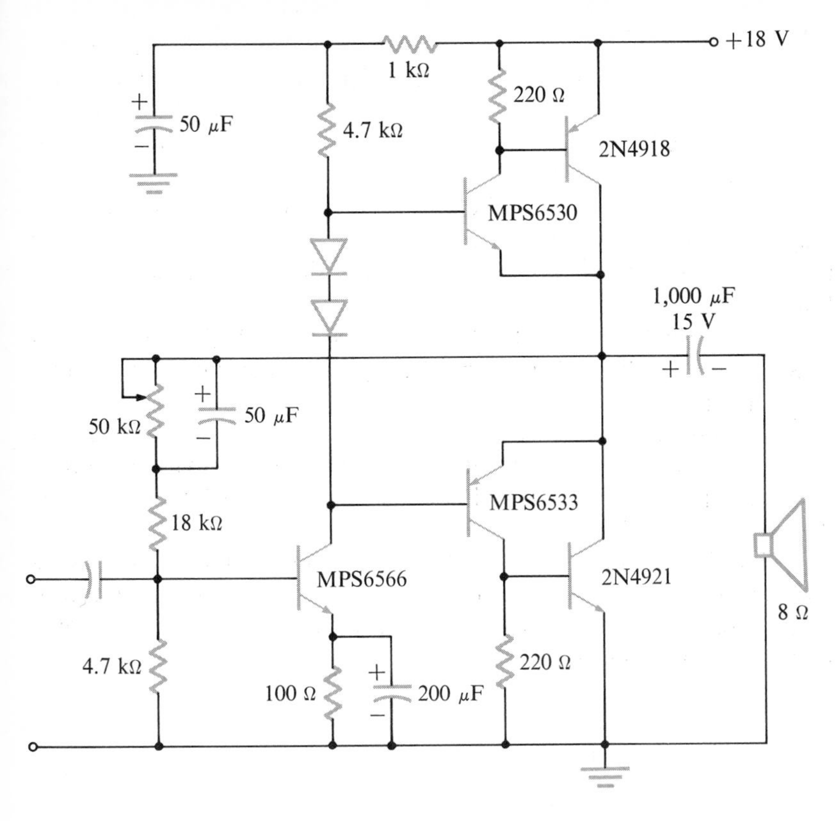

FIG. 15-10 A direct-coupled power amplifier with Darlington connected drivers but complementary-symmetry output transistors.

The Darlington pair output serves several functions. It allows direct coupling and a very high power gain. It has a good frequency response. It gives a much higher input impedance to the voltage amplifier. It takes the signal current delivered by the voltage amplifier and delivers sufficient current to the output stage to drive it with no difficulty.

The input stage to the amplifier of Fig. 15-10 must supply the voltage gain for the amplifier. Its output is supplied to both Darlington drivers, which act as their own phase inverters. The two driver bases are separated by the forward voltage across the two silicon diodes, about 1.3 V. These diodes provide sufficient forward bias to each driver base to turn the output transistors slightly on and operate class AB. They also provide stabilization, since their forward voltage decreases as the temperature increases. This reduces the forward voltage at the base of each output driver, and hence to the output transistors.

In order to function as temperature stabilizers the two diodes should be placed on the same heat sink as the output transistors. The output transistors are plastic-cased units, but with a metal contact for heat transfer. This metal contact is connected internally to the collector of the transistor and should therefore be isolated from the heat sink by a mica or plastic washer. A good grade of silicone grease should be used between the heat sink and the power transistors to assure good heat transfer. Do not try to operate a power amplifier without using heat sinks on the output transistors. Clip-on heat sinks may be needed on the driver transistors if the amplifier operates much into the class AB region or is driven by a sine wave for any lengthy time.

The output speaker may be either a 4- or an 8-Ω speaker. More power can be delivered from the 4-Ω speaker, but most high-quality speakers are 8 Ω.

The power amplifier of Fig. 15-10 should deliver several Watts of rms power when driven by either a ceramic or crystal phonograph cartridge. It may also use a phono preamp or a tuner input that can deliver at least 0.5 V of signal.

The input stage is a class A voltage amplifier that uses dc feedback from the output stages. The base-current-limiting resistance must be adjusted so that the dc voltage across each output transistor is one-half the supply voltage.

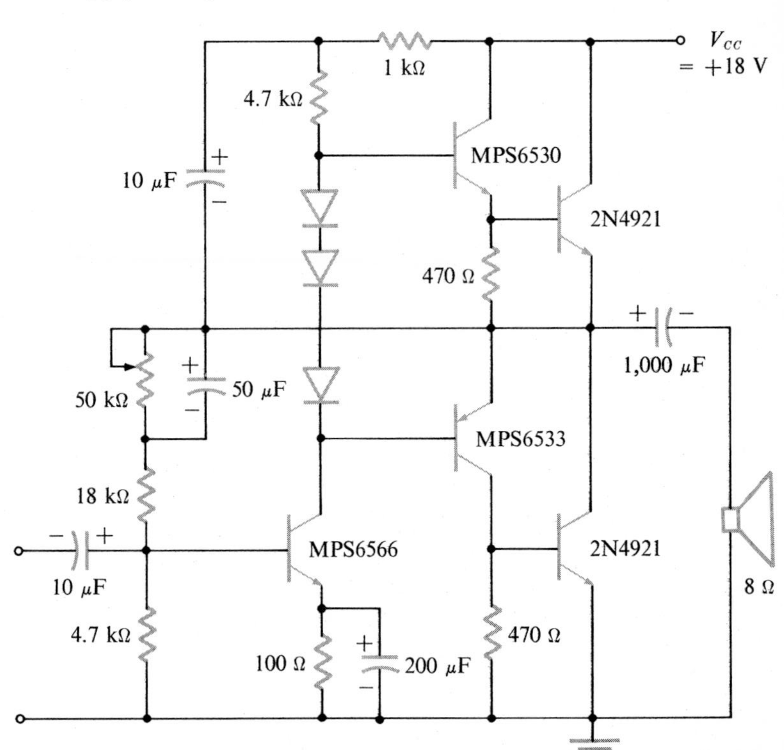

FIG. 15-11 A direct-coupled power amplifier with similar output transistors and complementary-symmetry Darlington drivers.

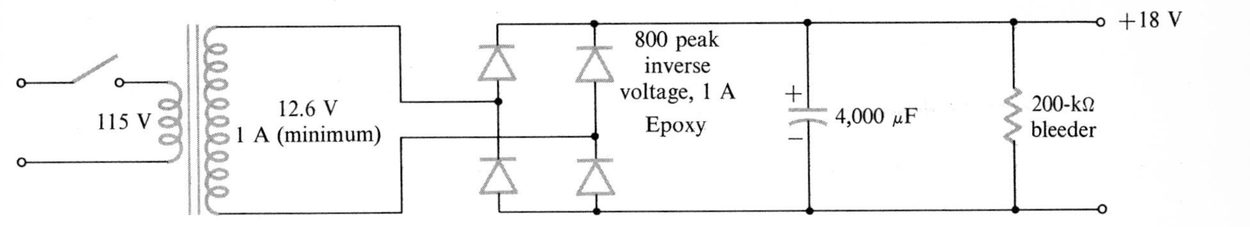

FIG. 15-12 A full-wave bridge 18-V power supply for powering the circuits of Figs. 15-10 and 15-11.

Figure 15-11 gives a variation of the direct-coupled class AB power amplifier. It differs from the circuit of Fig. 15-10 in that the output transistors are both *NPN* power transistors with complementary Darlington drivers instead of having complementary symmetry in both driver and output stages. Otherwise the two circuits function almost identically. Both may be powered from the brute-force type of nonregulated power supply shown in Fig. 15-12. Both should put out about 3 W at 18 V. However, this can be extended to about 5 W at 24 V for an 8-Ω speaker. A 40-V supply should give about 15 W at full drive.

For class AB operation three diodes may be required between the bases of the driver transistors. This can be checked by placing an ammeter in the power line. If very little current is flowing during quiescent conditions, the circuit is still operating as class B. Another diode may be required to offset the four junction potentials in Fig. 15-11.

A QUASI-COMPLEMENTARY POWER AMPLIFIER

When only the driver transistors in a power amplifier are in complementary symmetry, and the output transistors are of the same type, the amplifier is known as *quasi-complementary*. There are advantages, of course, in both full and quasi-complementary configurations.

Various methods have been devised to overcome the inherent faults of each basic circuit design. Texas Instruments, for example, suggests the design shown in Fig. 15-13 (page 386), which takes advantage of the low cost and high quality of plastic-encapsulated devices.

Note the absence of the usual diodes between the bases of the driver transistors. They have been replaced by a single transistor properly biased and stabilized by voltage-divider bias to give the proper base-voltage differential. The diodes in the emitters of the output transistors provide thermal stabilization for the transistors. A very high input impedance to the power amplifier is provided by the FET input stage. Last of all, the driver transistors are manufactured to be electrically equivalent as a complementary pair.

TYPES 2N5449, 2N5450, 2N5451
N-P-N EPITAXIAL PLANAR SILICON TRANSISTORS

TYPES 2N5449, 2N5450, 2N5451
BULLETIN NO. DL-S 6810924 MAY 1968

SILECT† TRANSISTORS

Encapsulated in Plastic for Such Applications as Medium-Power Amplifiers, Class B Audio Outputs, and Hi-Fi Drivers

- **Electrically Equivalent to 2N3704, 2N3705, and 2N3706**
- **For Complementary Use with 2N5447 and 2N5448**
- **Rugged, One-Piece Construction Features Standard 100-mil TO-18 Pin Circle**

mechanical data

These transistors are encapsulated in a plastic compound specifically designed for this purpose, using a highly mechanized process‡ developed by Texas Instruments. The case will withstand soldering temperatures without deformation. These devices exhibit stable characteristics under high-humidity conditions and are capable of meeting MIL-STD-202C method 106B. The transistors are insensitive to light.

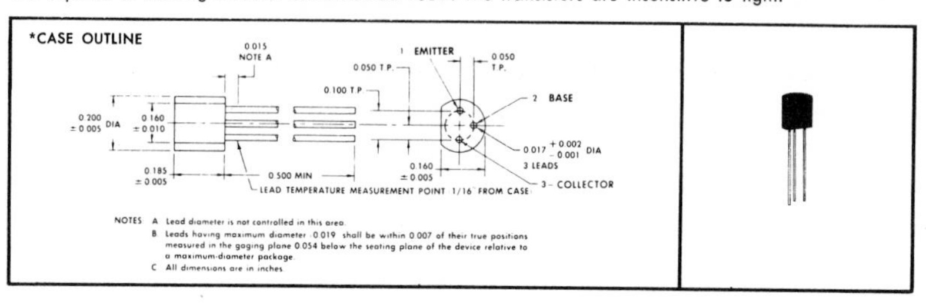

*absolute maximum ratings at 25°C free-air temperature (unless otherwise noted)

	2N5449 2N5450	2N5451
Collector-Base Voltage	50 V	40 V
Collector-Emitter Voltage (See Note 1)	30 V	20 V
Emitter-Base Voltage	5 V	5 V
Continuous Collector Current	←800 mA→	
Continuous Device Dissipation at (or below) 25°C Free-Air Temperature (See Note 2)	←360 mW→	
Continuous Device Dissipation at (or below) 25°C Lead Temperature (See Note 3)	←500 mW→	
Storage Temperature Range	−65°C to 150°C	
Lead Temperature 1/16 Inch from Case for 10 Seconds	←260°C→	

NOTES: 1. These values apply when the base-emitter diode is open-circuited.
2. Derate linearly to 150°C free-air temperature at the rate of 2.88 mW/deg.
3. Derate linearly to 150°C lead temperature at the rate of 4 mW/deg. Lead temperature is measured on the collector lead 1/16 inch from the case.

†Trademark of Texas Instruments
‡Patent pending
*Indicates JEDEC registered data

Types 2N5449, 2N5450, 2N5451 (*courtesy Texas Instruments Incorporated*).

TYPES 2N5449, 2N5450, 2N5451
N-P-N EPITAXIAL PLANAR SILICON TRANSISTORS

*electrical characteristics at 25°C free-air temperature

PARAMETER		TEST CONDITIONS	2N5449		2N5450		2N5451		UNIT
			MIN	MAX	MIN	MAX	MIN	MAX	
$V_{(BR)CBO}$	Collector-Base Breakdown Voltage	$I_C = 100\ \mu A, I_E = 0$	50		50		40		V
$V_{(BR)CEO}$	Collector-Emitter Breakdown Voltage	$I_C = 10$ mA, $I_B = 0$, See Note 4	30		30		20		V
$V_{(BR)EBO}$	Emitter-Base Breakdown Voltage	$I_E = 100\ \mu A, I_C = 0$	5		5		5		V
I_{CBO}	Collector Cutoff Current	$V_{CB} = 20$ V, $I_E = 0$		100		100		100	nA
I_{EBO}	Emitter Cutoff Current	$V_{EB} = 3$ V, $I_C = 0$		100		100		100	nA
h_{FE}	Static Forward Current Transfer Ratio	$V_{CE} = 2$ V, $I_C = 50$ mA, See Note 4	100	300	50	150	30	600	
V_{BE}	Base-Emitter Voltage	$V_{CE} = 2$ V, $I_C = 100$ mA, See Note 4	0.5	1	0.5	1	0.5	1	V
$V_{CE(sat)}$	Collector-Emitter Saturation Voltage	$I_B = 5$ mA, $I_C = 100$ mA, See Note 4		0.6		0.8		1	V
$\lvert h_{fe} \rvert$	Small-Signal Common-Emitter Forward Current Transfer Ratio	$V_{CE} = 2$ V, $I_C = 50$ mA, f = 20 MHz	5		5		5		
C_{cb}	Collector-Base Capacitance	$V_{CB} = 10$ V, $I_E = 0$, f = 1 MHz, See Note 5		12		12		12	pF

NOTES: 4. These parameters must be measured using pulse techniques. $t_p = 300\ \mu s$, duty cycle $\leq 2\%$.
5. C_{cb} is measured using three-terminal measurement techniques with the emitter guarded.

*Indicates JEDEC registered data

TYPICAL CHARACTERISTICS

2N5450
STATIC FORWARD CURRENT TRANSFER RATIO
vs
COLLECTOR CURRENT

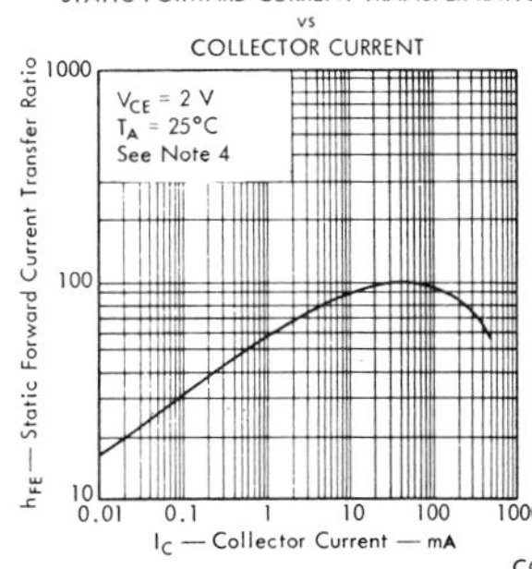

BASE-EMITTER VOLTAGE
vs
COLLECTOR CURRENT

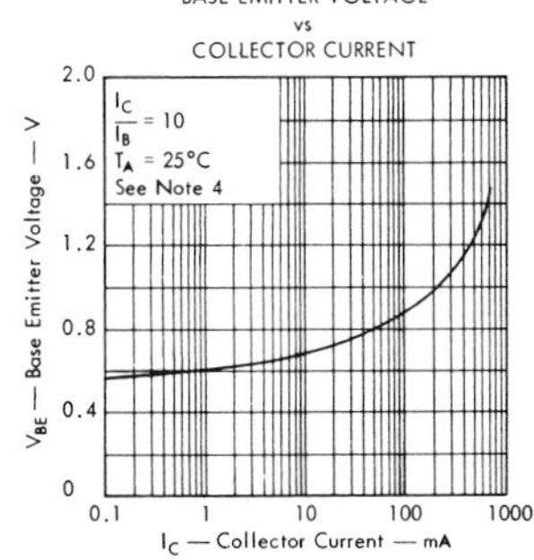

COLLECTOR-EMITTER SATURATION VOLTAGE
vs
COLLECTOR CURRENT

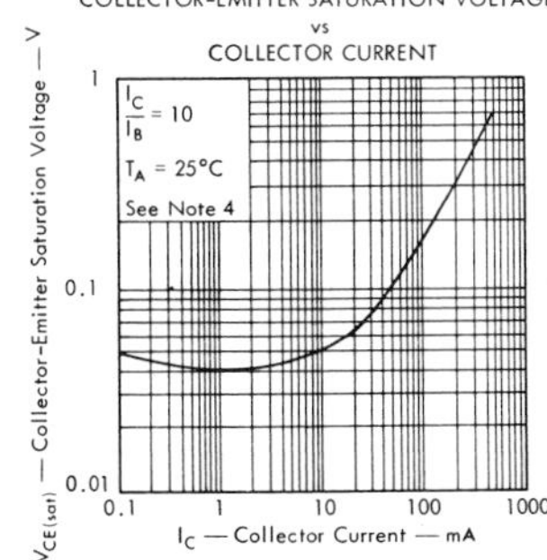

FIG. 15-13 A 15-W quasi-complementary power amplifier (*courtesy Texas Instruments Incorporated*).

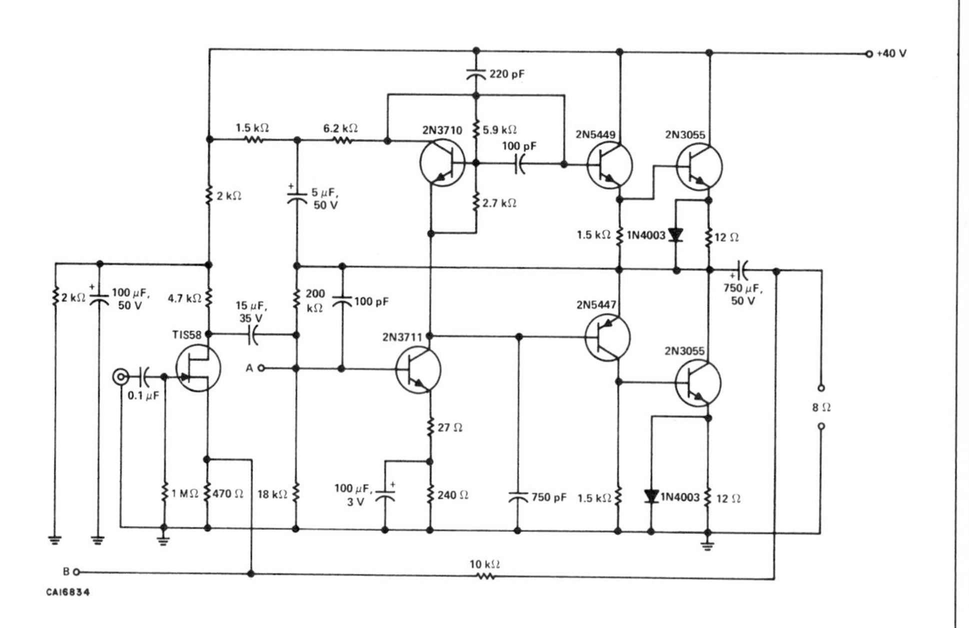

TYPICAL PERFORMANCE SPECIFICATIONS	
Continuous Output Power	15 W @ 0.15% THD
Power Bandwidth @ 7.5 W	20 Hz – 20 kHz
Frequency Response ± 0.5 dB	10 Hz – 50 kHz
Total Harmonic Distortion @ 7.5 W	0.06%
Intermodulation Distortion @ 7.5 W	0.15%
Sensitivity @ 15 W	850 mV
Input Impedance	1 MΩ
Hum and Noise: "C" Weighting	
Input Shorted	−95 dB
Input Open	−85 dB
Damping Factor	48

A COMPLEMENTARY POWER AMPLIFIER WITHOUT BIAS

In an attempt to produce a good power amplifier at low cost with inexpensive but high-quality parts, Fairchild developed a power amplifier that operates without bias at zero signal conditions. This amplifier, shown in Fig. 15-14, is very efficient and yet has no crossover distortion while operating as class B. The nominal power output at less than 1 percent distortion is about 5 W.

This circuit should be constructed with the components indicated, although some substitution is allowable. The input transistor has a current gain in excess of 500, but at 0.1 mA. This low current allows extremely high input impedance without any bootstrapping arrangement. Thus the decoupling capacitor C_2, which filters power supply hum, can be very small, a disk capacitor of 0.01 μF. Transistors T_1 and T_2 act as voltage amplifiers, and T_3 and T_4 act as current or power amplifiers.

The main points of difference between this circuit and others is the manner of using both dc and ac feedback. Capacitor C_3 supplies ac feedback from the output stage to the emitter of T_1 through R_6; dc feedback is obtained from the collector circuit of T_2 through R_5 and R_6. The dc feedback eliminates the need for current in the output transistors at zero signal. The open-loop gain is large enough to provide sufficient feedback to operate the output transistors without bias. The feedback is sufficient to compensate for nonlinear devices.

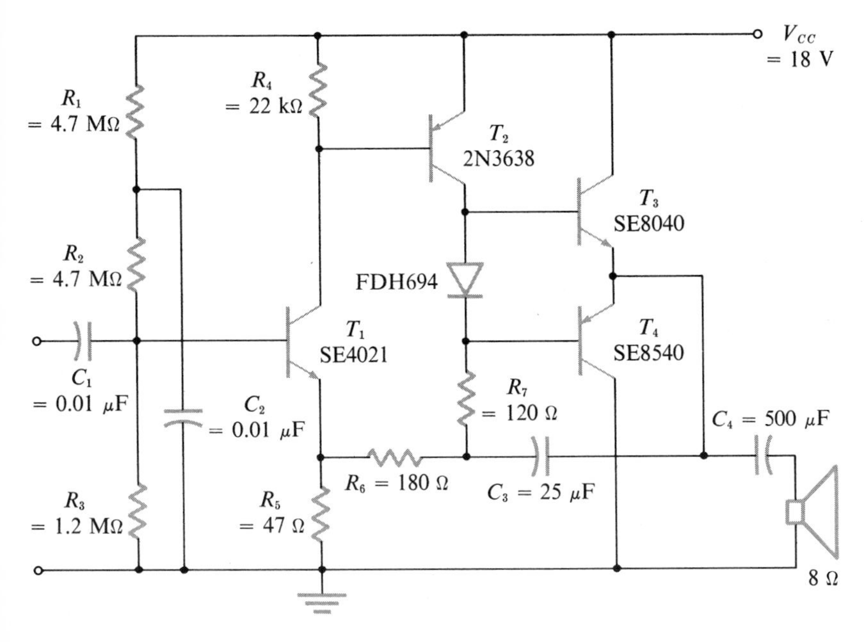

FIG. 15-14 A complementary power amplifier without bias yet with low distortion and high efficiency (*courtesy Fairchild Semiconductor*).

INTEGRATED-CIRCUIT POWER AMPLIFIER

A number of companies market power amplifiers in an integrated-circuit package. General Electric has a line of both preamplifiers and power amplifiers that operate to form a stereo high-fidelity amplifier capable of 10 W rms power, or 5 W per channel.

The PA 237, a 2.5-W unit, was modified to produce 5 W by additional heat-sinking capability and a higher-voltage power supply. The PA 246, a 5-W unit, has a frequency response of 30 Hz to 100 kHz at ±3 dB. The sensitivity for full 5 W output is 180 mV. The PA 246 is on a silicon chip measuring 50 by 60 mils. It is mounted in a modified 14-pin DIP integrated-circuit package, as shown in Fig. 15-15.

The PA 246 requires several external components in order to function according to specifications. These involve the biasing for the input differential amplifier and frequency compensation and feedback. When it is properly connected the unit puts out 5 W into a 16-Ω load (two 8-Ω speakers in series) with less than 1.0 percent distortion at an efficiency of 57 percent. The unit can operate into an 8-Ω load with a little less efficiency.

The circuit design for the PA 246 is much the same as those we have discussed, as are the power-amplifier integrated circuits developed by other companies such as Motorola, RCA, and Sinclair.

Whether the power amplifier is a discrete-device unit or an integrated-circuit unit, the same principles and techniques concerning oscillation, thermal stability, and power supply must be observed. The underlying concepts are in all cases the same.

FIG. 15-15 (*a*) The circuit layout for the General Electric integrated circuit power amplifier PA 246, including the external components for bias and feedback; (*b*) the PA 246 package showing the two large heat-sink tabs; (*c*) distortion curves, with intermodulation distortion measured for 60-Hz and 6-kHz signals mixed 4:1 and THD measured at 1 kHz. Reprinted (by special permission from the November 25 issue of *Electronics:* © 1968 by McGraw-Hill, Inc., New York, N.Y. 10036.)

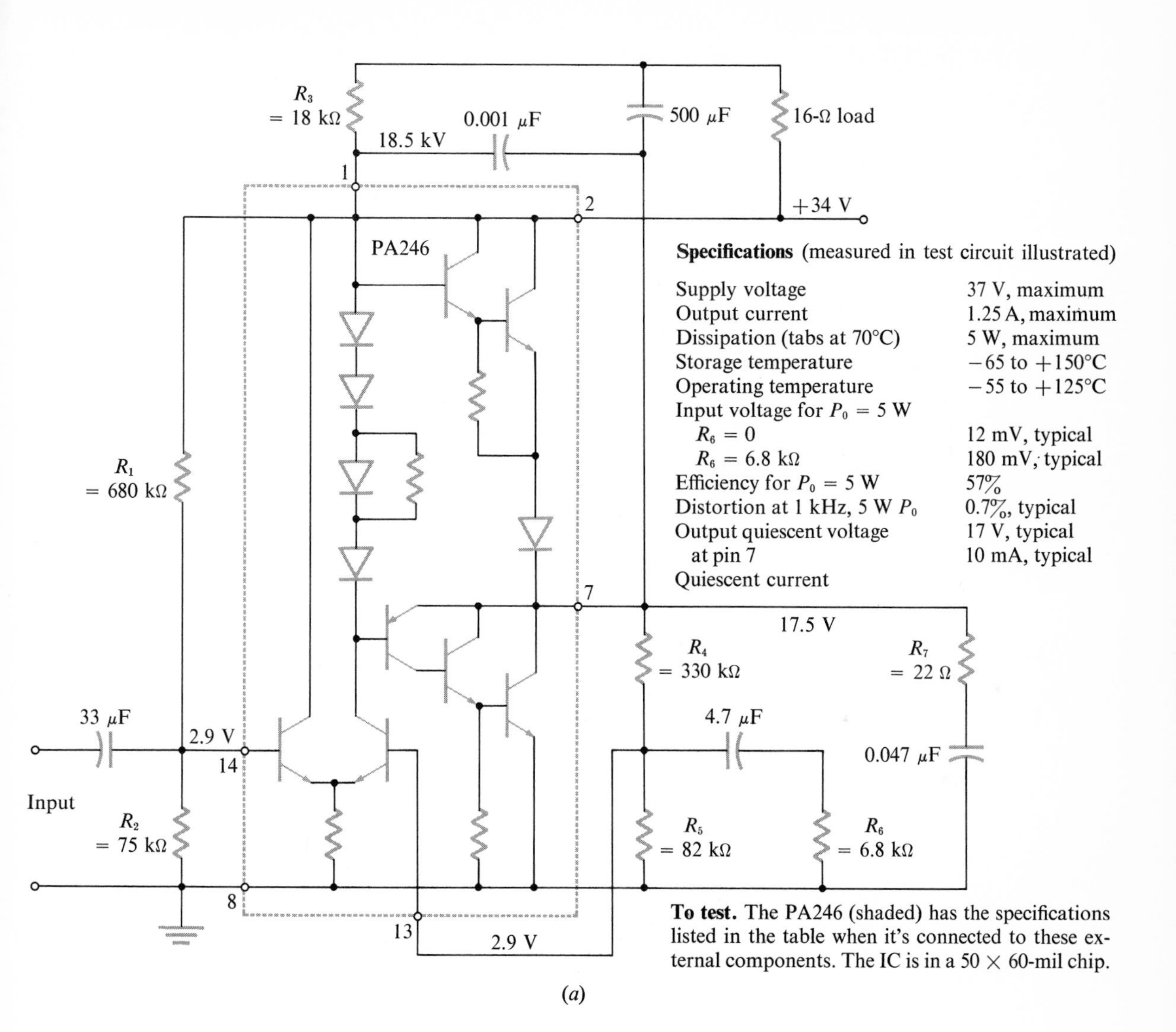

Specifications (measured in test circuit illustrated)

Supply voltage	37 V, maximum
Output current	1.25 A, maximum
Dissipation (tabs at 70°C)	5 W, maximum
Storage temperature	−65 to +150°C
Operating temperature	−55 to +125°C
Input voltage for P_0 = 5 W	
$R_6 = 0$	12 mV, typical
R_6 = 6.8 kΩ	180 mV, typical
Efficiency for P_0 = 5 W	57%
Distortion at 1 kHz, 5 W P_0	0.7%, typical
Output quiescent voltage at pin 7	17 V, typical
Quiescent current	10 mA, typical

To test. The PA246 (shaded) has the specifications listed in the table when it's connected to these external components. The IC is in a 50 × 60-mil chip.

(*a*)

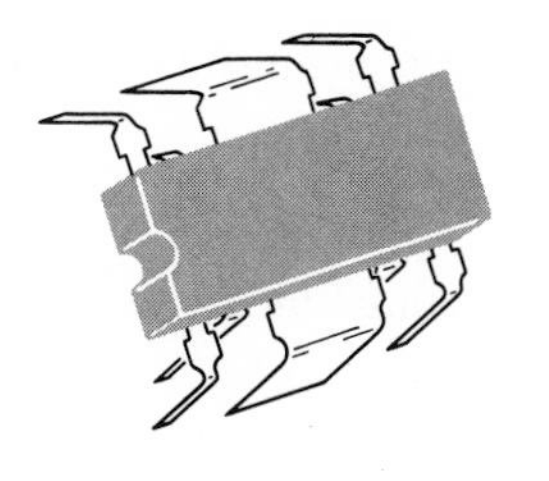

(*b*)

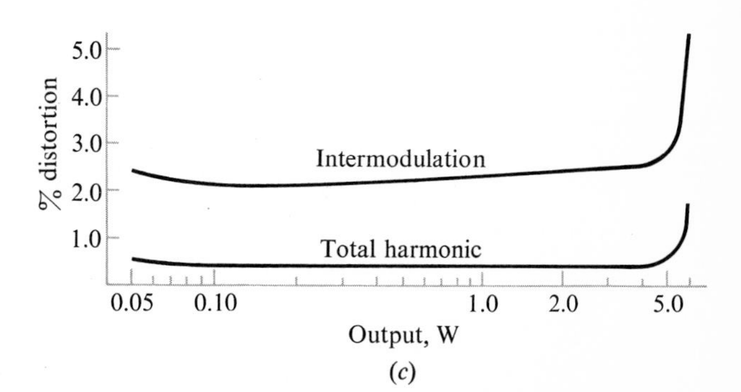

(*c*)

SUMMARY OF CONCEPTS

1. An amplifier that has high current gain and high voltage gain has high power gain.
2. A power amplifier generally must deliver a large amount of signal power to a load.
3. A single-ended amplifier must operate as class A if it is to function with low distortion.
4. Transformer coupling is generally used for class A single-ended power amplifiers.
5. The transformer primary impedance acts as the load impedance for the output transistor and is the reflected impedance from the secondary load.
6. Class A operation requires that current flow during the entire cycle.
7. A class A transformer-coupled power amplifier has only 50 percent efficiency.
8. Efficiency is the ratio of power delivered to the load to power delivered by the source.
9. The dc load line for a transformer-coupled amplifier goes almost straight up, and so the ac load line from the reflected impedance to the primary must be used.
10. Power transistors must be derated for power dissipation at a rate of so many milliwatts per degree of rise in temperature. Although power dissipation may be high at a case temperature of 25°C, it is usually low at a free-air temperature of 25°C because the case cannot dissipate the heat to the air rapidly.
11. Resistive bias stabilization is normally required when transformer input coupling is used.
12. Push-pull amplifiers usually operate as class AB or class B and use two transistors of similar ratings.
13. Class B amplifiers are very efficient but usually have a great deal of crossover distortion at low signal levels.
14. The maximum efficiency of the class B amplifier is 78.5 percent.
15. A center-tapped input and output transformer is required for push-pull transformer-coupled amplifiers.
16. Class B amplifiers are biased at cutoff.

17. Even harmonic distortion is removed from the amplifier output of the push-pull amplifier by the action of the output transformer.
18. Direct-coupled power amplifiers generally use two output transistors operating in an emitter-follower arrangement, with the speaker attached by means of capacitor coupling to the effective emitter of the upper power transistor.
19. Phase inversion is required for power amplifiers operating as push-pull devices. This requires two signals of equal amplitude but opposite phase.
20. Phase inversion is usually achieved in direct-coupled power amplifiers by means of complementary-symmetry transistors either on the output or in the driver stages.
21. Diodes are generally used between the bases of the output transistors of a direct-coupled power amplifier to give sufficient forward bias for class AB operation. They provide the correct offset voltage and also yield good thermal stabilization of the output.
22. Oscillation may occur in the power amplifier when there is feedback through too many stages, causing the total circuit capacitance to shift the phase too far.
23. Feedback may be negative at some frequencies and positive at others.
24. A Darlington connected driver amplifier gives very high power gain and improved input impedance.
25. Crossover distortion may be removed with the right amount of negative feedback and loop gain.

GLOSSARY

α crowding A high-current condition in which a high charge density in the base begins to repel carriers from the emitter region. This results in nonlinear amplification, since equal increments of base current do not produce equal increments of collector current.

Class A operation An amplifier circuit biased in the middle of the transfer characteristic curve.

Class AB operation An amplifier circuit operating just above cutoff, usually in a push-pull arrangement.

Class B operation An amplifier circuit biased at cutoff, a condition allowing increased device efficiency.

Closed-loop gain The gain of an amplifier circuit with feedback; normally less than open-loop gain.

Complementary symmetry A circuit arrangement in which two transistors, one *NPN* and one *PNP*, having similar characteristics are operated together, as in a direct-coupled output stage of a power amplifier.

Crossover distortion The odd-harmonic distortion introduced in a push-pull class B amplifier as the signal crosses from one transistor to the other.

Efficiency The ratio of signal power delivered to the load to power used by the circuit.

Oscillation The sustained generation of an ac voltage as a result of positive in-phase amplifier feedback.

Phase inversion Development of a signal 180° out of phase with the original signal, usually used in a push-pull power amplifier requiring two signals opposite in phase to drive the output stage.

Power amplifier An amplifier designed specifically to deliver large amounts of power. It generally operates with a large amount of current and has a very high current gain.

Push-pull amplifier An amplifier with two output transistors that operate alternately, one on the positive half-cycle and the other on the negative half-cycle, so that the signal seems to be alternately pushed and pulled to the load.

Single-ended amplifier An amplifier designed to drive the load from a single device. It is usually transformer coupled and operated as class A.

REFERENCES

Brazee, James G.: *Semiconductor and Tube Electronics*, Holt, Rinehart and Winston, Inc., New York, 1968, pp. 350–377.

Campbell, David L., and Richard Westlake: *Design Concepts for Low-power Amplifiers*, Application Briefs 72, 58, Fairchild Semiconductor, Mountain View, Calif., 1968.

Cutler, Phillip: *Semiconductor Circuit Analysis*, McGraw-Hill Book Company, New York, 1964, chaps. 7–8.

Horowitz, Mannie: *Practical Design with Transistors,* Howard W. Sams & Co., Inc., Indianapolis, 1968, chaps. 6, 11.

"I-C Audio Amplifier Puts Out 5 Watts," *Electronics,* vol. 41, no. 24, p. 111, Nov. 25, 1968.

Lenert, Louis H.: *Semiconductor Physics, Devices, and Circuits,* Charles E. Merrill Books, Inc., Columbus, Ohio, 1968, chap. 8.

"Linear I–C Applications," *Popular Electronics,* vol. 27, no. 6, pp. 49–56, December, 1967.

Meyer, Daniel: "L'il Tiger Stereo Power Amplifier," *Popular Electronics,* vol. 27, no. 6, December, 1967.

Preferred Semiconductors and Components from Texas Instruments, Catalog CC202, Texas Instruments Inc., Dallas, Tex., 1970, pp. 1702A, 16117–16120.

The Semiconductor Data Book, 4th ed., Motorola Semiconductor Products, Inc., Phoenix, Ariz., 1969, pp. AN16–AN29.

Surina, T., and Clyde Herrick: *Semiconductor Electronics*, Holt, Rinehart and Winston, Inc., New York, 1964, chap. 6.

Thornton, R. D. et al.: *Handbook of Basic Transistor Circuits and Measurements,* John Wiley & Sons, Inc., New York, 1966, chap. 2.

Transistor Manual, 7th ed., General Electric Co., Syracuse, N.Y., 1964, chaps. 3, 11.

REVIEW QUESTIONS

1. What factors are used to classify a power amplifier as class A?

2. Draw a transistor transfer curve and indicate where the amplifier should be biased for class A operation.

3. Describe a single-ended power amplifier.

4. How does the input impedance of a power amplifier compare to the normal voltage amplifier?

5. What factors contribute to the value of the ac load line used in a power amplifier?

6. In what way is the reflected impedance different for a push-pull transformer than for a single-ended transformer?

7. How is amplifier efficiency determined, and what is its importance?

8. What would happen to a power amplifier whose ac load line passed through the curve of maximum power dissipation?

9. What is meant by a thermal derating factor? Give an example.

10. What is a heat sink, and why is it needed for power transistors?

11. What are some of the advantages of transformer input coupling over *RC* coupling?

12. In what ways are push-pull amplifiers different from single-ended amplifiers?

13. How does class B amplification differ from class A amplification?

14. What are some disadvantages of class B amplifiers?

15. Why is class AB operation preferable in most cases to class B?

16. What is the difference in maximum efficiency between class A and class B amplifiers?

17. If the output signal current for a class B amplifier is not known, how can the maximum power to the load be determined if the amplifier is transformer coupled?

18. What is a major advantage of the direct-coupled power amplifier over the transformer-coupled one?

19. What is the function of the diodes often used between the bases of the output transistor drivers in a direct-coupled class AB amplifier?

20. Why should the voltage amplifiers of the power amplifier be supplied by a voltage that is decoupled from the power supply?

21. Why is feedback necessary, and how can it cause oscillation?

22. Describe the action of a Darlington configuration in a direct-coupled power amplifier.

23. What is the advantage of using complementary symmetry in the output stages of a power amplifier, and how does this effect the need for phase inversion?

PROBLEMS

1. If the supply voltage in Fig. 15-1 is doubled, what will happen to the maximum power output?

2. If the turns ratio of an output transformer is 3.5:1, what is the reflected load impedance to the primary winding if the secondary load is 4 Ω? If it is 8 Ω?

3. If a push-pull output transformer has a turns ratio of 10:1, with a primary center tap, what is the turns ratio used by each transistor? What load impedance is seen by each transistor if the secondary load is 8 Ω?

4. From the curves of Fig. 15-2, what supply voltage would be required to place the Q point and ac load line tangent to the maximum power dissipation curve?

5. If a class A power amplifier is using 500 mW of power, how much power is being dissipated by the power transistor? How much power can be delivered to the load?

6. If a power amplifier load is using 3 W of power and the circuit is consuming 7 W, what is the efficiency?

7. If a class B amplifier has a supply voltage of 20 V and an output transformer with a primary load of 150 Ω, what is the power consumed at zero signal? What is the maximum power that load can deliver?

8. If a power transistor can dissipate 40 W at a case temperature of 25°C and the derating factor is 30 mW per degree, how much power can it dissipate at 65°C?

9. If the transistor in Problem 8 can dissipate only 5 W at 25°C free-air temperature, derated at 20 mW per degree, what power can it dissipate at 65°?

10. From the curves of Fig. 15-5a, what would be the peak current if the peak input base current were 300 μA?

11. In Prob. 10, what would be the peak output voltage from one of the transistors? From both transistors?

12. In Fig. 15-8, what would be the approximate voltage difference between the bases of the output transistors if three silicon diodes were used?

13. If the feedback ratio used for a certain power amplifier is 0.02 and the voltage gain is −500, what is the closed-loop gain?

14. If the input signal to a power amplifier is 0.4 V p-p and the output voltage is 20 V p-p, what is the overall voltage gain?

15. In Prob. 14, if the current gain is 10,000, what is the power gain?

ANSWERS TO ODD NUMBERED PROBLEMS

Chapter 1
No problems

Chapter 2
1. 10 N
3. 10 W
5. 3 A
7. 5 W
9. 0.05 S or 50 mS
11. 0.005 F or 5000 μF
13. 1.66 mA
15. 10 mA
17. 5 K
19. 20 V
21. 14.67 kΩ
23. 13.7 V
25. 0.5 mA

Chapter 3
1. 50 V rms = 141.4 V *p-p* = 70.7 V peak
3. $V_C = 9.5$ V
5. $I = 0.6$ mA at $T = 0$, $I = 0$ mA at $T = 3.75$ s
7. $I_{\max} = 0.05$ A, $I_{1TC} = 0.0316$ A, $I_{5TC} = 0.05$ A
9. $X_C = 53$ kΩ
11. $X_C = 132.5$ kΩ
13. $Z = 132.8$ kΩ
15. $\omega = 6{,}280$ rad/s
17. Linear from 0 Ω to 50.4 kΩ at 4 kHz
19. $X = -742.8$ Ω
21. $Z = 897$ Ω at $-56°04'$
23. $f = 11.25$ kHz
25. $V = 500$ V

Chapter 4
1. 0.1 V
3. 6,666 Ω/V
5. 10 kΩ
7. 1.01 Ω
9. 1.0 V
11. 5 kHz
13. 55 μS, 18.2 kHz

Chapter 5
1. 40 μA at 40°C, 80 μA at 50°C
3. 15,000 cm/s

Chapter 6
1. $V_L \cong 9.35$ V
3. 50 Ω
5. 15 Ω
7. 17.2 V
9. 8.9 V peak, 6.3 V dc
11. 4.5 percent

Chapter 7
1. 0.68 V
3. 40 μA
5. $h_{FB} = 0.97$, $h_{FE} = 33.3$
7. 3.85 V
9. $h_{FB} = 0.992$

Chapter 8
1. $V_R = 7.5$ V, $V_{CE} = 2.5$ V
3. $\Delta I_C = 1$ mA
5. $V_{CC} = 10.5$ V
7. 1,000 nA = 1 μA

Chapter 9
1. 16.6 percent
3. 33.3 μA *p-p*
5. $h_{ib} = 30$ Ω, $h_{fe} = 72$
7. 8 V *p-p*, signal may decrease 4 V and so may increase 4 V without distortion
9. 8
11. 1 mA *p-p*, $A_i = 100$
13. 16.4 V
15. 1.29 mA

Chapter 10
1. 100 nA
3. 155 μA
5. $h_{ib} \cong 25$ Ω, $Z_{\text{in}} \cong 15.4$ kΩ
7. $V_{R_L} = 7$ V, $V_{CE} = 7.2$ V, $V_B \cong 1.4$ V
9. $h_{FE} = 133$, $Z_{\text{in}} \cong 55.2$ kΩ
11. $A_v \cong -11$
13. $V_{be} = 0.16$ V, $A_v = -37.5$
15. $R_L = 5.4$ kΩ
17. $R_L = 1.9$ kΩ, $V_{R_L} = 1.52$ V, $V_B = 1.6$ V, $R_B = 625$ kΩ, $R_{E,\ \text{bypassed}} = 1{,}100$, $R_{E,\ \text{unbypassed}} = 150$ Ω

Chapter 11

1. $R_f = 940$ kΩ
3. $Z_{in} = 1.77$ kΩ
5. $A_i \cong 69$
7. $I_{R_L} = 1.01$ mA
9. $R_E = 780$ Ω, $R_L = 4.3$ kΩ, $R_1 = 10$ kΩ, $R_2 = 56$ kΩ
11. $Z_{in} \cong 7.7$ kΩ
13. $Z_{in} \cong 6.2$ kΩ
15. $V_e = 0.943$ V *p-p*

Chapter 12

1. $h_{FE} = 99$, $h_{FC} = 100$
3. $V_{out} = 1.49$ V *p-p*, $A_v = 0.994$
5. $Z_{in} = 236$ kΩ
7. $R_L = 8.2$ kΩ, $Z_{out} \cong 8$ kΩ
9. $A_v = 250$
11. $Z_{out} \cong 8.9$ kΩ

Chapter 13

1. $I_D = 0$ at $V_{GS} = -V_P = -4$ V
3. $Z_{in} \cong 470$ kΩ
5. $R_L = 8.2$ kΩ
7. $Y_{fs} = 1.6$ mS
9. $R_{DS} = 3.67$ kΩ
11. $Y_{fs} = 1.62$ mS
13. $R_L = 5$ kΩ
15. $Y_{fs} = 0.775$ mS
17. $V_G = 0.47 \times 10^{-3}$ V or 0.47 mV

Chapter 14

1. $A_v = 750$
3. $C = 6.1$ μF
5. $R_{E,\text{ unbypassed}} = 630$ Ω
7. $Z_p = 32$ kΩ
9. $A_v = -4.4$
11. $V_{R_{L1}} = 7.5$ V, $V_{R_{L2}} = 4.75$ V
13. $V_{CM} = 2.1$ V
15. $f = 15.9$ Hz
17. $A_p = 3$ dB, $A_p = 6$ dB

Chapter 15

1. Increase by a factor of 2^2
3. 5:1, $Z_L = 200$ Ω
5. $P_{tr} = 250$ mW, $P_L = 250$ mW
7. $P_{\text{output signal}} = 0$ W, $P_L = 1.3$ W
9. $P_{\text{dissipation}} = 4.1$ W
11. $V = 4$ V peak, $V = 8$ V *p-p*
13. $A_v = -45.5$
15. $A_p = 500{,}000$

INDEX